'The book completes its mission to show that climate change is a driver of conflict, an obstacle to conflict resolution, and a creator of strategic shocks, tensions, opportunities, and risks... A must read for anyone interested in defence, security, or climate change.'

**General Tom Middendorp**, *International Military Council on Climate & Security (former Chief of Defence of The Netherlands)*

'Climate change is the single most significant long-term defence issue. Its consequences will shape threats, challenge governments (as competition for food, water increases) undermine military capabilities, and drive economic instability. The book offers insights that are at once timely, important, and alarming.'

**Tobias Ellwood MP**, *Chair of the UK House of Commons Defence Committee*

'The climate crisis is the top geopolitical issue of our time. From the High North to equatorial Africa, global heating will be the major driver of the conflicts and threats we will face in the coming decades. This vital and timely book challenges us to marshal a much better response, urgently.'

**Tom Fletcher**, *University of Oxford (former UK Foreign Policy Adviser to 3 UK Prime Ministers)*

'There are uncomfortable gaps between knowledge, practice, and reality when it comes to how climate change is shaping human and hard security. This book is an important contribution to closing that gap, showing how to stop worrying about securitizing climate and start climatizing security.'

**Hon. Sharon E. Burke**, *Ecospherics (Former US Assistant Secretary of Defense for Operational Energy)*

'Climate Change and Insecurity is a clarion call for the military and security sectors. This exceptional collection shows how climate change and its impacts are reshaping the world and how nations must prepare for what is to come.'

**Captain Dr Andrea Cameron**, *US Naval War College*

# CLIMATE CHANGE, CONFLICT, AND (IN)SECURITY

This book offers a multidisciplinary exploration of how climate change is impacting conflicts, contention, and competition in the world.

The volume examines how climate change is creating and exacerbating insecurities for millions of people globally and how states, intergovernmental bodies, and others are attempting to meet challenges today and in the near and medium term. It shows that climate change insecurity is relevant to a battery of security areas, including warfighting, stabilisation, human security, influence, resilience, and capacity building. The volume provides insights into how climate change has and will impact security at different scales and in different localities, including national and ethnic tensions, food and water security, resource competition, mass displacement, and even the recruitment profiles and operations of violent and extremist organisations. With contributions from pioneering researchers and practitioners, the book discusses shifting operational requirements and responsibilities and the need for clarity around the size and shape of capacity gaps.

In addition to practitioners and policy-makers working in these areas, the book will be of significant interest to researchers and students of defence studies, peace and conflict studies, climate change and environmental security, and international relations.

**Timothy Clack** is the Chingiz Gutseriev Fellow at the University of Oxford, UK. He is co-editor of various titles including *Cultural Heritage and Armed Conflict* (2022) and *The World Information War* (2021).

**Ziya Meral** is a Senior Associate Fellow at the Royal United Services Institute.

**Louise Selisny** is a Research Associate at the University of Oxford, UK.

# Routledge Advances in Defence Studies

**Series editors:** Timothy Clack, *University of Oxford, UK,* and Oliver Lewis, *University of Southern California, USA*

**Advisory Board:** Tarak Barkawi *London School of Economics, UK* Richard Barrons *Global Strategy Forum, UK* Kari Bingen-Tytler *Center for Strategic and International Studies, USA* Ori Brafman *University of California, Berkeley, USA* Tom Copinger-Symes *British Army, UK* Karen Gibsen *Purdue University, USA* David Gioe *West Point, USA* Robert Johnson *Oxford University, UK* Mara Karlin *John Hopkins University, USA* Tony King *Warwick University, UK* Benedict Kite, *British Army, UK* Andrew Sharpe *Centre for Historical and Conflict Research, UK* Suzanne Raine *Cambridge University, UK*

*Routledge Advances in Defence Studies* is a multi-disciplinary series examining innovations, disruptions, counter-culture histories, and unconventional approaches to understanding contemporary forms, challenges, logics, frameworks, and technologies of national defence. This is the first series explicitly dedicated to examining the impact of radical change on national security and the construction of theoretical and imagined disruptions to existing structures, practices, and behaviours in the defence community of practice. The purpose of this series is to establish a first-class intellectual home for conceptually challenging and empirically authoritative studies that offer insight, clarity, and sustained focus.

**Cultural Heritage in Modern Conflict**
Past, Propaganda, Parade
*Edited by Timothy Clack and Mark Dunkley*

**How Wars End**
Theory and Practice
*Edited by Richard Iron and Damien Kingsbury*

**Understanding UK Defence Exports**
The International Trade in Defence Capabilities
*John Louth*

**Climate Change, Conflict, and (In)Security**
Hot War
*Edited by Timothy Clack, Ziya Meral, and Louise Selisny*

For more information about this series, please visit: https://www.routledge.com/Routledge-Advances-in-Defence-Studies/book-series/RAIDS

# CLIMATE CHANGE, CONFLICT, AND (IN)SECURITY

## Hot War

*Edited by Timothy Clack, Ziya Meral, and Louise Selisny*

LONDON AND NEW YORK

Cover image: © British Army's Long Range Reconnaissance Group (LRRG) in central Mali in 2021 (credit: UK PJHQ; Crown copyright)

First published 2024
by Routledge
4 Park Square, Milton Park, Abingdon, Oxon OX14 4RN

and by Routledge
605 Third Avenue, New York, NY 10158

*Routledge is an imprint of the Taylor & Francis Group, an informa business*

*British Library Cataloguing-in-Publication Data*
A catalogue record for this book is available from the British Library

*Library of Congress Cataloguing-in-Publication Data*
Names: Clack, Timothy, editor. | Meral, Ziya, editor. | Selisny, Louise, editor.
Title: Climate change, conflict, and (in)security : hot war / edited by Timothy Clack, Ziya Meral, and Louise Selisny.
Description: Abingdon, Oxon ; New York, NY : Routledge, 2024. | Series: Routledge advances in defence studies | Includes bibliographical references and index.
Identifiers: LCCN 2023030301 (print) | LCCN 2023030302 (ebook) | ISBN 9781032455808 (hardback) | ISBN 9781003377641 (paperback) | ISBN 9781003377641 (ebook)
Subjects: LCSH: Climatic changes. | Security, International. | National security--Climatic factors.
Classification: LCC QC903 .C55234 2024 (print) | LCC QC903 (ebook) | DDC 304.2/8--dc23
LC record available at https://lccn.loc.gov/2023030301
LC ebook record available at https://lccn.loc.gov/2023030302

ISBN: 978-1-032-45580-8 (hbk)
ISBN: 978-1-032-45579-2 (pbk)
ISBN: 978-1-003-37764-1 (ebk)

DOI: 10.4324/9781003377641

Typeset in Times New Roman
by MPS Limited, Dehradun

## Pyrrhic Wisdom

Do you know:
How the mountains melted?
When the seas rose and claimed the shores?
Why the tempest-tossed found no harbour, found no open doors?
How the poisoned air filled the lungs of the little ones?
What made the bees all fly away?
When the tomorrows were lost to the mistakes of yesterday?
Where nothing is and nothing grows?
How ascent of man was chastened by the sun?
I know; it is a pyrrhic wisdom.

*– Louise Selisny*

# CONTENTS

# CONTRIBUTORS

**Pauline Baudu** is a Research Assistant with the Polar Institute and Environmental Change and Security Program of the Woodrow Wilson International Center for Scholars and a Research Assistant with the Center for Climate and Security and works with the French National Asylum Court. With extensive experience in human rights and migrations, her current research focuses on environmental migrations and on the security implications of climate change in the Arctic. She holds an MA in Defense, Security and Crisis Management from the French Institute for International and Strategic Affairs and an MA in Crisis Analysis and Humanitarian Action from Université Savoie-Mont-Blanc. She holds a BA in Applied Linguistics from Université de Tours.

**Tim Benton** is Director of the Environment and Society Programme and a Research Director of Emerging Risks at Chatham House. He joined Chatham House in 2016 as a distinguished visiting fellow, at which time he was also dean of strategic research initiatives at the University of Leeds. From 2011 to 2016, he was the 'champion' of the UK's Global Food Security Programme, which was a multi-agency partnership of the UK's public bodies. He has worked with the UK government, the EU, and G20. He has been a global agenda steward of the World Economic Forum and is an author of the IPCC's 'Special Report on Food, Land and Climate' (2019) and the UK's 'Climate Change Risk Assessment' (2017, 2022).

**Elizabeth Boulton** is an Eco-Military Theorist with Destination Safe Earth and a Research Affiliate with the Climate Change and (In)Security Project. She investigates what happens if climate and ecological crisis are prioritised as the central threat. Her background is in emergency logistics, as an officer in the Australian Army and humanitarian NGO sector. After completing a Master's in climate policy at the University of Melbourne in 2007, she moved into climate policy and risk communication work, including at the Australian National Climate Centre, before completing a PhD at the Australian National University in 2020.

**Cynthia Brady** is a Global Fellow with the Wilson Center's Environmental Change and Security Program. Previously, she served as a Senior Peacebuilding and Conflict Advisor with the U.S. Agency for International Development's Bureau for Democracy, Conflict and Humanitarian Assistance, where she spearheaded the Agency's technical work on environment, conflict and fragility and led the field support division of the Office of Conflict Management and Mitigation. Her applied research has focused on the nexus of international development with conflict, fragility and peacebuilding.

**Oli Brown** is an Associate Fellow in the Energy, Environment and Resources Programme at Chatham House. Between 2014 and 2018, he was based in Kenya where he coordinated the UN Environment's work to minimise the risks and impacts of disasters, industrial accidents, and armed conflicts. Between 2010 and 2012, he managed a UN Environment programme in Sierra Leone, and before that, he was a Senior Researcher and Programme Manager with the International Institute for Sustainable Development. He has specialist knowledge in environmental management, peacebuilding and conflict analysis, trade policy, migration policy, climate change, and extractives.

**Tristan Burwell** was commissioned into the British Army's 9th/12th Royal Lancers and deployed to Afghanistan as an Afghan Army Mentor in 2011 and Brigade Reconnaissance Force Liaison Officer in 2013. Later, he was an Exchange Officer in the Italian Army's Acqui Division HQ. He is currently a Research Associate of the Climate Change and (In)Security Project. He has an MA in Global Diplomacy from SOAS and a Master's in International Strategic Military Studies from the University of Turin. Awarded the British Army's Farmington Fellowship, he researched the role of non-state actors in processes of security sector reform at the University of Oxford.

**Timothy Clack** is the Chingiz Gutseriev Fellow at the School of Anthropology and Museum Ethnography (SAME), University of Oxford, UK. He is also an Associate Fellow of the British Army's Centre for Historical Analysis and Conflict Research (CHACR). His anthropological work has focused primarily on issues of conflict, identity, and heritage in the Horn of Africa, Central Asia, and the South Atlantic. Keen to apply theory to practice, he spent a decade in UK government service, including with the Cabinet Office and Foreign and Commonwealth Office. He is the Director of the Climate Change and (In)Security Project, a collaboration between Oxford and CHACR. He is also general editor of the Routledge Advances in Defence Studies (RAiDS) book series and co-editor, with Mark Dunkley, of *Cultural Heritage and Armed Conflict* (2022), and, with Robert Johnson, of *The World Information War* (2021), *Before Military Intervention* (2018), and *At the End of Military Intervention* (2015).

**Tom Deligiannis** is an Instructor in Global Studies at Wilfrid Laurier University. He also teaches in the Peace Studies Program at McMaster University, Hamilton. He received his PhD from the University of Toronto in 2020. He holds an MA in International Relations History (Toronto) and BA in History and Political Studies (Guelph). Between 2008 and 2017, he was a part-time faculty member in the Department of Political Science at Western University. Between 2005 and 2007, he was a resident Assistant Professor at the UN-mandated University for Peace in Costa Rica.

**Thammy Evans** is a Non-Resident Senior Fellow at the GeoTech Center of the Atlantic Council and a member of its Global China Hub. She is also a member of the Council for Security Risks Alliance for Ecological Security and a Senior Research Associate of the Climate Change and (In)Security Project. Having worked for NATO, the UN, UK MOD, private sector, and various NGOs, her career spans systems sustainability, security sector reform, gender and security, and public information and political advising. She has worked around the world and published with *Carnegie Europe, Chatham House Journal, British Army Review, Modern Asian Studies*, and *Small Wars Journal*.

**Rachel Fleishman** is a Senior Fellow for Asia-Pacific at the Center for Climate and Security where she focuses on the Asia-Pacific region. She also delivers Asia-Pacific liaison at the International Military Council on Climate and Security (IMCCS). She started her career in national security policy, working in nuclear arms control at SAIC, the Pentagon, and NATO. She currently advises businesses and non-profit institutions on climate change and circular economy issues at Insight Sustainability. She holds a BA from Tufts University, an MA from the University of Maryland, and an MBA from Northwestern University.

**Sherri Goodman** is the Secretary General of the International Military Council on Climate and Security. She is also the Chair of the Board at the Council on Strategic Risks, a Senior Strategist and Advisory Board member at the Center for Climate and Security, and a Senior Fellow at the Woodrow

Wilson International Center's Polar Institute and Environmental Change and Security Program. Previously, she served as the US' first Deputy Undersecretary of Defense (Environmental Security) from 1993 to 2001. As Senior Vice President and General Counsel of CNA (Center for Naval Analyses), she founded the CNA Military Advisory Board where she published a number of landmark reports, including *National Security and the Accelerating Risks of Climate Change*.

**Thomas Homer-Dixon** is a multi-award-winning academic author and teacher, policy consultant, and speaker. As well as being the Director of the Cascade Institute, he holds a University Research Chair in the Faculty of the Environment at the University of Waterloo, Canada. His current work focuses on threats to global security in the 21$^{st}$ century, including economic instability, resource scarcity, and climate change. Since receiving his PhD in Defence and Conflict Theory from the Massachusetts Institute of Technology (MIT) in 1989, he has produced over three decades of ground-breaking research and analysis that has illuminated the field of Environmental Security and informed the judgements of senior decision makers in the United States, Canada, and the United Kingdom. Adding to his interdisciplinary work, he has most recently published *Commanding Hope: The Power We Have to Renew a World in Peril* in 2020.

**Samuel Jardine** is a geopolitics specialist and historian. He specialises in the geopolitics, policy, and security of the Polar regions and space, with a specific interest in climate conflict, mineral politics, and multipolar competition. He holds an MA in Modern History from King's College London and a BA (Hons) in History from the Open University. He is the current Head of Research at *London Politica*, a global political risk consultancy. He is also a Research Associate of the Climate Change and (In)Security Project, a military sciences 'Rising Star' at the Royal United Services Institute (RUSI) and a former 2021 Fellow of the Arctic Geopolitics Program, working on the militarisation and national security working group, jointly hosted by the Arctic Institute, Ecologic Institute Berlin, and the Migrations in Harmony Research Coordination Network.

**Katarina Kertysova** is a Policy Fellow at the European Leadership Network (ELN) in London, a Global Fellow with the Wilson Center's Polar and Kennan Institutes, and a Research Associate with the Climate Change and (In)Security Project (CCIP). She writes on Arctic geopolitics, NATO's climate security agenda, military innovation and decarbonisation, and the climate-nuclear nexus. She contributed to the NATO 2030 Reflection Process and authored the climate chapter of the NATO 2030 Young Leaders Report. She holds degrees from the University of St Andrews and Sciences Po Paris.

**Ana Maria Kumarasamy** is a PhD Student in politics at Lancaster University. Her thesis focuses on environmental inequalities, identity politics, and governance in Beirut. She is also a PhD fellow at the Sectarianism, Proxies, and De-sectarianism (SEPAD) project and an associate fellow at the Middle East and North Africa Programme at the Swedish Institute of International Affairs.

**Gary Lewis** is a Former UN Resident Coordinator in Iran (2013–2018). During a nearly 35-year career in the UN, he has served in director-level positions in UNODC (drugs/crime), UNEP (environment), and UNDP (development). A native of Barbados, his work has taken him to duty stations in Switzerland, Kenya, Iran, Thailand, India, South Africa, Pakistan, Afghanistan, Austria, and Barbados. He is a published author on Barbadian history and has written and contributed to academic articles related to drug control and security. Currently, he mainly presents on ecological security issues. His main academic interests are paleo-anthropology and evolutionary psychology.

**Simon Mabon** is Professor of International Relations in the Department of Politics, Philosophy and Religion at the University of Lancaster. His research sits at the intersection of Middle East Studies, international relations, and political theory. He is the Director of the Richardson Institute, the UK's oldest peace and conflict research centre, and the Sectarianism, Proxies & De-sectarianism (SEPAD) Project. His books include *Houses Built on Sand: Violence, Sectarianism and Revolution in the Middle East* (Manchester University Press), *Sectarianism in the Contemporary Middle East* (Routledge), and *Saudi Arabia and Iran: Soft Power Rivalry in the Middle East* (IB Taurus).

**Kimberly Marten** is Professor of Political Science at Barnard College, Columbia University, specialising in international relations, international security, Russia, and environmental politics. She is a faculty member of Columbia's Harriman Institute for Russian, Eurasian, and East European Studies and Saltzman Institute of War and Peace Studies. She is a founding member of PONARS-Eurasia and a member of the Council on Foreign Relations and the International Institute for Strategic Studies. Her books include *Warlords: String-Arm Brokers in Weak States* (2012); *Enforcing the Peace: Learning from the Imperial Past* (2004); *Weapons, Culture, and Self-Interest: Soviet Defense Managers in the New Russia* (1997); and *Engaging the Enemy: Organization Theory and Soviet Military Innovation* (1993).

**Ziya Meral** is a Senior Associate Fellow of the Royal United Services Institute, a Visiting Fellow at the Royal Navy's Strategic Studies Centre, and a Senior Associate Fellow of the European Leadership Network. He is also a Senior Research Associate and Co-Founder, with Timothy Clack, of the Climate Change and (In)Security Project and a member of the Advisory Board of the British Institute in Ankara. He was previously the Director for Research and Programmes at the Centre for Historical Analysis and Conflict (CHACR), the British Army's civilian and military think tank providing independent analysis and research on defence and security based at the Royal Military Academy Sandhurst.

**Richard Milburn** is Director of Research of the Marjan Study Group at King's College London (KCL). His research studies armed conflict and the environment. He completed his PhD within the Department of War Studies at KCL, exploring the links between wildlife conservation, security, peacebuilding, and development, conducting fieldwork in eastern Democratic Republic of Congo. He acts as the UK representative of the Pole Pole Foundation, an award-winning Congolese conservation charity based in Bukavu. He is also co-founder of a social enterprise, Tunza Games, which uses gamification to engage people in conservation and raise funds to support global conservation projects.

**Neil Morisetti** is the Vice Dean (Public Policy) for the Faculty of Engineering Sciences and a Professor of Climate and Resource Security in the Department of Science, Technology, Engineering and Public Policy at University College London. Formerly, Commander UK Maritime Forces and Commandant of the Joint Services Command and Staff College. He was UK Government Climate and Energy Security Envoy and the Foreign Secretary's Special Representative for Climate Change (2009–2013). He is an Associate Fellow of Chatham House, a Member of the Military Advisory Board for the US Think Tank CNA (Washington DC), and an International Fellow of the Australian Security Leaders Climate Group.

**Lieutenant-General (Retired) Richard Nugee** is currently the Climate Change and Sustainability Non-Executive Director at the UK Ministry of Defence. He was previously the Climate Change Policy and Sustainability Lead for the Ministry of Defence until his recent retirement. His career saw service as Assistant Chief of Defence Staff (Personnel and Training) and Defence Services Secretary; Chief of Staff to the ISAF Joint Command in Kabul; Army Director General Personnel; Director of Manning (Army); and the Chief of Defence People at the UK MOD. In addition to being appointed Member of the British Empire (MBE) in 1998, Commander of the Order of the British Empire (CBE) in 2012, Commander of the Royal Victorian Order (CVO) in 2016, and Companion of the Order of the Bath (CB) in 2020, he was awarded the US Legion of Merit for his service in Afghanistan in 2014.

**Matthew Paterson** is Professor of International Politics at the University of Manchester and the Director of the Sustainable Consumption Institute. His research focuses on the political economy, global governance, and cultural politics of climate change. His latest book is *In Search of Climate Politics* (Cambridge University Press) and previous publications include *Climate Capitalism: Global Warming and the Transformation of the Global Economy* (with Peter Newell, Cambridge University Press) and *Transnational Climate Change Governance* (with Harriet Bulkeley and others, Cambridge University Press). He was a Lead Author for the IPCC's Fifth Assessment Report.

**Lauren Risi** is the Director of the Environmental Change and Security Program at the Wilson Center. She works with policy makers, practitioners, donors, and researchers to tackle security challenges related to environmental change and natural resource management. She is also the managing editor of *New Security Beat*. She has authored and edited a number of reports, including 'On the Horizon 2021: Environmental Change and Security'; '21st Century Diplomacy: Foreign Policy is Climate Policy'; and 'Climate, Migration, and Conflict in a Changing World'.

**Louise Selisny** is a Senior Research Associate of the Climate Change and (In)Security Project and an Expert Participant of the International Military Council on Climate Security (IMCCS). She is also a Research Associate at the School of Anthropology and Museum Ethnography at the University of Oxford. A qualified barrister, with experience across the UK MOD, the United Nations, and Amnesty International, she currently supports policy development within a number of defence and security forums, including the All-Party Parliamentary Group for Climate and Security where she is the Secretariat Coordinator. She has a wide range of local governance and stakeholder relations experience in Eastern Africa and Central Asia. Whilst her practical expertise relates mainly to human security and conflict, she maintains a distinct focus on promoting gender equity.

**Alex Tasker** is an ESRC Policy Fellow in International Relations and National Security and a Lecturer in Human Ecology at University College London. He is also a Research Associate at King's College Conflict and Health Research Group (CHRG). He is a dual-trained veterinary surgeon and development anthropologist whose research focuses on One Health explorations of human–animal–environmental relationships in complex settings, including forced migration, climate uncertainty, and conflict.

# FOREWORD

Climate change is an immediate global security threat and one that national governments must take steps to address. Around the world, the indications are clear that climate change is leading to food and water insecurity and risks escalating the number, length, and complexity of armed conflicts, accelerating and multiplying disasters, triggering and scaling migration, exacerbating inequalities and grievances, and weakening international law and governance. Adaptations and mitigations in these areas are, in turn, crucial for the development of climate resilience.

Our collective ability to respond to climate change threats is underpinned and potentially frustrated by a complex web of political, economic, and social realities at home and abroad. The World Bank has estimated that in response the world will need to make significant investment in infrastructure – around US$90tn by 2030. Meanwhile, the UN's Intergovernmental Panel on Climate Change (IPCC) has reported that an annual investment of US$2.4tn is needed globally in the energy system alone through to 2035 to keep temperature rise below 1.5°C from pre-industrial levels. Also relevant is resistance from countries dependent on hydrocarbon exports, a public discourse often characterised by climate misinformation and disinformation, inter-state competition over the minerals required to scale the transition to renewable energy technologies, and the political difficulties of meeting climate ambitions without compromising diplomatic priorities, trade, and supply chains.

The geopolitics of the near-term future will be further complicated by tensions around cross-border water access and management, for example, linked to the Euphrates, Indus, Mekong, Nile, Omo, and Tigris rivers. The escalating acquisition of the world's agricultural lands and food production infrastructure by certain overseas nations will secure access for some at the expense of others. There will also likely be tensions between countries feeling the worst and immediate effects of climate change and those most responsible in historical and present terms for the emissions driving it. This will be associated with tensions between countries that do take action to reduce carbon emissions and those that do not.

Mass migration caused by climate change also has the potential to generate strategic shocks, and, as was seen in 2021 when Belarus propelled migrants into Poland and Lithuania in an attempt to get EU sanctions lifted, it may become increasingly weaponized. There is also a risk that growing climate activism, which to this point has remained largely peaceful, could become violent.

No nation is immune to the security effects of climate change. As Chair of the UK's parliamentary Environmental Audit Committee and All-Party Parliamentary Group for Climate and Security, I'm conscious that the international character of many climate risks holds the potential for driving cascading effects across borders and sectors. The changing oceanic temperatures driving the migration of global fish stocks, for example, will have not only implications for access to and cost of fish, but also transform ecosystems, create wider economic harm, and increase competition for resources. The potential for conflict will also increase as nations – those who previously relied on the fish, those who find the fish in their territorial waters, and those who need to diversify their economic and subsistence options – seek to protect access to diminishing stocks.

As part of the *Routledge Advances in Defence Studies* (RAiDS) series, *Climate Change and (In)Security* offers insight into a range of important thematic areas, including systemic risk, climate intelligence, military preparations and mitigation efforts, and frameworks of entangled and ecological security. It also offers a geographical focus on the polar regions, sub-Saharan Africa, the Levant, and Central America. Together, the contributors illuminate the scale and breadth of the security threats nations and global humanity now face due to climate change. Armed with greater understanding, the right responses can be developed and delivered.

*Climate Change and (In)Security* makes clear that climate change must be seen as a priority security and defence issue. The book is timely and important and demands a broad readership not only in academe but also amongst practitioners and policy makers.

Rt Hon Philip Dunne MP
House of Commons, 2023

# PREFACE

This book has come together during a period of heightening military awareness as to the implications of climate change for defence and security, particularly in the West. Through the Climate Change and (In)Security Project (CCIP), a collaboration between the University of Oxford and the British Army's Centre for Historical and Armed Conflict Research (CHACR), we have taken the opportunity to explore the insecurities created by climate change and how to respond to them and created ongoing conversations between researchers, policy makers, and decision makers.

CCIP has been fortunate that, in the main, the attitude of relevant stakeholders has been at once positive and pragmatic about engagement. For most, the evidence of anthropogenic climate change is abundant and clear. The implications for the threat environment are being registered and considered. Importantly, the conversation, in places, has now turned to the challenges of what to do and how to do it.

It is important to remain mindful that the focus of those in the policy, security, and military sectors can be diverted, understandably, by other priorities and world events. The Russia–Ukraine War has, of course, shifted the dial on overseas commitments, force composition, defence procurement, and much else. Yet the threats from climate change – some baked into and exacerbating current conflicts, including in Ukraine – still require the affordance of significant capacity. As many of the chapters in this book make clear, the clock is ticking. If the first duty of government is to keep its citizens safe and the country secure, then it is crucial to also recognise that the window to mitigate and prepare is diminishing and, as the scale of the challenge grows, costs associated with the response are rising.

Responding to the threats from climate change requires cooperation. Even the world's largest polluters and militaries will be unable to restrain global heating and mitigate attendant threats unilaterally. This book is a testament to a newfound awareness of climate security issues and the value of a cooperative posture. The collaborative attitude of assorted militaries, services, government departments, and NGOs, as well as researchers, is captured in the pages here. As such, we thank the contributors for their insights and commitment to the project. Many of the authors have also assisted CCIP in other ways, including sharing their insights at our conferences and institutional briefing sessions.

We must thank the CCIP team for their efforts. What started out as a small undertaking funded by the University of Oxford has grown into a significant enterprise in no small part due to their collective efforts. Amongst others, the project has also worked with, and supported, the British Army, Defence Science and Technology Laboratory (DSTL), Global Interagency Security Forum (GISF), International Military Council on Climate Change (IMCCS), NATO, Peers for the Planet, Princeton University's Global Systemic Risk Group, Royal College of Defence Studies, Royal Navy, UK Ministry of Defence, UN, and the Westminster Energy Forum. Since December 2022, CCIP has delivered the Secretariat for the UK's All-Party Parliamentary Group for Climate and Security.

We would also like to offer particular thanks to Richard Brewin, Programme Manager of NATO's Climate and Energy Security Section, General Tom Middendorp, Chair of the IMCCS, Major-General (Retired) Andrew Sharpe, Director of the CHACR, Erin Sikorsky, Director of the US Center for Climate & Security, and colleagues at the University of Oxford in the School of Anthropology and Museum Ethnography, Environmental Change team at Reuben College, and Changing Character of War Centre at Pembroke College for their collegiality and support.

SIPRI and Chris Hodder kindly gave their permission for the reproduction of Figure 0.1, IPCC Secretariat for Figures 10.1–10.3, and PJHQ, Sergeant Simon Coakley, and Lance Corporal Joshua Simms for Figure 12.2. Vicki Herring deserves thanks for redrawing the illustrations in the introduction and Chapters 8, 14, and 15. This book has enjoyed the support of Andrew Humphrys and Devon Harvey at Routledge.

The CCIP website and other resources can be found at: https:// cciproject.uk.

Timothy Clack, Ziya Meral, and Louise Selisny
Oxford, 2023

# INTRODUCTION

## Climate change and (in)security

*Timothy Clack, Ziya Meral, and Louise Selisny*

### Introduction

Attention in politics is often afforded to the urgent at the expense of the important. Moreover, defence and security policies and practices tend to focus on the most immediate threats a nation is facing and how those threats can be addressed. Longer-term trends that map across national boundaries might feature in 'horizon scanning' exercises and forecast studies that look at the shaping of the world a few decades ahead, but they tend to be seen as issues 'out there', which might or might not actualise. Thus, insights and recommendations about the future are often set aside in the pursuit of high-level priorities articulated by governments. The security of the present trumps that of the forecasted future.

To some degree, this is at once understandable and rational. There are limits to the resources any nation can deploy; therefore, a state will always use a triage to assess risks and decide on how to structure and use its defence and security capabilities to meet the most immediate, significant, and likely risks. While this is a 'cost saving' exercise in some sense by seeking to ensure precious and finite resources are not spent injudiciously, it is also a 'cost occurring' weakness. A defence and security policy and structure that limits its thinking to today will almost always fail a nation by leaving it unprepared for the challenges and contexts of tomorrow. As threats that were not taken as 'important' yesterday become the dominant threats of today, those responsible for defence and security rush to make sense of what is happening; often assuming a reactionary posture in search of rapid fixes. These 'late in the day' responses are often described by practitioners as 'stable door', 'sticking plaster', and 'impermanent' solutions, but they are, nevertheless,

DOI: 10.4324/9781003377641-1

accepted on the bases of being 'good enough', 'where we are', and within the 'art of the possible'. No charge is directed at practitioners at the sharp end for such delivery but it has long been noted both anecdotally and in Western military doctrine that 'prevention is better than cure' (Houghton 2015; Johnson and Clack 2019, pp. 1, 18–19). Preparedness and management of resource are crucial in order to navigate national security threats and this means thinking about tomorrow, today.

The competition between the urgent and important in the security arena is a long-standing issue. The pattern has been clear across 20th-century defence thinking. In the United Kingdom, the United States, and France, the lead-up to both World War I and World War II, for example, was characterised by dismissals from most political and military leaders of what was ahead and then a frantic rush to adapt and respond to conflicts of unprecedented scale and threat (Kupchan 2020; Stedman 2011; Wapshott 2015). In the 2000s, a similar rush was evident to meet the challenges posed by international terrorism, with substantial mistakes committed in failures to understand the nature of the threat and ways to address it (Cook 2020; Cox and Rosoux 2002; Kilcullen 2016). Since 2014, the invasion of Crimea and then Eastern Ukraine by Russia has borne witness to another historic shift in defence and security posture: the return of geopolitics, inter-state rivalry, brutal invasions, territorial annexations, and the potential for escalatory peer-on-peer conflicts. Within a comparatively short amount of time, the West has gone from a world that was, in historical perspective, stable with a low likelihood of inter-state wars to a world characterised by defence budgets rises, investment into emerging warfare technologies, a new generation of nuclear weapons, and calls from elected officials and military leaders for increases in conventional forces and equipment. Amongst other things, Western publics see wars and conflicts unfolding 27/4 on their TV screens and social media feeds, witness the displacement of millions of people from their homes in Ukraine and in other parts of the world, and face shortages, inflation, and cost of living increases in their daily lives. The relative peace and prosperity that had been taken for granted since the end of the Cold War are now distant. A telos the West assumed was given and shared for all now seems to be abandoned.

But what if the West and others are committing the mistake by allowing the clear defence and security risks posed in the current geopolitical space to consume all of our thinking, planning, and adaptation. Is adequate attention also being given to a threat that is already here and altering our futures? And what if the said threat is not the kind that we can meet through the usual 'late in the day' decision making and action which we have seen characterises defence and security thinking in crises, i.e. a rush to respond only when it becomes the most dominant, present threat? And what if this threat is not only to our defence and security but 'our way of life' and is, actually, the only truly 'existential threat' we have ever faced and one that simply dwarfs any threat that we have hitherto been defending ourselves from?

For many in defence and security policy and practice, climate change might still feel like a topic well outside of their profession, domain, and interests, left optimally to activists, aid, and development agencies, and broader government structures. Such a position, however, would reflect a profound lack of understanding of what climate change is, how it is impacting our world, and what it all means for defence and security. It is these understandings, impacts, and implications that this book foregrounds for both academic and practitioner awareness.

## Impacts of climate change

There is a large body of scientific studies and climate forecasts available. The UK Met Office, for example, leaves no space for ambiguity when it comes to global warming. It notes that '[c]limate change is already having visible effects on the world. The Earth is warming, rainfall patterns are changing, and sea levels are rising ... average global temperatures have risen by more than 1°C since the 1850s ... [and] 2015, 2016, 2017, 2018, 2019 and 2020 were the hottest years ever recorded' (UK Met Office 2022). It further warns that 'the ice in the Arctic is melting fast. It is already 65% thinner than it was in 1975. Late summer Arctic Sea ice area is currently the smallest in at least 1,000 years ... [W]e could see ice-free summers in the Arctic by the middle of this century. When ice sheets and glaciers melt, freshwater flows into the sea'. As well as making the sea-level rise, freshwater also reduces the salinity (saltiness) of the water, which can slow or change ocean currents (UK Met Office 2022). The outcomes of these shifts vary across the world from flood risks in coastal and riverine areas to heavy rains and more frequent storms in some localities and droughts, prolonged extreme heat waves, and forest fires in others. The security implications of these climate outcomes are at once significant and diverse.

The death toll and economic losses caused by natural disasters in the world have increased substantially between 1970 and 2019 and compared to all preceding years on record. According to the World Meteorological Organisation (WMO) and the UN Office for Disaster Risk Reduction (UNDRR) weather, climate, and water extremes have accounted for, '50 per cent of all disasters, 45 per cent of all reported deaths and 74 percent of all reported economic losses. There were more than 11,000 reported disasters attributed to these hazards globally, with just over two million deaths and [US]$3.64 trillion in loses. More than 91 per cent of the deaths occurred in developing countries' (UN 2022). The WMO and UNDRR also note that in the 1970–2019 period, 'of the top 10 disasters, droughts proved to be the deadliest hazard ... causing 650,000 deaths, followed by storms that led to

577,232 deaths; floods, which took 58,700 lives; and extreme temperature events, during which 55,736 died' (UN 2022).

UN Refugee Agency (UNCHR) reports that 'hazards resulting from the increasing intensity and frequency of extreme weather events, such as abnormally heavy rainfall, prolonged droughts, desertification, environmental degradation, or sea-level rise and cyclones are already causing an average of more than 20 million people to leave their homes and move to other areas in their countries each year' (UNHCR 2022). Droughts, heavy rains, and unreliable and fast-changing climate patterns create a further fundamental challenge: food security. The UN World Food Programme estimates that a '2°C rise in average global temperature from pre-industrial levels will see a staggering 189 million additional people in the grips of hunger' and reports that, 'large swathes of the globe, from Madagascar to Honduras to Bangladesh, are in the throes of a climate crisis that is now a daily reality for millions. The climate crisis is fuelling a food crisis' (WFP 2022).

The fact that these outcomes are set to impact some of the world's poorest states with limited capacity to respond to such grave humanitarian conditions within their borders only adds further difficulties for hundreds of millions of people. It also creates and galvanises deep cleavages and fault lines within countries and between communities. Furthermore, as the International Committee of the Red Cross (ICRC 2020) points out, while climate change is not seen as a primary causal driver of conflicts, 14 of the 25 states assessed to be most vulnerable to climate change are mired in conflict, which restricts their ability to mitigate risks posed by climate change and limits their opportunities to receive humanitarian aid. Thus, for such states, the complex relationship between conflicts and climate change forms a vicious cycle: climate change adds additional stresses, challenges, and suffering to people facing violent conflict, which, in turn, hinders their response to, and contributes to, climate change. As the International Crisis Group puts it, '[t]he relationship between climate change and deadly conflict is complex and context-specific, but it is undeniable that climate change is a threat multiplier that is already increasing food insecurity, water scarcity and resource competition, while disrupting livelihoods and spurring migration. In turn, deadly conflict and political instability are contributing to climate change – including through illegal logging' (ICG 2022).

Here it must be noted that climate change is creating liability risks for certain nations, particularly those that are the largest carbon out-putters and/or were the earliest to industrialise. On the basis of advances in attribution science, countries, particularly those in the Global South, are increasingly likely to seek damages from others for loss and damage resulting in economic, physical, and cultural harms (Minnerop and Otto 2020; Otto 2022). Some claims are already reaching court (see Lloyd and Shepherd 2021) whilst the merits of many others feature repeatedly in political and diplomatic discourses.

There is also increasing awareness of how climate change is contributing to conditions that might supplement the emergence of, and exacerbate, violent extremism and terrorism. The Wilson Center's Africa Program has released a study on climate change and its impact on extremism in the Lake Chad Basin, for example, which notes:

> In the 1960s, Lake Chad had an area of more than 26,000 $km^2$. By 1997, it had shrunk to less than 1,500 $km^2$, and further to 1,350 $km^2$ by 2014. Decades of depletion due to climate variabilities have contributed to fuelling insecurity in local communities whose economic livelihoods depend on the lake, leading to a humanitarian crisis across the Lake Chad Basin. Countries surrounding the Lake Chad Basin, including Cameroon, Chad, Niger, and Nigeria, have been confronted with violent extremism and terrorism, in addition to ethnic, religious, and farmer-herder conflicts.
>
> (Frimpong 2020)

Climate change can be an indirect contributor to terrorism, including as an ideological driver and as a means to control communities (Benjaminsen and Ba 2018; Boyd and Henkin 2022). Furthermore, recruitment into Violent and Extremist Organisations (VEOs) under conditions of climate-related deprivation and/or conflict has been indicated in many studies (see Adelphi 2017; Chaturvedi and Doyle 2015; King 2023; Vestby 2014). The Malthusian logic which underpins such models is that deteriorating environmental conditions and sustained population growth are exacerbated by climate change which, in turn, fuels competition over resources and drives recruits into radical groups and political violence. Even if direct causal links between climate change and terrorist recruitment remain debated (see Telford 2020; Raineri 2020), it is clear that conditions of environmental scarcity shape contexts and do not promote stability. The abundance of resources relevant to climate adaptation found in certain areas, for example in the DRC, has been called a 'green curse' due to the associated security challenges around access and extraction (Vinke et al. 2023, pp. 13–15).

The point should be made, in line with Selby et al. (2022, pp. 57–58), that 'scarcity' and 'abundance' are imprecise and relational concepts. Moreover, conflicts can also be driven by local political forces and strategies rather than shifts in availability. The so-called 'blood diamond' conflicts, for example, in Angola and Sierra Leone were not triggered by changes in either supply or demand but were, in large part, caused by local responses to global price fluctuations (Selby et al. 2022, p. 57).

Various VEOs seemingly recognise these links – or at least the emotive resonance of the issues – for climate change and environmental conditions feature in the propaganda outputs of various terrorist and insurgent groups.

In 2002 and 2010, for example, Osama Bin Laden criticised the United States for not signing the Kyoto Protocol and the scale of its 'industrial waste and gases' (The Guardian 2002). After the Abbottabad raid, an open letter for future distribution was located which described the 'catastrophic consequences' of global warming and encouraged people in the United States to 'rise up' (Landay 2016). There are many other attempted mobilisations of the climate/ environment in VEO propaganda. In 2019, Al Qaeda in the Arabian Peninsula (AQAP) repaired wells to win hearts and minds and achieve a propaganda effect in various parts of Yemen (Bodetti 2019), in 2018, al Shabaab put a prohibition on plastic bags and the logging of certain trees which is still in force in the territories the group governs today (Binding 2018) and, in 2017, the Taliban encouraged Afghan farmers to plant more trees because of their 'important role in environmental protection' (BBC 2017).

There are myriad security issues relating to climate change, including food insecurity, large-scale displacement, natural disasters, and violent conflicts. Given the limited ability, or will, of most states to significantly mitigate the risks of climate change, the world is highly likely to be entering an unprecedented era of instability at a global scale, with a deepening rift of conditions and living standards between those in wealthier nations and those living in the Global South. This poses substantial moral questions and duties, as while we share the same planet facing the same conditions, respective experiences and capacities to respond to them and maintain meaningful futures differ dramatically (Brooks 2022; O'Brien et al. 2022). Even if the forecasted climate contexts and ability to adapt and mitigate are comparatively better for wealthier nations, it must be recognised that the security problems 'out there' in time and space seldom stay distant. As such, even when they might not feel so now, ultimately, they will in some form reach all of us. Understanding and preparation will be crucial.

## Implications for defence

The breakdown of outcomes noted above will have already given the reader a sense of why climate changes pose substantial questions for defence and security. Breaking down areas of defence and security helps to further contextualise implications.

Defence planning and operations describe 'three levels of warfare' – strategic, operational, and tactical – which 'provide a framework within which to rationalise and categorise military activity' (UK MOD 2014, p. 19). The strategic level refers to policy-level decisions taken by a government to establish goals and allocate resources in their pursuit. The operational level refers to the level of military activity which plans, prepares, and coordinates military operations in pursuit of strategic aims. The tactical level refers to the front line of military operations: the level at which military units operate in their missions, may it be fighting against an adversary, undertaking humanitarian activities, or a mission to support local authorities in national emergencies. These levels reflect both an analytic framework and an organisational one in how defence thinking processes and responds to challenges and acts on policies set by a national government. Unsurprisingly, given the enormity and complexity of the challenge, climate change has direct implications for all levels of warfare.

### Strategic climate

At the strategic level, climate change intensifies the trends that have been observed for a decade. Within a context of renewed geopolitical competition, particularly involving the world's major and medium powers, shifting geographical conditions present opportunities as well as direct national risks. Discussions over the control of previously inaccessible mineral and energy resources in, for example, the polar regions, as well as the opening up of new trade and transportation routes have been continuing for well over a decade and have intensified in light of the speed of global heating (Jardine 2022; Mager 2009; Ostreng et al. 2013; Stephenson et al. 2013). Seizing such opportunities, or denying them to another country, will always be part of the state's strategic calculus and informed by competition, cooperation, and contexts. However, whatever one sees as a positive emerging from physical changes in the world, the insecurities and vulnerabilities escalated by changes to climate far outweigh them.

As a recent United States assessment makes clear geopolitical tensions are rising over how to accelerate reductions in greenhouse gas emissions, control of resources and transitional technologies, and climate financing (US ATA 2023, p. 22). This is in addition to the strategic impacts of climate change, which include, for example, the degrading of economies, agricultural regimes, and food production capacity (2023, pp. 22–23). The risk of capital flight is also highly relevant. Financial regulation, for example, is increasingly forcing banks and insurers to undertake impairment-based risk modelling in order to understand the stress on their solvency margins over time. This exercise is already leading to a review of existing capital allocation liabilities and future capital allocation decisions. Climate impacts are, in short, shaping lending decisions, lending rates, and insurance coverage. It is highly likely that due to climate risks, adequate capital may not be allocated to certain locations in the future. This will have a destabilising impact on countries and locations over time. It is also likely that, in some contexts, financial decisions will create and compound insecurities before the physical impacts of climate change become obvious.

There has already been an increased focus on preparing militaries for disaster relief operations. The UK MOD's 'Disaster Relief Operations Oversea: The Military Contribution' doctrine, for example, starts with this important recognition:

Over the next 30 years urbanisation, population growth and climate change are all likely to contribute to greater numbers of people inhabiting areas that will be at significant risk of environmental disaster. This is particularly so in areas susceptible to volcanic and seismic activity and in low-lying coastal regions where extreme weather events, such as tropical cyclones and flooding are likely to become more prevalent. Droughts and heatwaves are also likely to increase in intensity, duration and frequency. Some of these events could precipitate natural disasters which, because of the interdependencies enabled by globalisation, may have consequences far beyond the site where the disaster occurs (for example, the spread of disease due to poor sanitation). These circumstances will almost certainly result in an increase in humanitarian crises throughout the world, the response to which may require the UK military's involvement.

(UK MOD 2016, p. 3)

In fact, the UK military is utilised increasingly in disaster relief missions. In 2017, Hurricane Irma, a Category 5 storm caused substantial damage across the Caribbean. The UK response involved two Royal Navy ships, nine helicopters, and, 'at its height, 2000 military personnel were deployed to the region alongside 40 UK aid experts, 50 UK police officers, 40 prison officers and 80 FCO staff' (House of Commons 2018, p. 5). In 2020, the United Kingdom sent Royal Fleet Auxiliary (RFA) ship Argus to support efforts to respond to Hurricane Eta (GOV.UK 2020) and United Kingdom forces in Belize for an exercise altered their deployment in order to support relief efforts in the country (British Army 2020).

The strategic implications of issues caused by climate change do not end in the utilisation of military assets in disaster relief. Intensified conflicts across the world, displacement, irregular migration, and disruptions to trade and food production no matter how distant their locality can and do have direct security impacts on the West. Such situations create more human suffering, deepen instabilities, amplify political division and pressure, and demand financial, practical, and diplomatic commitments.

There are already well-established links reported between so-called 'ungoverned spaces' that see extremist groups flourish amidst dire economic and environmental conditions, with global security risks caused through international networks and resonance of ideological messaging (above). One can imagine scenarios in which there will be a new generation of UN-lead peace-keeping missions to parts of Africa, Asia, the Middle East, South America, and other 'climate conflict hot spots' (Clack and Nugee 2022), for example, if not further examples of powerful nations seeking to address terror risks posed by extremist groups by conducting unilateral military operations. Given the recent experience of the COVID-19 pandemic (Hameiri 2021; Ide 2021), it is important to recognise the risk that conflict

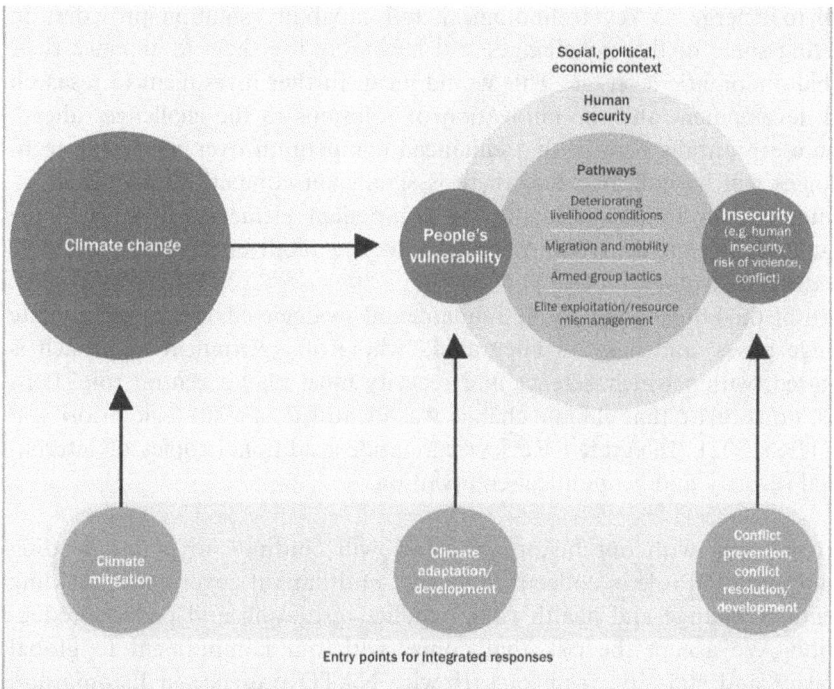

**FIGURE 0.1**  Entry points for integrated responses.

*Credit:* SIPRI and Christophe Hodder.

actors seek to escalate their activities in such contexts of scarcity and that increased food insecurity and droughts will usher conflicts and tensions over supply chains and resources beyond the precious commodities that might or might not be possible to extract in the Arctic, Antarctic, and elsewhere.

For governments around the world, the core strategic challenge with climate change will be to mitigate these risks, support efforts to address them, and find resources for the structures and technologies required to respond to them. For all, this is a strategic challenge of mitigation and adaptation. The UN's climate security work in Somalia, for example, has identified four climate security 'pathways' and three strategic 'entry points' for integrated responses (Figure 0.1).

Interventions in the four pathways – livelihood deterioration, migration and mobility, armed group tactics, and elite exploitation – can take the form of climate mitigation, climate adaptation, and conflict prevention and resolution (Broek and Hodder 2022, pp. 5–6). Thus, economics, political structures, and human agency are all viable and deployable strategic instruments.

One strategic competition linked to climate change that we can perhaps welcome as a positive will be how increased numbers of wealthier nations will seek to emerge as key technological and capability solution providers in meeting some of these challenges and inevitably use them to enhance their global diplomatic leverage. This would mean further investment in research and development and the innovation of solutions to the challenges ahead. (There are pitfalls here in that enhanced competition over renewable technologies will be indicated and there is significant competitive and environmental impact in the extraction of component elements. Moreover, the singular focus on decarbonisation can reduce biodiversity and negatively stress the hydrological cycle.)

All of the above point to the fundamental strategic challenge that climate change poses, and how an integrated, 'whole of government' approach is required, within which defence and security must play a central role. It is, thus, no surprise that climate change was identified as a strategic priority in the UK's 2021 'Integrated Review' alongside traditional topics of international security and geopolitical competition:

> In keeping with our history, the UK will continue to play a leading international role in collective security, multilateral governance, tackling climate change and health risks, conflict resolution and poverty reduction. We accept the risk that comes with our commitment to global peace and stability, from our tripwire NATO presence in Estonia and Poland to on-the-ground support for UN peacekeeping and humanitarian relief.
>
> (UK Government 2021, p. 11)

It is important to note here that the 2023 refresh of the 'Integrated Review', a process catalysed by the security landscape having been dramatically altered after the Russian invasion of Ukraine, maintained a focus on mitigating the multiplicator threats from climate change:

> [C]limate change and biodiversity loss are important multipliers of other global threats, and are guaranteed to continue to worsen over the next decade ... The consequences are both acute and chronic, resulting in a sharp increase in global migration and the number of people in need of immediate humanitarian assistance.
>
> (UK Government 2023, p. 10)

> [W]e will continue to strengthen the UK's resilience to the range of interlinked risks associated with *climate change and environmental damage*. This is firmly linked to the international agenda on climate change, biodiversity loss and sustainable development.
>
> (UK Government 2023, p. 47, emphasis in original)

### *Operational climate*

At the operational level, climate change poses substantial questions for the preparation and management of military operations and facilities. US Secretary of Defense, Lloyd J Austin III, noted that 'the Defense Department has been impacted directly in just the past few years by extreme weather caused by climate change' (Vergun 2021). He also provided a series of examples, including the damage from Hurricane Michael on Florida's Tyndall Air Force Base, damage caused by the flooding of the Missouri River on Nebraska's Offutt Air Force Base, repeat evacuations of military installations in California due to wildfires, and Typhoon Wutip interruption of a joint exercise involving the United States, Australia, and Japan in Guam (Vergun 2021).

Rising temperatures and declining air density will have a direct impact on the design of military planes, their load capacity, and the runway lengths they need to be able to take off and land (Coffel et al. 2017). Without adequate fixed-wing capabilities, modern militaries simply cannot function or operate as they serve as a basis for logistic support in operations in remote locations, offer overwatch protection, operate as signals and surveillance platforms, and deliver kinetic effect at reach against an adversary. Either the planes and the infrastructure that facilitates their use are redesigned or a significant loss of capability will be evident in the future. A similar revision of capability challenge can be recognised linked to naval vessels. Not only will more frequent and stronger storms make operating more arduous and conditions harder to forecast, but there have also been concerns over warming waters and viability/performance of ships' engines (Norton-Taylor 2016).

Naval infrastructure, given its largely coastal positioning, is directly vulnerable to climate change. Indeed, the US Navy in Alaska, for example, 'is being forced to rebuild and relocate roads, buildings, and airfields as the permafrost melts, and it might eventually have to relocate some of its bases' (Reinhardt and Toffel 2017). Given that the US Navy is estimated to have, '111,000 buildings and structures on bases and other installations, located on 2.2 million acres around the world', the economic costs of such upgrades and relocations could amount to hundreds of billions of US dollars (Reinhardt and Toffel 2017). Such upgrades would also take time and political and military focus and consume human and equipment resources.

Climate change also has implications for the navigation, communication, and tracking systems used by militaries. This has been recognised as being particularly relevant in the polar regions where technical challenges are multiplied and an intensification of military tempo will require an enhanced degree of precision (see Naval Studies Board 2011, pp. 94–113).

The examples listed above highlight the substantial costs and challenges on the horizon for militaries. Without adequate and safe equipment, infrastructure, supply chains, and transportation, there is simply no way defence can deliver optimally any national strategic priority, support global humanitarian assistance missions, or maintain deterrence against adversaries. Climate change is going to demand more from defence operationally than the current budget and planning frameworks can accommodate.

### Tactical climate

At the tactical level, military personnel will struggle in the face of certain conditions, particularly over longer timeframes. Western militaries are required, of course, to operate to deliver a complex series of missions, including warfighting, projection of force, holding of territory, deterrence, stabilisation, and peace-keeping operations, security of key infrastructure, humanitarian and other evacuations, and the transportation and distribution of aid. To deliver this mission set, military personnel are often tasked to live in remote localities with extreme weather conditions. This presents challenges in relation to maintaining military skills and equipment, not least when prolonged gaps between resupply or support are likely.

The physical health of military personnel facing increased temperatures and extreme weather conditions cannot be overlooked (Hasemyer 2019). Future operating conditions will affect what personnel wear, what they eat, where they are based, when, and how, they operate, and what logistical, medical, and wellbeing support systems are in place for them. The possibility of new pandemics and rapid epidemiological contagion, with the potential to impact large numbers of personnel swiftly and with consequences for force size and readiness, should not be overlooked. Military installations, whether at home station or deployed, are often crowded and personnel have frequent access to domestic and global populations. Given the efficiency of transport and operational reach, militaries can be unwitting carriers of diseases around the world and any failure to operate will hinder warfighting outcomes and the crucial role militaries play in the response to large humanitarian crises (e.g. US Army and Spanish Flu; see Byerly 2005).

Personnel and others may also face increased risks from poorly transported, stored, and maintained ordnance, likely exacerbated if they are engaging with militaries and partners with obsolete equipment or operating in theatres replete with older mines and ammunition stocks (Schwartzstein 2019). Climate change conditions also require adaptability of military and security forces to a broad range of skills and preparation for a broad range of operations. The skills, training, and professionalisation that are needed to meet a humanitarian crisis with vast numbers of people in need of shelter, food, and relocation clearly differ from those required to conduct a peace-keeping mission in a theatre where communities are in conflict over resources or deter an adversary that might be threatening national assets and boundaries. On this basis, one can legitimately argue that states need new rapid response structures beyond militaries to be ready for the long-term challenges posed by climate change. Indeed, it can reasonably be questioned as to whether seeing militaries as 'swiss army knives' meeting every challenge is a sustainable or appropriate governmental impulse (Lindsay and Lischer 2003; Rolfe 2011; cf Lischer 2007).

## Human security

Given the renewed focus of Western militaries on human security in recent years (Stoltenberg 2021; UK MOD 2021c), the significant threat climate change poses to human security must be emphasised. The concept of human security is also used by the United Nations Intergovernmental Panel on Climate Change (IPCC) in order to integrate relevant socioeconomic and development factors into climate change security considerations (Adger et al. 2014). The UN's 'human security approach' presents five fundamental principles that should steer decision making. Security considerations should be: people-centred; comprehensive; context-specific; prevention-orientated; and facilitate protection and empowerment.

According to UK military doctrine, 'human security is an approach to national and international security that gives primacy to human beings and their complex social and economic interactions' (UK MOD 2019, p. viii). Around the world, human security will almost certainly be affected by climate-related threats to agricultural regimes, including increased pest and disease presence, spikes in food prices, and shocks to food production and food logistics (see Wellesley and Benton 2022). The recalibration of diplomatic alliances, displacement and dispossession of peoples, border disputes, endemic famine, and warfare will also, of course, precipitate violations of human security.

The threats from climate change to human security are not only faced disproportionately in certain parts of the world but also within societies. In certain communities, where there is a lack of access to education, economic opportunity, and land ownership, the domestic workload of women and girls is amplified by climate change, for example in collecting water and provision of agricultural labour. Climate change amplifies gender inequality, which is strongly associated with inter-state and intrastate instability and conflict (Hudson et al. 2021). It has also been pointed out in this regard, that globally, climate change and security institutions tend to be male-dominated environments, with consequences for framing and interventions (Magnusdottir and Kronsell 2021).

A number of studies have highlighted importantly how climate change insecurity exacerbates gender-based violence (e.g. Harville et al. 2010; Nguyen 2018; Thurston et al. 2021; Van Daalan et al. 2022). The realities of travelling further to get water, for example, make women more vulnerable to sexual assault. Moreover, there is evidence of girls being traded for resources between households during times of peak scarcity, thereby increasing their vulnerability to sexual exploitation and trafficking. There are other negative implications felt most keenly by women. In water-scare and impoverished environments, girls are also married off at increasingly younger ages so as to transfer the 'burden of provision'. Van Daalan et al. (2022) report spikes in early marriage in Bangladesh coinciding with the floods of 1998 and 2004. In many contexts of subsistence stress, girls also tend to be the first to be withdrawn from education to engage in domestic and agricultural chores, such as firewood collection. Women also have less agency in defining adaptation strategies, with customary roles excluding them from subsistence decisions, e.g. in cattle-herding communities, men manage the cattle and their movement whilst women do the milking.

The blame for natural disasters, failed harvests, and other climate effects is often directed at specific demographics, e.g. women for perceived immorality, youth for a failure to perform customary rituals, and those from other ethnicities or clans for misappropriating resources. It would be a mistake to consider this to characterise only traditional societies. After Hurricane Katrina hit New Orleans in 2005, for example, some described the disaster as a 'divine punishment', with the gay community in their sights. As Van Daalan et al. (2022) note some same-sex couples and transgender people were refused aid from the Federal Emergency Management Agency, prevented from entering relief centres, and experienced physical and sexual violence in shelters. These incidents highlight the greater human security risks that will be faced by certain demographics as climate change grips. The human security capabilities of militaries will need to be conscious and prepared for such realities.

## Causation and complexity

The intersection of the myriad national and human security risks amplifies the scale and complexity of the climate change threat. This and other linkages and consequences can be illuminated through the climate security chain of causation (Figure 0.2). This begins with the 'climate' system itself and works through the 'causes of climate change', the 'effects of climate change', 'climate change hazards', and then on to the 'primary (natural systems) impacts' and the 'secondary (human systems) impacts'. The chain then moves to 'compound cascade risks', which have the potential to interact with and amplify each other, resulting in political and then defence and security implications. In short, climate change constitutes a global crucible that reshapes and redefines risk so as to multiply, magnify, and intensify threats. Whilst the chain implies a unidirectionality, it must be stressed that feedbacks, amplifications, and tipping points are also evident. In the cascade, the

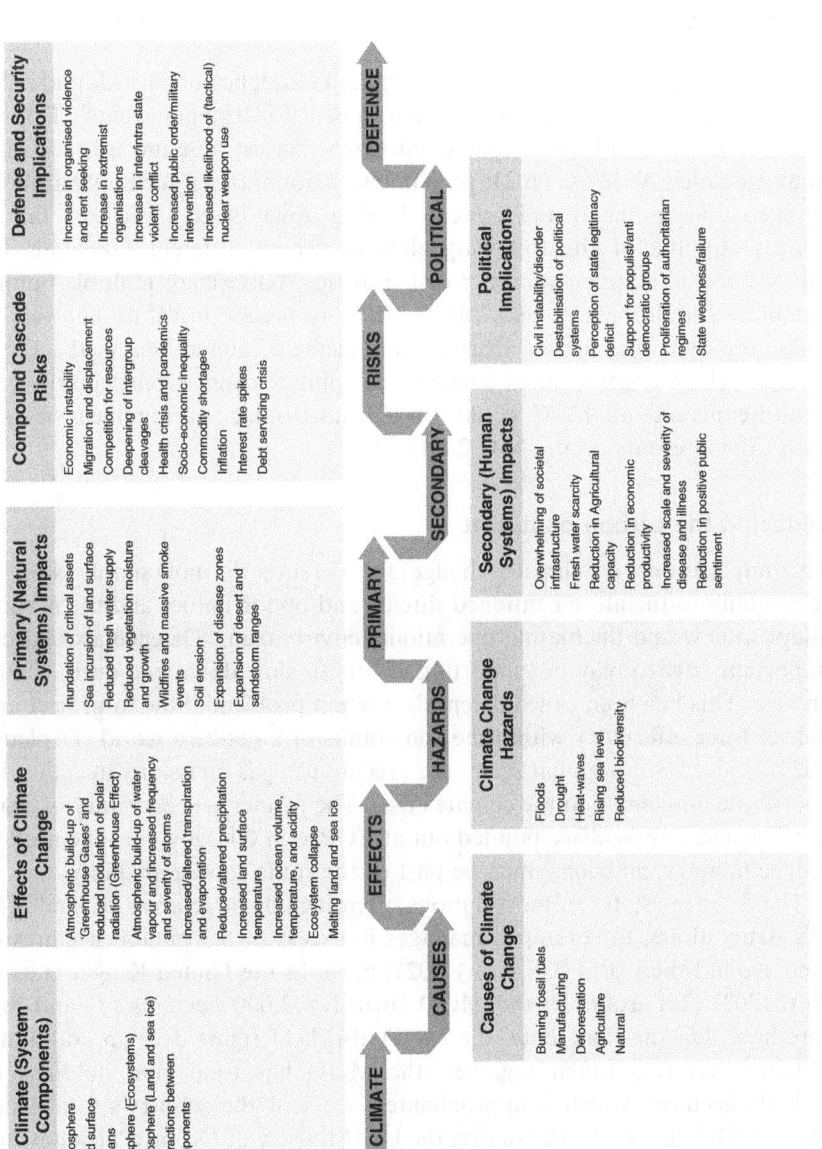

**FIGURE 0.2** The climate security chain of causation.

*Source:* Authors' creation.

frequency of hazards, shocks, and humanitarian disasters increases. This means that there is less time between one crisis and the next, and this, in turn, serves to further undermine societal infrastructure, overwhelm any systemic resilience, and compromise other human systems' necessities, such as agriculture.

As well as increasing the scale, severity, and frequency of hazards and risks, climate change creates the perfect environment for 'tipping events'. Tipping events can be broadly categorised into two phases: 'domino cascade' and 'joint cascade'. As Klose (2021, p. 1) notes, 'a domino cascade may hardly be stopped once initiated and critical slowing down-based indicators fail to indicate tipping of the following element. These different potentials for intervention and anticipation across the distinct patterns of multiple tipping dynamics should be seen as a call to be more precise in future analyses of cascading dynamics arising from tipping element interactions in the Earth system'. The US government concurs that climate change, particularly temperature increase after 2°C, could cause, 'catastrophic and nonlinear' risks if such tipping events occur (NIC 2021).

## Reducing the carbon bootprint

The implications of climate change for defence do not simply stop at responding to the aforementioned threats and opportunities and the need to adapt to new and fluctuating operational environment. Defence also has an important part to play in supporting efforts to slow down and curb climate change. This has been called defence's 'carbon problem': how to project and deliver force effectively within the constraints of a net-zero world (Depledge 2023). Indeed, on a global scale, the greenhouse gas emissions of militaries contribute immensely to the climate crisis. The former UK Secretary of State for Defence, Ben Wallace pointed out at COP26 in Glasgow that the need to reduce military emissions must be part of the route to sustainability.

The defence estates of most nations often include large areas of land. The US Army alone, for example, manages in excess of 5.3 million hectares of land around the world (US Army 2022, p. 8). In the United Kingdom as of April 2021, for example, the MOD owned 233,000 hectares of land and foreshore. On the same date, the MOD also held rights over an additional 111,200 hectares. Taken together, the MOD has total land holdings of 344,200 hectares, which is approximately 1.5% of the country's total land-mass (MOD 2021b). Furthermore, the UK Ministry of Defence 'accounts for 50% of the UK central Government's emissions' (UK MOD 2021a, p. 6). With land holdings and emissions of such scale, defence has a moral and practical duty to incorporate climate change calculations into how it manages and runs its infrastructure, and what kinds of agreements it enters with farmers and other land users.

Beyond the question of defence estates, militaries consume supplies at great scale. These supplies, including everything from rations and clothing to fuel and equipment, are purchased from the global marketplace. Existing in a feedback loop, climate change impacts military supply chains, and these supply chains impact climate change. Consequently, the US Army has noted that it is time to ask 'suppliers to further reduce both embodied emissions and the impact that supply chain activities have on the climate' (US Army 2022, p. 13). In adopting a 'Buy Clean' procurement posture, the US Army intends to 'lower embodied carbon emissions from manufacturing, transportation, installation, maintenance, and disposal sub-processes' in order to reach net zero in its procurements by 2050 (2022, p. 13).

In some respects, the defence sector has taken some important steps. The British Army, for example, has invested in prototype electric hybrid armoured, reconnaissance, and logistic vehicles, with significantly reduced emissions and improved performance (British Army 2021). The electric trucks that transport a field hospital can now supply power for up to 12 hours, providing the equivalent of nine diesel generators. New buildings on the UK's military's training estate are also net negative, supplied from renewable sources such as anaerobic digesters and solar farms (DIO 2021). The other military services in the United Kingdom have innovated similarly in this space. The Royal Air Force recently achieved a world-first flight powered by 100% synthetic fuel (Chuter 2021), authorised the use of 50% sustainable aviation fuel in all its aircraft (UK MOD 2020), and plans to order electric-powered planes for training (Rufford 2021). The Royal Navy, meanwhile, is incorporating alternative-fuelled sustainability into new ship design. There is still much that is required, of course, and for some commentators, the (limited) progress to date has been described as 'military grade greenwashing' (Kitchlew 2022). Reputation and legitimacy are also at stake.

## Reputation

There are reputational risks from climate change as countries respond, or do not, to crises and the demands of their allies, partners, and publics. In addressing their carbon bootprint, militaries enhance their role in sustainable security. However, they can also become agents of 'climate diplomacy' insomuch as they can influence positive change in other nations (overseas) and government departments (domestically). This might take the form of soft power influence with dividends for reputational security. Cull (2022) makes the point that the dual drivers of the COVID-19 pandemic and the return of Great Power competition have jolted the Western world back into a focus on reputational and collaborative statecraft. Climate shocks and concomitant responses are likely to offer similar 'jolts' and precipitate further collaboration.

At home and, at times, abroad, those militaries and governments that are seen to be acting responsibly and urgently to climate considerations are going to be viewed more positively, with implications for support in other lines of activity. These enhancements to the military's public legitimacy can become exaggerated amongst certain parts of the population. The lens of human security has highlighted how certain demographics face a higher psychological and physiological burden linked to climate change (above). Moreover, younger people are indicated to have greater sensitivity to climate change concerns, with global youth movements demanding of urgent climate action (Murphy 2021; Sanson and Bellemo 2021; Von Storch et al. 2021). Most prevalent in younger generations, a trend of eco-anxiety has even been described which in its most extreme forms has been linked to depression, socio-ethical paralysis, and loss of wellbeing (Panu 2020; Verplanken et al. 2021). This matters most directly in the areas of military recruitment and retention. To attract recruits, NATO militaries focus on brand management and target youth populations at institutions of further and higher education (see NATO 2007). The relevance of climate in this context should not be overlooked.

Whilst narratives and evidence of military leadership and action on climate issues are likely to produce positive dividends, the absence of such generates risks. In the era of subthreshold and information warfare, narratives and resonance are instrumentalised to create cohesion and influence, but also division, discord, and political paralysis (Clack and Johnson 2021). The West has been the target of a lot of content which portrays it as the perpetrator of past and present climate crimes (Daily Mail 2012; Dickie 2007; Nelson 2009). Whether these are accurate or not is moot if the narrative is compelling and resonates an 'emotional truth' to audiences. The more visible militaries are in their response to climate change, the less traction will be afforded to alternative representations.

### Securitisation

The climate conflict nexus is now accepted as a threat by a range of militaries and governments, including the US Biden administration. Amplifying this focus with section 103(c) of Executive Order 14008, the US Secretary of Defense is required to develop, 'an analysis of the security implications of climate change (climate risk analysis) that can be incorporated into modelling, simulation, wargaming, and other analyses' (US DOD 2021, p. 2). However, for the reader approaching this book from outside of the research, policy, and practice of defence and security, thinking through climate change from a defence lens might feel uncomfortable. After all, there is a track record that can be pointed to as to how the securitisation of a humanitarian or political issue has resulted in misfocused debates and unconstructive security outcomes (Foley 2010; Kivimäki 2019; Kuperman 2001).

Securitisation is a concept particularly deployed in critical security perspectives to capture how the political utilisation of an issue as a security threat can be for illegitimate reasons and can result in second-order effects and unintended consequences. Moreover, security frameworks and resources can be hijacked for other agendas, all rights and norms suspended, and security language used to promote fear and political mobilisation. Studies have shown how political agendas that evoke security responses can actually create more insecurities, lead to deterioration in human rights standards, marginalise vulnerable communities and individuals, and hinder finding long-term solutions. One can posit here a myriad of examples across the world of 'moral panics' over refugees, asylum seekers, minority religions, ethnic conflicts, political dissent, and pandemics represented as terrorism and existential threats (e.g. see Caballero and Emmers 2006; Futak-Campbell 2021; Gozdecka 2021).

The securitisation of climate change remains a contentious issue and has been criticised heavily (e.g. Buhaug et al. 2015; Nordas and Gleditsch 2007; Salehyan 2008). The semantics of key terms, such as 'scarcity' and 'resource', also continue to engage scholarly focus (e.g. Fairhead, 2000; Gleditsch 1998). Buxton and Eade (2021) outline a range of issues with the climate security framework, including the linkages between the state, militaries, and the oil and extractive economies, profiteering across the military-industrial complex, shaping of other agendas such as food and water management, and problematic assumptions around climate-induced conflict and migration. Furthermore, Crawford (2022) has described how the US economy and military, in particular, have fashioned a long-term cycle of growth, fossil fuel use, and dependency. In her anti-securitisation approach, Crawford offers the provocation that, in light of the primacy of the threat from climate change, the United States should rethink its grand strategy and, in doing so, reduce the size and operations of its military (Crawford, 2022, pp. 286–290).

Interestingly, whilst complaints are directed at the military over the securitisation of the climate change response, some senior military leaders have expressed concern about the effectiveness and readiness of military forces being reduced because they have had to – or at least be prepared to – respond to environmental emergencies. This is understandable in light of anxieties about the size and shape of current and future forces and the diversification of their roles often with fewer available resources.

Defence must remain mindful of the issues of securitisation and capacity. However, discussing the implications of climate change for defence and security, highlighting the need for defence and security policies to incorporate climate change outcomes into their planning and preparations, and urging defence and security structures to contribute to efforts to curb emissions should not be seen as 'securitisation' in the pejorative sense. The fact that there is relatively little scholarly attention to this area should cause us to pause and reflect. Defence and security apparatuses are not the solution to climate change nor can they change the human activity that leads to it. However, it is undeniable that as central and sizeable facets of modern states, they contribute to the problem and offer opportunities for the state to reduce its carbon footprint. Moreover, defence and security have, or can develop, the tools to ameliorate negative impacts through conflict and disaster response. There will be other roles as well and dialogue between researchers from assorted disciplines and defence and security policy-makers and practitioners is needed more than ever. Put simply, climate change is a direct threat to human existence and brings with it substantial challenges that no state, or no single mechanism, no single campaign, no single approach, and no single area of focus can address in isolation.

### The structure of the book

This book intends to start a conversation between researchers and the defence and security sectors. Further, it intends to build bridges between scholarly work and practitioner requirements and help defence and security thinking in the West and globally to move beyond the 'here and now' of immediate threats and look at, and prepare for, the future. This future is already being felt in certain parts of the world, including in the West and countries of Western strategic interest. Understanding, preparation, and collaboration are vital – the challenges are beyond the capabilities of any state to singularly address.

The book is divided into three sections. Composed of seven chapters, climate security contexts are the focus of the first section. In Chapter 1, Tim Benton, Neil Morisetti, and Oli Brown describe cascading risks through the lens of food security and displacement. The connectivity and co-dependencies of factors in what might be called the climate threatscape are illuminated. The authors posit that building resilient trade and its governance offer optimal prospects of reducing vulnerability and exposure.

Chapter 2 is written by Kimberly Martens and Chapter 3 by Samuel Jardine and Timothy Clack. These chapters explore the shifting geopolitical and security situations in the polar regions linked to climate change. In both areas, climate change is happening alarmingly fast. In the case of the High North, all eight Arctic states (Canada, Denmark/Greenland, Finland, Iceland, Norway, Russia, Sweden, and the United States) are impacted, and the ambitions of two authoritarian governments have been fuelled: Russia, which hopes to use its increasingly ice-free Northern Sea Route (NSR) for industrial and military purposes; and China, which hopes to insert itself into the region through its Polar (Ice) Silk Road. Martens describes three political themes common across the Arctic. Firstly, Arctic states share an interest in upholding norms of state sovereignty. This is likely to limit both regional territorial aggression and China's external ambitions. Secondly, the region faces investment uncertainty associated with 'stranded assets': products and their associated infrastructure that no one will want to possess in a net-zero future. This may limit regional oil and gas development and also curtail the grand strategic plans of various nations. Thirdly, indigenous peoples are facing threats and opportunities. Some (as in Greenland or in Russia's Norilsk) are becoming empowered, while others (as on Russia's Yamal Peninsula) face increasing repression, with implications for livelihoods and human security.

Similarly, in the Antarctic and South Atlantic, the prospects of increasingly cost-efficient and greater-scale human and economic activity loom ever larger. As the Antarctic Treaty opens for renegotiation, Jardine and Clack assess that this will exacerbate the tensions facing an increasingly fragile diplomatic status quo with states seeking to reconfigure the regulation of physical presence, mining, bioresources, tourism, and other uses. Longer-term, the potential for larger-scale persistent human habitation along the Antarctic coast will further complicate the situation. It is thus highly likely that the state of frozen competition over historic, but maintained, claims will end.

Chapter 4 by Ana Kumarasamy and Simon Mabon considers the security politics of climate change in the Levant, a geographical context which is experiencing an increased tempo and intensity of droughts, desertification, forest fires, and sandstorms, with consequences including internal displacement and rising violence. The authors describe the state responses to these issues and how they tend to be characterised and shaped by corruption by governmental officials and the regulation of the inhabitants rather than attempts to address the underlying environmental and human security implications. In concert with other challenges, such as economic deterioration, COVID-19, and political exclusion within the region, this has resulted in mass protests, for example, in Basra in Iraq in 2018 due to an acute water crisis and in Beirut and Mount Lebanon in Lebanon in 2019 where the protests were initially ignited by forest fires. The chapter argues, compellingly and with implications for other areas, that the devastating repercussions of climate change across the Levant pose an array of challenges to regional political life and organisation both within and between states and that the intersectional nature of climate change disproportionately affects those already living in precarious conditions.

Tom Deligiannis, in Chapter 5, describes the important differences between sudden-onset and slow-onset climate risks and contends that slow-onset climate risks are not suited to military intervention but are, in practice, better left to development agencies, national bureaucracies, and state-building assets. This is on the grounds that the solution to slow-onset risks is the alleviation of the vulnerabilities faced by households, communities, and states through improvements to livelihoods and resilience. Deligiannis asserts that climate change needs to be decentred from our analysis of human–environmental conflicts, and contextualised within the range of processes and transformations people impose on nature. He also outlines an adaptive learning approach to research and policy which deals with the slow-onset consequences of climate security risks.

There are parallels in Deligiannis' argument concerning response to slow-onset climate risks with criticisms directed at (Western) militaries for their assorted state-building, governance, and development projects to win 'hearts and minds' linked to, for example, the recent counterinsurgency wars in Iraq and Afghanistan (Egnell 2010; Murtazashvili 2016). That noted, the case for military involvement in disaster and humanitarian relief – sudden-onset climate risks – is, in general, stronger due to the military's reach and rapid response capabilities. However, it could also be argued that, even here, there are issues, such as the strain on assets. There is a case to be made for the creation and maintenance of rapidly deployable, non-military disaster response organisations and capabilities.

Chapters 6 and 7 also concern social and humanitarian risks. In Chapter 6, Lauren Risi emphasises scale through her discussion of cascading impacts, compound risks, and tipping points. She notes that in light of unprecedented scales and risks, long-established institutional frameworks for addressing threats to national security and global stability are no longer fit for purpose. In response, Risi proposes a new framework developed by the Wilson Center in partnership with the National Oceanic and Atmospheric Agency (NOAA) that provides a common conception of the security threats posed by weather and climate-related disruptions. Through an exploration of regional and country-specific cases where the connections between climate disruptive events and security challenges are evident, her chapter lays out four sets of vulnerabilities – or 'tipping points' – through which climate impacts can create cascading impacts and compound risks, and how applying this framework can reveal new entry points for more effective climate and security responses.

In Chapter 7, Matthew Paterson forces the recognition that climate change is a social as well as a natural phenomenon. The conceptualisation of climate in terms of either collapse (in social organisation, governance systems, and food systems, for example) or transformation (in the energy, agricultural, and construction sectors, for example) both come with significant attendant security considerations. The framing of collapse, of course, captures the existential nature of the threat and forecasts a future of large-scale food scarcity and mass displacement (see Centeno et al. 2023). Often under-theorised is that the transformation frame also has deep implications for security institutions given their current widespread dependence on fossil fuels for operational effectiveness, as well as the historical importance of the military to economic and technological innovation across societies.

In relation to Paterson's contribution, sociological work on implicatory denial is worth highlighting (e.g. Cohen 2001; Norgaard 2019; Wullenkord 2022). This is a 'coping strategy' of sorts, which effects individuals and groups alike, and involves the recognition of a problem but the denial of its consequences. The denial of ramifications is of short-term convenience in situations such as climate change and, insidiously, can cause paralysis of policy and response. The climate security challenge is enormous, the costs vast, and effects at once complex and frightening for both decision makers and publics. Moreover, the ability of states to dampen the effects of climate change and, in turn, mitigate some of the security upshots may be limited and, in some forms, possibly even illusory. It is partly for these reasons that climate change (and the more limited climate security) discussions often carry political overtones or are seen as being of marginal or activist interest. Irrespective of how states and their publics conceptualise the issues, however, climate change is going to be a central thread tying together and shaping a complex web of security challenges in the decades to come.

Six chapters comprise Section II, entitled 'Defence and Security Implications'. Katarina Kertysova, in Chapter 8, details NATO's energy efficiency and mitigation efforts over the past decade. Her chapter first provides an overview of NATO's efforts to reduce reliance on fossil fuels and enhance the efficiency of alliance armed forces to date, including through the Climate Change and Security Action Plan, adopted at NATO's Brussels Summit in June 2021. Her chapter also describes the climate-related decisions that were taken at the Madrid Summit, in June 2022, and considers what more NATO can do to contribute to global mitigation efforts and support its members in their military emissions reduction efforts. In Chapter 9, Richard Nugee, a retired senior general of the British Army, and Timothy Clack discuss why, and how, Western militaries should prepare to respond to climate change threats now and over the next 30 years. Drawing on the example of the UK military, they posit strongly that if defence were not to adopt an approach to sustainability and ignored the effects of climate change, there is the potential for significant disadvantage, operationally, politically, socially, economically, and legally.

A focus on climate change impacts on the maritime domain is provided by Sherri Goodman, Pauline Baudu, and Rachel Fleishman in Chapter 10. In their contribution, the implications of two climate scenarios are considered – less than and above 2°C – as well as the roles of mitigation (emissions reduction and active carbon sequestration), adaptation, and conservation in maritime response to climate and biodiversity issues. In Chapter 11, Alex Tasker illuminates how worsening climate change intersects with other multidimensional factors such as unprecedented migration and political instability to drive the creation of feedback loops across the globe, amplifying, and sustaining volatile situations. By drawing on evidence from the ongoing Syrian conflict, he shows how these dynamic relationships form a complex tapestry woven from ecological, social, and biological connections that drive new conflicts in a rapidly changing world. His analysis also shows how climate-led disruptions to established networks can rapidly accelerate existing tensions, which require new thinking and new responses from security actors. These analyses highlight how impacts on human–animal–ecological relationships can shift the balance of power in an increasingly uncertain world, and how climate shapes both the direction and speed of change.

Louise Selisny, Timothy Clack, Tristan Burwell, and Richard Nugee argue in Chapter 12 that hyper-localised climate change and insecurity data are required to inform appropriate state responses and resource allocations across contexts of regional, national, and subnational variability. Their chapter describes the situation in parts of Mali, where temperatures are increasing at a rate 1.5 times higher than the global average, frequency of droughts has doubled between 2015 and 2020, flooding is common, and 20% of the country's population (3.6 million) are already food insecure. In this theatre, ongoing conflict, political instability, and weak government institutions – each exacerbated by climate change – worsen the security situation. The chapter uses insight from personnel who were part of a recent British Army Long Range Reconnaissance Group deployment to the UN Multidimensional Integrated Stabilisation Mission in Mali to advocate for the collection, collation, and distribution of climate intelligence (CLINT).

In Chapter 13, Richard Milburn examines the role of military support to conservation activities globally, particularly focused on efforts to reduce illegal wildlife and timber trades. His chapter examines the impact of biodiversity loss on climate change and the strategic and tactical level drivers of these trades and the role militaries have played in counteracting those threats to date. With a focus on Operation CORDED, the British Army's counter-illegal wildlife trade operation, Milburn discusses performance, lessons, and best practices. He argues that the military has a unique set of skills that can have a significant impact to improve nature conservation activities if utilised in the right way, and that those activities can be a useful component of efforts to achieve net zero and develop the capabilities to respond to the growing environmental security threats faced around the world.

The final section of the book, Section III, concerns 'framings and reflections'. National responses to climate-related hazards such as heat waves, wildfires, floods, and drought through military assistance to civil authorities (MACA) are the focus of Chapter 14 by Thammy Evans and Gary Lewis. Their contribution covers prediction, preparedness, and potential and notes that the current 'zeitgeist of climate change' masks the totality and enormity of the risks. Military preparations for MACA roles are increasing through national risk registers, local resilience forums, and annually renewed military call-out orders. They lament, however, that intelligence assets have focussed little on climate and ecological security. Given the breadth of scenarios militaries will need to respond to in the short-term future, the authors assert that a switch in focus is urgently required to include mitigation, adaptation, resilience, and regeneration. They also suggest that there is a need for progress in this space to be made publicly available to facilitate whole of society preparedness.

In Chapter 15, Elizabeth Boulton outlines a new theoretical approach to security and proposes that together, the climate and environmental crises, are not merely scientific or economic governance problems, but rather constitute a new form of violence and harm. She casts this as the hyperthreat and applies to it military-style threat analysis and strategic planning methods. In doing so, two major insights appear. Firstly, that humanity has far more response options that hitherto realised and, secondly, based upon defunct industrial era logic, current security strategy and threat posture are now profoundly out of step with an altered security environment. Boulton's analysis finds that there is now no longer any path to security from conventional warfare methods. As such, the chapter introduces an alternate politico-military response, PLAN E, which is a climate and ecologically centred security paradigm.

In the final contribution, Chapter 16, some reflections on 30 years of climate conflict research are provided by Thomas Homer-Dixon. As one of the pioneers of the field of environmental security studies, Homer-Dixon has engaged in long-term conceptual research in the areas of complexity science, climate change, food and water scarcity, and conflict (Homer-Dixon 1994, 1999; Homer-Dixon and Blitt 1998). In reviewing the past 30 years, Homer-Dixon asserts that it is clear that climate change is now a powerful factor influencing economic productivity, agricultural capacity, and human livelihoods around the world. Indeed, he notes that climate change will increasingly contribute to tipping events, where economic stagnation and food insecurity combine with and amplify other stressors to produce sharp outbreaks of social instability and violence. In sum and as a final assessment, Homer-Dixon makes clear that climate change, for decades considered a security 'threat multiplier' or 'shaping threat', must now be recognised as a direct threat to global security in itself.

## References

Adelphi 2017. Insurgency, terrorism and organised crime in a warming climate: analysing the links between climate change and non-state armed groups. *Adelphi Climate Diplomacy Report* https://www.adelphi.de/en/publication/insurgency-terrorism-and-organised-crime-warming-climate (accessed 1 May 2023).

Adger, W. N., J. M. Pulhin, J. Barnett, G. D. Dabelko, G. K. Hovelsrud, M. Levy, Ú. Oswald Spring, and C.H. Vogel, 2014. Human security. (In) C. Field, V. Barros, D. Dokken, K. Mach, M. Mastrandrea, T. Bilir, M. Chatterjee, K. Ebi, Y. Estrada,

R. Genova, B. Girma, E. Kissel, A. Levy, S. MacCracken, P. Mastrandrea and L. White (eds) *Climate Change 2014: Impacts, Adaptation, and Vulnerability. Part A: Global and Sectoral Aspects. Contribution of Working Group II to the Fifth Assessment Report of the Intergovernmental Panel on Climate Change.* Cambridge: Cambridge University Press, pp. 755–791.

BBC 2017. Taliban leader urges Afghans to plant more trees. *BBC News* https://www.bbc.co.uk/news/world-asia-39094578 (accessed 1 May 2023).

Benjaminsen, T. A. and B. Ba 2018. Why do pastoralists in Mali join jihadists groups? A political ecological explanation. *The Journal of Peasant Studies* 46(1). 10.1080/03066150.2018.1474457

Binding, L. 2018. Al Shabaab terror group bans single-use plastic bags. *Sky News* https://news.sky.com/story/al-shabaab-terror-group-bans-single-use-plastic-bags-11427892 (accessed 1 May 2023).

Bodetti, A. 2019. Climate change expands the terrorist threat. *YaleGlobal Online* https://archive-yaleglobal.yale.edu/content/climate-change-expands-terrorist-threat (accessed 1 May 2023).

Boyd, M. and S. Henkin 2022. A climate of terror? Approaches to the study of climate change and terrorism. *National Consortium for the Study of Terrorism and Responses to Terrorism* https://www.start.umd.edu/research-projects/climate-terror-approaches-study-climate-change-and-terrorism (accessed 1 May 2023).

British Army 2020. British Army assists Belize with disaster relief in the wake of Hurricane Eta. https://www.army.mod.uk/news-and-events/news/2020/11/british-army-assists-belize-with-disaster-relief-in-the-wake-of-hurricane-eta/ (accessed 1 May 2023).

British Army 2021. Army hybrid vehicles power forward. https://www.army.mod.uk/news-and-events/news/2021/07/army-hybrid-vehicles-power-forward/ (accessed 1 May 2023).

Broek, E. and C. Hodder 2022. *Towards an Integrated Approach to Climate Security and Peacebuilding in Somalia.* Stockholm: SIPRI.

Brooks, T. 2022. *Climate Change Ethics for an Endangered World.* London: Routledge.

Buhaug, H., T. A. Benjaminsen, E. Sjaastad, and O. Theisen. 2015. Climate variability, food production shocks, and violent conflict in Sub-Saharan Africa. *Environmental Research Letters* 10(12): 125015. https://iopscience.iop.org/article/10.1088/1748-9326/10/12/125015

Buxton, N. and D. Eade 2021. Primer on Climate Security: The Dangers of Militarising the Climate Crisis. *TNI.* https://www.tni.org/en/publication/primer-on-climate-security (accessed 1 May 2023).

Byerly, C. 2005. *The Influenza Epidemic in the US Army During World War I.* New York: NYU Press.

Caballero, A. and R. Emmers 2006. *Non-Traditional Security in Asia: Dilemmas in Securitisation.* Aldershot: Ashgate.

Centeno, M. A., P. W. Callahan, P. A. Larcey, and T. S. Patterson 2023. *How World's Collapse: What History, Systems, and Complexity Can Teach Us about Our Modern World and Fragile Future.* London: Routledge.

Chaturvedi, S. and T. Doyle 2015. *Climate Terror: A Critical Geopolitics of Climate Change.* Basingstoke: Palgrave Macmillan.

Chuter, A. 2021. British air force hails first-ever test flight using only synthetic fuel. *Defense News* https://www.defensenews.com/global/europe/2021/11/17/british-air-force-hails-first-ever-test-flight-using-only-synthetic-fuel/ (accessed 1 May 2023).

Clack, T. and R. Johnson (eds). 2021. *The World Information War: Western Resilience, Campaigning and Cognitive Effects.* London: Routledge.

Clack, T. and R. Nugee 2022. Climate change is creating security threats around the world – and militaries are responding. *The Conversation* https://theconversation.com/climate-change-is-creating-security-threats-around-the-world-and-militaries-are-responding-173668 (accessed 1 May 2023).

Coffel, E. D., T. Thompson, and R. Horton 2017. The impacts of rising temperatures on aircraft takeoff performance. *Climatic Change* 144: 381–388.

Cohen, S. 2001. *States of Denial: Knowing about Atrocities and Suffering.* Cambridge: Polity.

Cook, J. 2020. *A Woman's Place: US Counterterrorism Since 9/11.* Oxford: Oxford University Press.

Cox, M. and V. Rosoux 2002. Paradigm shifts and 9/11: international relations after the twin towers. *Security Dialogue* 33(2): 247–255.

Crawford, N. C. 2022. *The Pentagon, Climate Change, and War: Charting the Rise and Fall of US Military Emissions.* Cambridge, MA: MIT University Press.

Cull, N. 2022. *From Soft Power to Reputational Security: Rethinking Public Diplomacy and Cultural Diplomacy for a Dangerous Age.* London: Routledge.

Daily Mail 2012. Iran VP claims country's drought is part of West's weather war on Islamic republic. *Daily Mail* https://www.dailymail.co.uk/news/article-2175110/Iran-VP-claims-countrys-drought-Wests-weather-war-Islamic-republic.html (accessed 1 May 2023).

Depledge, D. 2023. Low-carbon warfare: climate change, net zero and military operations. *International Affairs* 99(2): 667–685.

Dickie, M. 2007. China blames the west for global warming. *Financial Times* https://www.ft.com/content/e7301c1e-b5fc-11db-9eea-0000779e2340 (accessed 1 May 2023).

DIO 2021. Training estate delivery successes support armed forces. *GOV.UK* https://insidedio.blog.gov.uk/2021/09/27/training-estate-delivery-successes-support-armed-forces/ (accessed 1 May 2023).

Egnell, R. 2010. Winning 'hearts and minds'? A critical analysis of counter-insurgency operations in Afghanistan. *Civil Wars* 12(3): 282–303.

Fairhead, J. 2000. The conflict over natural and environmental resources. *War, Hunger, and Displacement* 1: 147–178.

Foley, A. M. 2010. Uncertainty in regional climate modelling: a review. *Progress in Physical Geography: Earth and Environment* 34(5): 647–670.

Frimpong, O. B. 2020. Climate change and violent extremism in the Lake Chad Basin: key issues and way forward. *Wilson Center Occasional Paper* https://www.wilsoncenter.org/publication/climate-change-and-violent-extremism-lake-chad-basin-key-issues-and-way-forward (accessed 1 May 2023).

Futak-Campbell, B. 2021. Facilitating crisis: Hungarian and Slovak securitization of migrants and their implications for EU politics. *International Politics* 59(3): 541–561.

Gleditsch, P. 1998. Armed conflict and the environment: a critique of the literature. *Journal of Peace Research* 35(3): 381–400.

GOV.UK 2020. UK aid and military support sent to Central America in wake of Hurricane Eta. *GOV.UK* https://www.gov.uk/government/news/uk-aid-and-military-support-sent-to-cental-america-in-wake-of-hurricane-eta (accessed 1 May 2023).

Gozdecka, D. A. 2021. Human rights during the pandemic: Covid-19 and securitisation of health. *Nordic Journal of Human Rights* 39(3): 205–223.

Hameiri, S. 2021. Covid-19: is this the end of globalization? *Canada's Journal of Global Policy Analysis* 76(1): 30–41.

Harville, E., C. Taylor, H. Tesfai, X. Xiong, and P. Buekens 2010. Experience of Hurricane Katrina and reported intimate partner violence. *Journal of Interpersonal Violence* 26(4): 833–845.

Hasemyer, D. 2019. Military fights a deadly enemy: heat. *ABC News*. https://www.nbcnews.com/news/us-news/military-s-climate-change-problem-blistering-heat-killing-soldiers-during-n1032546 (accessed 1 May 2023).

Homer-Dixon, T. 1994. On the threshold: environmental changes as causes of acute conflict. *International Security* 16(2): 76–116.

Homer-Dixon, T. 1999. *Environment, Scarcity, and Violence*. Princeton: Princeton University Press.

Homer-Dixon, T. and J. Blitt (eds) 1998. *Ecoviolence: Links among Environment, Population, and Security*. New York: Rowman & Littlefield.

Houghton, N. 2015. Interesting Times. General Sir Nick Houghton, Chief of the Defence Staff. *RUSI* https://www.gov.uk/government/speeches/interesting-times (accessed 1 May 2023).

House of Commons 2018. The UK's response to hurricanes in its Overseas Territories: Fifth report of session 2017–2019. *House of Commons Foreign Affairs Committee* https://publications.parliament.uk/pa/cm201719/cmselect/cmfaff/722/722.pdf (accessed 1 May 2023).

Hudson, V., D. Bowen, and P. Nielsen 2021. *Political Order: How Sex Shapes Governance and National Security Worldwide*. New York: Columbia University Press.

ICG 2022. Climate change and conflict. *International Crisis Group* https://www.crisisgroup.org/future-conflict/climate (accessed 1 May 2023).

ICRC 2020. Seven things you need to know about climate change and conflict. *International Committee of the Red Cross* https://www.icrc.org/en/document/climate-change-and-conflict (accessed 1 May 2023).

Ide, T. 2021. Covid-19 and armed conflict. *World Development 140*. https://www.sciencedirect.com/science/article/pii/S0305750X20304836?via%3Dihub

Jardine, S. 2022. On thin ice: the geopolitical impact of climate change in the Antarctic. *Climate Change & (In)Security Project* https://static1.squarespace.com/static/60800d20f65a1555173d7f03/t/628e0b321d9b7a3f35a648e3/1653476152084/Jardine.pdf (accessed 1 May 2023).

Johnson, R. and T. Clack 2019. Introduction: anticipating future stabilisation. *Before Military Intervention: Upstream Stabilisation in Theory and Practice*. Basingstoke, Palgrave, pp. 1–25.

Kilcullen, D. 2016. *Blood Year: Islamic State and the Failures of the War on Terror*. London: Hurst.

King, M. D. 2023. *Weaponizing Water: Water Stress and Islamic Extremist Violence in Africa and the Middle East*. Boulder: Lynne Rienner.

Kitchlew, I. 2022. Is super-polluting Pentagon's climate plan just 'military-grade greenwash'? *The Guardian* https://www.theguardian.com/us-news/2022/mar/10/pentagon-us-military-emissions-climate-crisis (accessed 1 May 2023).

Kivimäki, T. 2019. *The Failure to Protect: The Path to and Consequences of Humanitarian Interventionism.* Cheltenham: Edward Elgar.

Klose A. K., N. Wunderling, R. Winkelman, and J. F. Donges. 2021 What do we mean, 'tipping cascade'? *Environmental Research Letters* 16(12): 125011.

Kupchan, C. 2020. *Isolationism: A History of America's Efforts to Shield Itself from the World.* New York: Oxford University Press.

Kuperman, A. J. 2001. *The Limits of Humanitarian Intervention: Genocide in Rwanda.* Washington: Brookings Institution Press.

Landay, J. 2016. Bin Laden called for Americans to rise up over climate change. *Reuters.* https://www.reuters.com/article/us-usa-binladen-climatechange/bin-laden-called-for-americans-to-rise-up-over-climate-change-idUSKCN0W35MS (accessed 1 May 2023).

Lindsay, J. and S. Lischer 2003. Humanitarian aid is not a military business. *Harvard Kennedy School Belfer Center for Science and International Affairs* https://www.belfercenter.org/publication/humanitarian-aid-not-military-business (accessed 1 May 2023).

Lischer, S. 2007. Military intervention and the humanitarian multiplier. *Global Governance* 13: 99–118.

Lloyd, E. and T. Shepherd 2021. Climate change attribution and legal contexts: evidence and the role of storylines. *Climatic Change* 167(28). 10.1007/s10584-021-03177-y.

Mager, D. 2009. Climate change, conflicts and cooperation in the Arctic: easier access to hydrocarbons and mineral resources? *International Journal of Marine and Coastal Law* 24(2): 347–354.

Magnusdottir, G. and A. Kronsell (eds) 2021. *Gender, Intersectionality and Climate Institutions in Industrial States.* London: Routledge.

Minnerop, P. and F. Otto 2020. Climate change and causation: joining law and climate science on the basis of formal logic. *Buffalo Journal of Environmental Law* 27: 1–22.

Murphy, P. 2021. Speaking for the youth, speaking for the planet: Greta Thunberg and the representational politics of eco-celebrity. *Popular Communication* 19(3): 193–206.

Murtazashvili, J. B. 2016. Afghanistan: a vicious cycle of state failure. *Governance* 29(2): 163–166.

NATO 2007. *Recruiting and Retention of Military Personnel.* Salisbury: DSTL-AGARD. https://www.nato.int/issues/women_nato/recruiting%20&%20retention%20of%20mil%20personnel.pdf (accessed 1 May 2023).

Naval Studies Board 2011. *National Security Implications of Climate Change for US Naval Forces.* Washington: The National Academies Press.

Nelson, D. 2009. Manmohan Singh blames West for India's climate change problems. *The Telegraph* https://www.telegraph.co.uk/news/worldnews/asia/india/5768824/Manmohan-Singh-blames-West-for-Indias-climate-change-problems.html (accessed 1 May 2023).

Nguyen, H. T. 2018. Gendered vulnerabilities in times of natural disasters: male to female violence in the Philippines in the aftermath of Super Typhoon Haiyan. *Violence Against Women* 25(4): 421–440.

Nordas, R., and N. Gleditsch. 2007. Climate change and conflict. *Political Geography* 26(6): 627–638.

Norgaard, K. M. 2019. Making sense of the spectrum of climate denial. *Critical Policy Studies* 13(4): 437–441.

Norton-Taylor, R. 2016. Destroyers will break down if sent to Middle east, admits Royal Navy. *The Guardian* https://www.theguardian.com/uk-news/2016/jun/07/destroyers-will-break-down-if-sent-to-middle-east-admits-royal-navy (accessed 1 May 2023).

Ostreng, W., K. Eger, B. Fløistad, A. Jørgensen-Dahl, L. Lothe, M. Mejlænder-Larsen, and T. Wergeland 2013. *Shipping in Arctic Waters: A Comparison of the Northeast, Northwest and Trans Polar Passages*. Berlin: Springer.

Otto, F. 2022. *Angry Weather: Heat Waves, Floods, Storms, and the New Science of Climate Change*. Vancouver: Greystone.

O'Brien, K., A. L. St Clair, and B. Kristoffersen (eds) 2022. *Climate Change, Ethics and Human Security*. Cambridge: Cambridge University Press.

Panu, P. 2020. Anxiety and the ecological crisis: an analysis of eco-anxiety and climate anxiety. *Sustainability* 12(7836). 10.3390/su12197836.

Raineri, L. 2020. Sahel climate conflicts? When 9fighting) climate change fuels terrorism. *European Union Institute for Security Studies* https://www.iss.europa.eu/content/sahel-climate-conflicts-when-fighting-climate-change-fuels-terrorism (accessed 1 May 2023).

Reinhardt, F. L. and Toffel, M. W. 2017. Managing climate change: lessons from the US Navy. *Harvard Business Review*. https://hbr.org/2017/07/managing-climate-change (accessed May 2023).

Rolfe, J. 2011. Partnering to protect: conceptualizing civil-military partnerships for the protection of civilians. *International Peacekeeping* 18(5): 561–576.

Rufford, N. 2021. Shocks away! RAF to fly electric planes. *The Times* https://www.thetimes.co.uk/article/shocks-away-raf-to-fly-electric-planes-tkdpn87vz (accessed 1 May 2023).

Salehyan, I. 2008. From climate change to conflict? No consensus yet. *Journal of Peace Research* 45(3): 315–326.

Sanson, A. and M. Bellemo 2021. Children and youth in the climate crisis. *British Psychological Bulletin* 45(4): 205–209.

Schwartzstein, P. 2019. Climate change may be blowing up arms depots. *Scientific American*. https://www.scientificamerican.com/article/climate-change-may-be-blowing-up-arms-depots/ (accessed 1 May 2023).

Selby, J. G. Daoust, and C. Hoffmann 2022. *Divided Environments: An International Political Ecology of Climate Change, Water and Security*. Cambridge: Cambridge University Press.

Stedman, A. D. 2011. A most dishonest argument? Chamberlain's government, anti-appeasers and the persistence of League of Nations language before the Second World War. *Contemporary British History* 25(1): 83–99.

Stephenson, S., L. Smith, L. Brigham, and J. Agnew 2013. Projected 21st-century changes to Arctic marine access. *Climatic Change* 118(3–4): 885–899.

Stoltenberg, J. 2021. A changing approach to security. *NATO*. https://www.nato.int/cps/en/natohq/opinions_181806.htm?selectedL (accessed 1 May 2023).

Telford, A. 2020. A climate terrorism assemblage? Exploring the politics of climate change-terrorism-radicalisation relations. *Political Geography* 79. 10.1016/j.polgeo.2020.102150.

The Guardian 2002. Full text: Bin laden's 'letter to America'. *The Guardian* https://www.theguardian.com/world/2002/nov/24/theobserver (accessed 1 May 2023).

Thurston, A., H. Stockl, and M. Ranganathan 2021. Natural hazards, disasters and violence against women and girls: a global mixed-methods systematic review. *BMJ Global Health* 6(4): e004377. 10.1136/ bmjgh-2020-004377

UK Government 2023. Integrated review refresh 2023: responding to a more contested and volatile world. *Her Majesty's Stationary Office.* https://www.gov.uk/government/publications/integrated-review-refresh-2023-responding-to-a-more-contested-and-volatile-world (accessed 1 May 2023).

UK Government 2021. Global Britain in a competitive age: the integrated review of security, defence, development and foreign policy. *Her Majesty's Stationary Office.* https://assets.publishing.service.gov.uk/government/uploads/system/uploads/attachment_data/file/975077/Global_Britain_in_a_Competitive_Age-_the_Integrated_Review_of_Security__Defence__Development_and_Foreign_Policy.pdf (accessed 1 May 2023).

UK MOD 2014. *Joint Doctrine Publication 0-01 (JDP 0-01): UK Defence Doctrine.* Shrivenham: Ministry of Defence. https://assets.publishing.service.gov.uk/government/uploads/system/uploads/attachment_data/file/389755/20141208-JDP_0_01_Ed_5_UK_Defence_Doctrine.pdf (accessed 4 July 2022).

UK MOD 2016. *Joint Doctrine Publication 3–52 (JDP 3–52): Disaster Relief Operations Overseas: The Military Contribution.* https://www.gov.uk/government/publications/disaster-relief-operations (accessed 1 May 2023).

UK MOD 2019. Joint Service Publication 1325 (JSP-1325): *Human Security in Military Operations.* https://assets.publishing.service.gov.uk/government/uploads/system/uploads/attachment_data/file/770919/JSP_1325_Part_1_2019_O.PDF (accessed 1 May 2023).

UK MOD 2020. Sustainable fuels to power RAF jets. *GOV.UK* https://www.gov.uk/government/news/sustainable-fuels-to-power-raf-jets (accessed 1 May 2023).

UK MOD 2021a. Ministry of Defence Climate Change and Sustainability Strategic Approach. *Ministry of Defence* https://assets.publishing.service.gov.uk/government/uploads/system/uploads/attachment_data/file/973707/20210326_Climate_Change_Sust_Strategy_v1.pdf (accessed 1 May 2023).

UK MOD 2021b. MOD Land Holdings: 2000 to 2021. *Ministry of Defence* https://www.gov.uk/government/statistics/mod-land-holdings-bulletin-2021/mod-land-holdings-2000-to-2021#:~:text=The%20Ministry%20of%20Defence%20(MOD,which%20operations%20can%20be%20instigated (accessed 1 May 2023).

UK MOD 2021c. Joint Service Publication 985 (JSP-985): *Human Security in Defence: Volume 1: Incorporating Human Security in the Way We Operate.* https://assets.publishing.service.gov.uk/government/uploads/system/uploads/attachment_data/file/1040257/20211209_JSP_985_Vol_1.pdf (accessed 1 May 2023).

UK Met Office 2022. Effects of climate change. https://www.metoffice.gov.uk/weather/climate-change/effects-of-climate-change (accessed 1 May 2023).

UN 2022. Climate and weather-related disasters surge five-fold over 50 years, but early warnings save lives – WMO report. *UN News* https://news.un.org/en/story/2021/09/1098662 (accessed 1 May 2023).

UNHCR 2022. Climate change and disaster displacement. https://www.unhcr.org/uk/climate-change-and-disasters.html#:~:text=The%20impacts%20of%20climate%20change,the%20world%20that%20host%20refugees (accessed 1 May 2023).

US Army 2022. Climate Change Strategy. *Department of Defense.* https://www.army. mil/e2/downloads/rv7/about/2022_army_climate_strategy.pdf (accessed 1 May 2023).

US ATA 2023. Annual Threat Assessment of the U.S. Intelligence Community. *Office of the Director of National Intelligence.* https://www.dni.gov/files/ODNI/ documents/assessments/ATA-2023-Unclassified-Report.pdf (accessed 1 May 2023).

US DOD (Department of Defense) 2021. Climate risk analysis. https://media.defense. gov/2021/Oct/21/2002877353/-1/-1/0/DOD-CLIMATE-RISK-ANALYSIS-FINAL.PDF (accessed 1 May 2023)

US NIC (National Intelligence Council) 2021. *National Intelligence Estimate.* https:// www.dni.gov/index.php/newsroom/reports-publications/reports-publications-2021/ item/2253-nationalintelligence-estimate-on-climate-change (accessed 1 May 2023).

Van Daalan, K., S. Kallesoe, F. Davey. S. Dada, L. Jung, L. Singh, R. Issa, C. Emilian, I. Kuhn, I. Keygnaert, and M. Nilsson 2022. Extreme events and gender-based violence: a mixed-methods systematic review. *Lancet Planetary Health* 10.1016/S2542-5196(22)00088-2.

Vergun, D. 2021. Defence Secretary calls climate change an existential threat. *DOD News* https://www.defense.gov/News/News-Stories/Article/Article/2582051/defense-secretary-calls-climate-change-an-existential-threat/ (accessed 1 May 2023).

Verplanken, B., E. Marks, and A. Dobromir 2021. On the nature of eco-anxiety: how constructive or unconstructive is habitual worry about global warming? *Journal of Environmental Psychology* 72. 10.1016/j.jenvp.2020.101528.

Vestby, J. 2014. Climate variability and individual motivations for participating in political violence. *Global Environmental Change* 56: 114–123.

Vinke, K., L. Campbell, D. Schirwon, K. Seyuba, F. Krampe, H. Maalim, and G. Mbungwal 2023. *Security in the Democratic Republic of Congo: Competing over Abundant Resources – Adapting to Change.* Berlin: German Council on Foreign Relations.

Von Storch, L., L. Ley, and J. Sun 2021. New climate change activism: before and after the Covid-19 pandemic. *Social Anthropology* 29(1): 205–209.

Wapshott, N. 2015. *The Sphinx: Franklin Roosevelt, the Isolationists, and the Road to World War II.* New York: W.W. Norton.

Wellesley, L. and T. Benton 2022. Welcome to a new age of food insecurity. *Prospect* https://www.prospectmagazine.co.uk/essays/welcome-to-a-new-age-of-food-insecurity-russia-ukraine (accessed 1 May 2023).

WFP 2022. Unprecedented hunger to follow in wake of the climate crisis, WFP calls for urgent action on World Food Day. *Relief Web* https://reliefweb.int/report/ world/unprecedented-hunger-follow-wake-climate-crisis-wfp-calls-urgent-action-world-food-day (accessed 1 May 2023).

Wullenkord, M. C. 2022. From denial of facts to rationalization and avoidance: ideology, needs, and gender predict the spectrum of climate denial. *Personality and Individual Differences* 193: 111616. 10.1016/j.paid.2022.111616.

**SECTION I**

# Climate security contexts

# 1

# CASCADING AND SYSTEMIC RISKS FROM ENVIRONMENTAL CHANGE

*Tim Benton, Neil Morisetti, and Oli Brown*

## Introduction

Almost daily, environmental events are in the news: floods, storms, drought, wildfires, heat, or even locust and disease outbreaks. The environmental headlines are not limited to climate change but include land-use change (e.g. deforestation) and degradation (e.g. from farming), and its associated biodiversity loss. Changing the use, or intensity of use, of land, may interact with climate change, impact indigenous livelihoods, and provide the materials underpinning the global economy. However, unless one is personally caught up in the events or places under discussion, the news of such environmental crises can all seem very distant: what can such events mean for the average citizen and their security?

This chapter aims to highlight some answers to exactly this question by discussing the way that events far away can have ripple effects that cascade across borders, geographies, and sectors and lead to significant outcomes across the world. In short, this is because we live in an increasingly inter-connected world, where the hazards – and the ability for their impacts to propagate spatially – are increasing in severity and frequency.

As an example of global embeddedness, take the United Kingdom. In 2022, the UK's GDP was about £2.2tn; the combined value of imports and exports was £1.72 tn, indicating how reliant the economy is on the flow of goods. UK society relies on a myriad of complex supply chains: building a car takes approximately 20,000 components, many of which will come from suppliers and manufacturers across the globe. Many of the ingredients in the majority of prepared foods (such as vegetable oil, starch, salt, and sugar) may have been imported. For some foods – like fruit and vegetables – at certain times of the year the United Kingdom is almost

DOI: 10.4324/9781003377641-3

totally reliant on its imports (over 80% at peak times), and increasingly they come from drought-prone countries (Scheelbeek et al. 2020). If one examines the local economy, sector by sector, almost all are reliant – to a greater or lesser extent – on products, finance, services, or people that come from abroad. The globalised connectivity and complexity of the web of supply are growing over time.

In addition to embeddedness, in the interests of efficiency gains, supply chain management has become increasingly 'lean' (Papadopoulou and Özbayrak 2005). Leanness depends on a range of factors such as (a) 'just in time' supply chains, reducing buffer stocks of components/products and instead relying on deliveries at the time they are needed, (b) centralising processing and distribution facilities to enjoy the economies of scale; and (c) supplier base rationalisation, shrinking the suppliers to a smaller number of bigger providers, again to reduce overheads and drive greater value per relationship. However, such lean operations are also potentially fragile, as leanness by definition removes both redundancy and diversity, both of which are important properties of resilience. If a just-in-time supply chain is interrupted because of unforeseen events, the product cannot be made; if there is a single supplier for key components that suffers an interruption in its business, then the users of those components cannot work. If there are centralised facilities that have to shut, the whole business closes. To give an example, following the Great East Japan Earthquake in March 2011, Fujita and Hamaguchi (2011) estimated that car assembly was down by 39% in Guangdong China in April, and by 48.5% in Thailand in May due to interruptions to lean supply chains. In the United States, COVID-19 resulted in the shutdown of centralised meat processing plants that reduced the availability of beef by between a quarter and a half in 2020 (Ijaz et al. 2021).

Following the financial crisis in 2008/2009, there was a focus on research on the way that failures occurring in one part of a complex system can, through contagion, create failures in other parts of the system, creating a risk cascade that can lead to overall system failure if enough parts of the system break (Haldane and May 2011; Battiston et al. 2012). Hence, two key concepts for this chapter are: *cascading risks* – the danger that an event leads to ripple effects that cascade through a range of transmission pathways across borders and result in impacts geographically divorced from the event; and *systemic risks* – where risk cascades can lead to system-wide risks that impact the way economies and societies function.

COVID-19 is a key example of both: the emergence of a new disease occurred in one place and spread around the world. Attempts to manage it required shutting borders, lockdowns, and shifts in the way governments and societies functioned. COVID-19 rapidly became more than a health issue, rather it became the biggest perturbation to society – limiting behaviour, affecting social cohesion and mental health – and the economy since the last world war. To indicate the system-wide nature of COVID-19, initial estimates were that the economic costs in 2020 were over 100× greater than the direct health-care-related costs of dealing with the disease (Challinor et al. 2021).

Risk cascades, by their very nature, transmit the impact of the original event from something directly affecting a localised population to something systemic, affecting people around the world. This amplification may be partly structural (e.g. a climate event causing a flood at semi-conductor manufacturing plants, as happened in Thailand in 2011, interrupting supply chains for many types of goods across multiple sectors with global reach), or it may be social. An example of the latter would be the 2007/8 food price spike where, in a tight market, the expectation of lower yields due to drought in Australia, helped fuel a run on the market, which was then amplified further by a combination of poor policy (export bans) and panic buying. The end result was the amplification of the price signal (by over 200%), far in excess of the magnitude of the yield shortfall (under 1%), with global impacts on food prices and people's access to food.

### Implications for human security

Risks to human security are never straightforward. Many factors contribute, some socially, economically or politically driven, others environmentally related. Climate change falls into this last category. The second- and third-order consequences can exacerbate existing stresses, contributing to increased instability and potential conflict. Many people appreciate this and understand that climate change can have significant impacts on human security, but the role of cascading impacts in propagating those consequences around the world at speed is perhaps under-recognised.

The 1994 UN Human Development Report defined the concept of human security based on two main criteria: '... [F]irst, safety from such chronic threats as hunger, disease and repression. And second ... protection from sudden and hurtful disruptions in the patterns of daily life – whether in homes, in jobs or in communities' (UNDP 1994, p. 23). It is clear from the paragraphs above – and described in more detail below – that cascading risks can be disruptive in both the senses encapsulated in the quote, particularly in the emergence and spread of diseases, contribution to food insecurity, and for the potential of disruptions to the patterns of daily life.

The current view of human security is people focused and with a concentration on, amongst other things, prevention and resilience building (in order to protect people from disruption) (UN 2016). This is the lens we use in this chapter. Furthermore, in this chapter, we describe the way that hazards can create risks that cascade across borders via a number of pathways. Subsequently, we focus on how risks are determined by a combination of hazards, exposure, and vulnerability, and how these are changing. We also describe some case studies and draw some conclusions about how to mitigate risks.

## Transmission pathways for cascading risks

There are multiple ways that environmental hazards (e.g. storms, droughts, or the emergence of pests or diseases) can create risks that transmit across borders and sectors (Challinor et al. 2021). These include causing:

- *Interruptions in the movement of goods.* Trade is a key route by which cascading impacts can occur following environmental hazards – like extreme weather events. Of particular urgency is the supply of food, as most countries rely on lean supply chains, and some may only have a few days of food within their own borders. Even in countries with large agricultural land holdings, the economic incentives of modern agriculture tend to lead to specialisation in a few products, exporting the supply not needed locally, and importing other foodstuffs that cannot be grown with comparative advantage locally. Interruptions to supply chains can arise through interruptions to production (e.g. agricultural drought, pests and diseases, and storms and floods impacting factories), transport, and logistics (e.g. reduced river flows, floods and storms affecting port infrastructure, blockage of critical canals and straits, buckling of railways due to heat, etc.).
- *Impacts on the movement of money.* Financial flows are as crucial as the flows of goods to the functioning of the global system. For example, finance for investment in building factories or infrastructure underpins production and transport facilities. In addition, insurance and re-insurance are typically necessary for facilities to function. Damaging environmental events can affect financial flows and thereby have a cascading impact on other sectors. As an example, the City of London covers about 8% (at about US$110bn) of the global insurance market (Carroll et al. 2020), so significant events elsewhere – Hurricane Katrina in 2005 cost over US $85bn – could potentially impact on the economy of the United Kingdom.
- *Impacts on the movement of fuel or energy.* Many countries use more energy than they produce and therefore have to import energy or fuel to underpin local industries. The least energy secure includes Ukraine, Thailand, Brazil, South Korea, and China (US Chamber of Commerce 2017). Climate hazards that block critical infrastructure like ports, therefore, can create energy insecurity for the countries that receive the fuel. Similarly, there are some international fuel pipelines (e.g. Nord Stream 1 between Germany and Russia) that when interrupted have significant consequences driving cascading impacts. Of course, cascades across space do not need to cross borders: the longest electricity transmission lines are thousands of

kilometres long (Future Power Technology 2020). For example, a hazard occurring in Para, Brazil, could interdict energy supply in Rio de Janeiro, over 2500 km distant. As markets are inter-connected, disruption to one fuel, such as gas, can affect global prices for other forms of energy.

- *Interruptions in the movement of information.* Just as finance is necessary for supply chains and economies to function, so is the movement of information. Information is increasingly carried via complex infra-structures involving undersea and overland cables, dedicated datacentres (Bhardwaj 2018), and satellite infrastructure. As with any infrastructure, environmental hazards can potentially impact. For example, in June 2021, a mudslide triggered by Congo River flooding severed undersea data cables (Livingstone 2021).

- *Increasing movement of people.* Disasters triggered more than three-quarters of the new displacements recorded worldwide in 2020 and more than 98% of these were the result of weather-related hazards such as storms and floods (IDMC 2021). Most human displacements are local (i.e. within a state, often rural to urban) and many are temporary, although the limited evidence we have suggests that many who flee are unable to return quickly to their home (Mosello et al. 2021). However, climate change is increasingly driving longer-term population movements due to environ-mental degradation. For example, as discussed below, changes in rainfall variability, as well as access to water and forage, can lead grazers to move into traditional pastoral land in the Sahel (Cepero et al. 2021). In some circumstances, as discussed below, an environmental trigger can lead to longer-term impacts eroding conditions sufficiently for significant outward migration from a region (e.g. drought leading to food price spike leading to destabilisation of economies – the Arab Spring – and incentivising migration out of North Africa into Europe, as happened in 2010/2011).

- *Impacts on governance, stability, and conflict.* Whilst climate hazards can lead to increased human displacement, in other cases communities become more susceptible to the influence of organised crime or violent extremist organisations (VEOs), thereby creating a new pathway for social, eco-nomic, or governance instability. These can create cascading risks in multiple ways. For example, increasing civil unrest as a result of climate-related reductions in food availability, or rise in prices, can destabilise governments, leading to knock-on consequences for the stability of a region: this might interrupt supply chains, require deployment of peace-keeping forces, or significant flows of aid, all of which can have global consequences far beyond the countries in question.

- *Impacts on health and wellbeing.* Physical health can be affected by risks cascading across geographies through, for example, reduced air quality from wildfires, movement of pests, diseases, or their vectors around the world, and interruptions to food supply, leading to food or nutritional insecurity. Populations in any part of the world can also suffer mental health impacts to do with the impact of information on environmental issues elsewhere. First, anxiety to do with families and friends at risk from natural disasters, particularly for people who have moved countries away from home communities that may be affected (Rees and Fisher 2020). Second, through a less specific anxiety about environmental crisis, so-called 'eco-anxiety' (Gifford and Gifford 2016; Panu 2020). In extremis, wellbeing or anxiety issues can lead to the rise of social movements that may themselves be destabilising.

• *Interacting transmission pathways.* The transmission pathways above are, of course, not independent of each other. As shown by COVID-19, the emergence of a new disease can interrupt the movement of people, goods, and flows of finance as well as physical and mental health. This interaction of risk transmission pathways contributes to creating the potential for economy-wide, systemic risk.

Figure 1.1 is a schematic encompassing some of major elements of cascading and systemic risks. A key element is whether, as the risk moves across borders, the risk attenuates or amplifies. This in part depends on the network – countries, institutions, or transport nodes and the way they are connected –

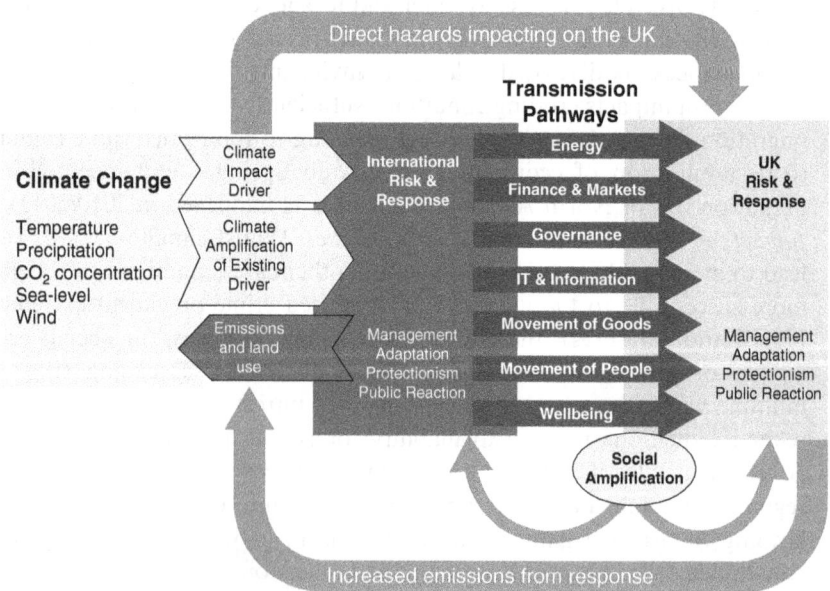

**FIGURE 1.1**  Schematic encompassing some of major elements of cascading and systemic risks.

*Source:* Challinor and Benton (2021).

and partly on the social response. For the former, if the hazard or cascade creates a shock to a node in the network (e.g. a country, or an institution) that is highly connected to, and influential on, other parts of the network, the shock is likely to amplify, quickly spreading across the network (Haldane and May 2011; Puma et al. 2015). Various social, policy, and market responses can likewise affect the way the risk propagates. For example, if an emerging hazard affects a crucial good important for national security (e.g. food or energy), for which it might be difficult to quickly obtain more because of the immediacy of constrained supply on a global market, the risk of not being able to have an assured supply as the hazard emerges may cause a market run as players seek to buy what they need as quickly as they can whilst supplies are still available. Such panic buying can amplify the price signal from the potential impact far in excess of the direct impact. Furthermore, such market dynamics can be further amplified by policy decisions (e.g. countries instituting export bans to ensure local supplies, but further tightening the international market). A key issue, in dealing with cascading and systemic risks is therefore to enhance the resilience of the system (see below).

There is a further amplifying feedback: times of high prices incentivise investment in new supply, which may lead to increased emissions and future climate risks through the system response (bottom blue arrow on figure). For example, commodity crop prices are significantly related to rates of deforestation globally (Berman et al. 2021), so a risk cascade leading to increased food prices plays through market drivers to rapidly increase agricultural outputs but which, ultimately, increases the collective risk. There is a similar potential in post-COVID-19 recovery to build back as fast as possible, rather than 'build back better/greener', locking in more climate change and systemic fragility.

## The risk equation: environmental hazards, exposure, and vulnerabilities

Risk is traditionally seen as the likelihood of an impact arising from a combination of hazard, exposure, and vulnerability. In traditional thinking about climate change, a hazard would be, for example, the occurrence of extreme rainfall in a particular place. Exposure would be related to whether, for example, a population is located at the base of a slope or on a floodplain that the extreme rainfall might destabilise or inundate. Vulnerability would be associated with any factors that might affect the severity of the impact, such as whether flood defences or afforestation of slopes existed when the hazard struck.

In the world of cascading risks, exposure and vulnerability take on quite different properties. Exposure relates to the degree to which you are integrated into the global economy, how reliant you are for cross-border flows of goods or services, and how porous your border might be to things a country may want to defend itself against (e.g. immigration control or surveillance for diseases). Vulnerability then equates to the absence of resilience to the way the risk may be transmitted: are there buffer stocks, a resilient supply infrastructure and diverse supply routes, significant credit and ability to control borders, and what proportion of people may be affected by any impacts, itself often related to inequality? What hazards arise from environmental change?

### Environmental hazards

Environmental change arises from anthropogenic drivers (such as emissions, land-use change, or pollution) that lead to changes to the natural world via climate change, habitat loss, fragmentation, and degradation and the knock-on impacts that arise from ecological change. Climate change can cause habitat loss or degradation (e.g. via drying out of wetlands or desertification) and can also cause ecological re-wiring in similar ways to direct land-use change or degradation. Environmental systems are, by their nature, complex bio-physical systems and are subject to non-linear behaviour, such as sudden changes in the way that they work (see tipping points below).

Below, we group environmental hazards into three main classes depending on whether they relate to (1) changes in the climate; (2) changes in the weather, particularly extremes; and (3) ecological changes driven by climate or land-use change and degradation.

#### Changes in the climate

'Climate' describes the weather conditions that an area may experience throughout a year, and is measured as the average across multiple decades. Hence, a key part of climate change is the way the average conditions change (such as the mean temperature, or average winter rainfall). So, one element of climate change is the way that an average year's weather may develop as time passes. Will the region get warmer or significantly hotter, will it dry out during the year or only at certain times (such as wetter winters and drier summers)? However, such climate change is, by definition, likely to be gradual, and it is the sense of climate change encompassed by targets like the 1.5 °C vision of the Paris Agreement. From a risk perspective, it is not the 1.5 °C that creates the risks in a changing climate, rather it is the way the weather may change as the climate changes.

#### Changing weather patterns, particularly extremes

Cascading risks are more likely to occur from high-impact weather events, which would typically be (current) climatological extremes, mainly because high-impact weather is typically a rare and acute event and therefore extreme. Extreme events include heat, drought, extreme rainfall (in terms of intensity, mm per hour, or duration), extreme storms (hurricanes, typhoons, and cyclones), tornadoes, and extreme cold. Owing to the rare nature of extreme events, they are often difficult to plan for and may create novel, unprecedented situations for which resilience and adaptation planning has not been properly accounted. The nature and pace of such events as we are seeing today are pushing our social systems outside historical ranges and, in doing so, are stressing managerial regimes at local, national, and international levels; regimes that have evolved to deal with only limited climate fluctuations.

Climate change, associated with Arctic warming, is thought to be contributing to changing configuration of the northern hemisphere jet stream (Mann et al. 2017). This can mean that when the weather is disruptive in one place, it is also disruptive elsewhere. For example, in June and July 2018, North America, Western Europe, and the Caucasus experienced extreme temperatures at the same time that South-East Europe and Japan experienced extreme rainfall and floods (Kornhuber et al. 2019).

From a cascading risks perspective, this implies that there is potential for, for example, simultaneously poor harvests in multiple breadbasket regions of the world (Kornhuber et al. 2020). Much of the conceptual framing of the need for resilience in global systems, like the food system, is based on framing risks as independent – if it is bad weather in one place it is likely to be good somewhere else. Instead, risks may be correlated in time and space, and thus compound the damage they are able to wreak (Carter et al. 2021).

### Ecological hazards

Climate change interacts with other anthropogenic processes that may create environmental hazards through disruption of the way that ecosystems function. Habitat loss and fragmentation lead directly to the loss of biodiversity by reducing the available habitat or breaking it up into pieces that do not support a viable population. Likewise, habitat degradation reduces the suitability of habitat for biodiversity and may typically occur through intensifying agriculture or pollution. Climate change also affects ecosystems by changing the habitat suitability for the naturally occurring animals and plants that live in a given area. Species respond differentially to changing climates: all things being equal, more mobile species can track climate change through shifting ranges more easily than less mobile species.

The combination of climate change and changes in the use of land creates new communities of species that have not interacted before. 'Ecological re-wiring' fundamentally changes the way that species interact, leading to the potential for ecological ripple effects to drive indirect changes in abundances. These include changes in the way species eat each other, compete with each other, and affect the environment for each other. Whilst many of the changes in species' distributions, and the rewiring of ecosystems, may be benign, some are not. Perhaps chief amongst these is that ecological rewiring leads to new mixes of pests, pathogens, reservoirs, hosts, and potential new hosts, perhaps contributing to the increasing emergence of infectious diseases, e.g. HIV, Ebola, SARS, MERS, and COVID-19) (Brooks et al. 2019), and the change in distribution of many existing vector-borne diseases (e.g. malaria re-emerging within Europe).

Unlike natural habitat, which has tended to be lost and fragmented, agricultural land has been a habitat type that has been rapidly growing, as more land is brought into farming. This provides more connected landscapes for crop pests and pathogens, allowing them to track climate change relatively easily and alter their geographical distribution. As a result, the pole-wards movement of agricultural pests and diseases is well documented (Bebber et al. 2013). Environmental change, as well as changes in the climate, contributes directly to biodiversity loss through land-use change and degradation, which can drive the emergence of new pests and diseases, particularly as urbanisation and people movement mean that humans interact with the biodiversity around them in new ways.

### Exposure and vulnerability

Beyond hazards, risks are determined by exposure and vulnerability. Exposure is about the degree to which a country or institution or individual is 'in harm's way'. For example, does a state or manufacturer rely on components or goods that come from areas where climate hazards are known to be increasing? Vulnerability is about the degree to which an impact from a hazard to which a country, institution, or individual will be affected. To continue with the supply chain example: is the supply chain exposed to climate hazards a just-in-time one, for which disruption would be critical to operations, or are there alternative suppliers, or significant buffering capacity (through storage) that means a disruption's impacts can be easily mitigated? Or, will a disruption in supply impact significantly on the poor or marginalised? Exposure and vulnerability to environmental hazards depend on many things:

- *Geography*. Many cities – for historical reasons – are built on lowland river plains, and often threatened by rising sea levels. The global population exposed to increasing sea-level rise is now up to 400 million, of which the majority are in the tropics; a figure that will only increase with the projected growth of global megacities (Hooijer and Vernimmen 2021).
- *Development status*. Wealthier countries may suffer exposure to hazards like poor countries but can afford to invest in infrastructural adaptation to reduce vulnerability, such as better flood defences and transport infrastructure, etc. This means that they are at less risk. On the other hand,

greater economic power also is associated with greater global integration and greater pressure on efficiency and leanness, perhaps increasing the exposure to impacts transmitted from elsewhere.

- *Inequality within countries.* Vulnerability to impacts cascading across geographies, such as a trade interruption, immigration, a new disease, etc., depends on the proportion of the population most exposed and vulnerable to the impacts. This is typically proportional to inequality within the country. Where more people are economically marginalised, they typically have less innate resilience to impacts and are therefore at greater risk.
- *The policy environment*, including the rule of law and protection of human rights. Policies related to the movement of goods, people, finance, and information, and the willingness to change these policies in times of crisis is a contributing factor to the way that risks can be magnified or attenuated, as well as the strength of legal protection for human security. For example, instituting barriers to the export of food from producer countries in a time of rising food prices – in the name of ensuring local food security – was a contributory factor for amplifying food price spikes in the past, leading to widespread impacts around the world (see below).
- *Market dynamics.* Traded commodities (and the financial flows that support them) are a key area where disruption can lead to market dynamics that amplify the impact of hazards on supply chains. For example, if there is an existing tight supply globally of an important good (for which there is little scope to meet needs in other ways) a disruption can contribute to a run on the market. This comes from panic buying as institutions and people try to fulfil their needs quickly, and can lead to price rises, further reinforcing the perception of limited supply, and fuelling more panic buying. Amplifying market dynamics can arise through factors that tighten supply (e.g. emerging alternative markets for goods, production systems already working at their limit with little scope for compensatory change, increases in demand from other causes, lack of transparency of market supply, and other barriers to trade).
- *Politics and conflict.* Poorly governed states, or states in conflict, may be more vulnerable to risks transmitted internationally than well-governed states in peace because of the latter's innate potential to plan for, and effectively mitigate, shocks. Furthermore, disputes and international conflicts (e.g. failure to engage with international rules-based cooperative approaches, trade wars, border closures, cyber attacks, and so on) all have the potential to increase vulnerability to risks transmitted from overseas.

The exposure and vulnerability parts of the risk equation are inherent properties of the socio-economic and political context of a country. Theoretically, even if the hazards are increasing (see above), the risks could decline through adaptation (e.g. resilience building) and mitigation (see below). And, whilst at a broad scale, we can predict the types – if not the timings – of hazards in the future (using earth-system models), the radical uncertainty of the contexts means that exposure and vulnerability of different populations and systems could increase or decrease (Kay and King 2020). On a global scale, it is clear states have increasingly become more deeply integrated in globalised systems; this may broadly increase risk through exposure to cascading impacts (Puma et al. 2015). Indeed, reducing exposure may require diversification away from highly concentrated supply chains rather than turning inwards and building more self-sufficiency. Furthermore, globally, inequality is rising (Goda and Torres García 2017; Court et al. 2021), which may also increase risk due to the amplified vulnerability of marginalised populations (or parts of populations).

## Case studies of cascading risks

There are an accumulating number of examples of risk cascades in the literature, and potentially these are a very small sample of the potential number of pathways that a hazard can lead to cascading risks with impact on distant societies (Carter et al. 2021). In many ways, the recent experience with COVID-19 illustrates many of the cascading and systemic risks highlighted in this paper. It may also have been made more likely through a combination of climate change and land degradation creating ecological rewiring. The hazard arose in China and it rapidly spread around the world through contagion, leading to significant economic impacts through border closures and local lockdowns changing demand for goods, with some impact on supply-side dynamics. Lockdowns suppressed demand, and economies shrank, leading to total economic costs that significantly dwarf the health costs alone. Post-COVID-19 economic recovery led to a demand/supply imbalance, and the growth of inflation that has made the unrolling and cascading impacts of the Ukraine War even more severe (Benton et al. 2022).

Cascading risks can, of course, come from hazards that are unrelated to environmental issues, such as violent conflict, as evidence by the 2022 invasion of Ukraine (Benton et al. 2022). However, conflicts – even soft-power conflicts – have the potential to add further stress to the demand-supply dynamics that can feed back to increase risks. For example, with the election of Donald Trump as US President in 2016, it was possible to sketch out the likely scenario whereby the US instituted a US–China trade war by increasing tariffs on Chinese steel followed by retaliatory Chinese tariffs on US soy, leading to incentives for production growth in Latin America and increasing deforestation in Brazil (Benton 2017). These scenarios increase greenhouse gas emissions and the potential for accelerating climate hazards globally. This prediction seems to have come to pass (Lu et al. 2020).

We illustrate further these points on cascade risk with two contrasting case studies: the 2010/11 global food price crisis and climate-amplified security risks in the Sahel.

### Case study: the 2010/11 food price spike

The 2010–2011 food price spike was initially triggered by extreme heat in the summer of 2010, impacting in Eastern Europe and particularly in Ukraine and Western Russia (Barriopedro et al. 2011; Watanabe et al. 2013; Hoag 2014). It was extreme in its temperature (over 40°C) and duration (lasting from July to mid-August). Wheat yields declined by about one-third (Wegren 2011; Philippe et al. 2016).

The thinness of global food markets means that interruptions in supplies can have big effects, because market shortfalls tend to produce market runs, which in turn, amplify the impact and cause policy responses – like export bans – that may relieve some problems locally but amplify the price spike. In 2010, Russia imposed an export ban in order to ensure Russian grain access. A range of countries responded in similar, and uncoordinated, ways as dictated by national self-interest and domestic politics (Jones and Hiller 2017). Combined, the yield shortfalls amplified by export bans led to a global food commodity price spike (Welton 2011) and may have been further amplified by financial speculation (Spratt 2013). This price spike was the largest during the modern era, only recently surpassed by the impacts on global food commodity markets by the illegal invasion of Ukraine – affecting both the same region and with similar consequences unfolding (Benton et al. 2022).

Across the world, the food price spike had impacts, particularly on low-income households in low and middle-income countries. Responses included increasing work to bolster earnings, accessing savings where available, reducing spending on 'non-essentials', and raising their voices in demonstrations. Rising food prices exacerbated feelings of powerlessness for economically marginalised groups, who identified collusion between powerful incumbents (such as between politicians and 'big business') alongside a disregard for their economic marginalisation (Hossain and Green 2011). The food price spike led in many places to a politicised response and to food-related civil protest, including riots, across a range of countries (Natalini et al. 2017).

In high-income countries, social impacts were more muted. Food inflation increased about fivefold in 2010, and as prices rose, across the board households bought less and traded down in quality, particularly in lower-income groups. Emergency food provision, exemplified by the food-bank charity, the Trussell Trust, saw a 50% increase in demand in 2010/11.

### *Case study: environmental stress, climate, and conflict in the Sahel*

In the coming years, the Sahel will face a variety of climate stresses, including more rainfall variability and more extreme events like droughts and floods. The region is also likely to see temperatures, particularly in the northern parts of the Central Sahel, rising up to one and a half times faster than the global average (Desmidt et al. 2021). Although the region has a wealth of natural resources, it also faces a daunting array of developmental challenges. Since the turn of the century, the population of the Sahel has roughly doubled and currently stands at 100 million people, or about 9% of Sub-Saharan Africa's population (excluding South Africa). The region is mostly rural with over 60% of employment in agriculture and less than one-third of the populace living in urban centres. Security challenges are rife – with nearly 3.5 million refugees or internally displaced people according to UNHCR data – and recent coups have taken place in Burkina Faso (in January 2022 and September 2022), Mali (in August 2020 and May 2021), Chad (in April 2021), and Niger (in July 2023), and a proliferation of violent extremist and armed opposition groups particularly in Northern Mali; the Liptako-Gourma region (border between Mali, Niger and Burkina Faso), and the Lake Chad Basin.

In this context, climate change impacts in the region have the potential to trigger a range of cascading risks. In particular, climate impacts such as prolonged droughts or increasingly unpredictable rainfall patterns will harm food security and agriculture-related livelihoods of rural communities, which are more vulnerable due to challenges in accessing services and markets. Pastoralist communities across the region will also be severely affected by shifts and shocks in the quantity and quality of available grazing resources (Desmidt et al. 2021). The effects of this are likely to lead to both increasing mobility and migration and also increased inter-communal tensions. For those who can move, adverse climatic change and its impacts on local communities could lead to further displacement, rural–urban migration, and also regional migration. For those who cannot move or escape, inter-communal – and state-community – conflicts could escalate, if systems of resource and conflict management fail. The historical marginalisation of pastoralist communities exacerbates this risk (Cepero et al. 2021). For example, by creating opportunities for armed groups to more easily recruit destitute farmers and pastoralists in search of food, money, or protection (Ba and Cold-Ravnkilde 2021).

The impacts of climate change are likely to continue to exacerbate and make resolution of the Sahel's myriad challenges more difficult. Military solutions to the security challenges in the region – particularly those driven by European states – have largely failed to improve local governance and institutions (Bisson 2020) and there is a general fatigue for both Sahelian and

European governments in sustaining a military presence in the region (Cepero et al. 2021). But how these peacebuilding and military interventions can move beyond an overly securitised approach, with often overlapping and incoherent initiatives focused on stabilisation rather than human security, development, and resilience will be crucial to meeting the climate challenges of the next decade head on.

## What to do?

The above discussions outline that cascading risks can be sparked by environmental hazards. These are increasing in frequency and severity. The risks also depend on the interaction of exposure and vulnerability which may also be increasing due to, amongst other things, increased global connectiveness across all aspects of society and economies, and growing inequality. Of course, there are two main courses of action that can mitigate the risks: (1) prevention of the hazards from getting worse (i.e. mitigation of the environmental causes) and (2) mitigate the impact of the hazards by decreasing exposure and vulnerability (i.e. adaptation to the risks).

### *Hazard mitigation*

Mitigating the climate hazards requires reducing greenhouse gas emissions – through both reducing reliance on fossil fuel-related energy and reducing the emissions from short-lived climate forcing gases (such as methane) and off-setting $CO_2$ emissions that are 'hard to abate' (such as aviation emissions) by increasing carbon dioxide removal from the atmosphere to compensate. Such off-setting can occur through, for example, planting trees or better agricultural management. Through the combination of reducing emissions and off-setting those that remain it is theoretically possible to achieve 'net zero' or even 'net negative' emissions, which is increasingly a key part of both governmental planning (as part of the long-term planning requirement of the Paris agreement) and institutional planning (to align with the Paris goals).

Despite the planning to decarbonise economies, currently, post-COP-27 pledges still put the world on course for well over 2°C of global warming. Given that the speed and severity of climate-related (and environment-related) hazards are increasing under today's existing environmental change, it is likely for the foreseeable future, hazards will continue to grow. Thus, mitigation – whilst essential – is not sufficient to manage the risks.

### *Adapting to the risks*

Given hazards, exposure, and vulnerability are, arguably, all increasing, managing the risks requires adapting to them by reducing the exposure and vulnerability to hazards that are foreseeable, as well as building resilience to shocks and disruptions that may be less foreseeable.

Standard approaches to risk management often involve qualitative or quantitative approaches to identifying hazards, assessing exposure, and reducing vulnerability. For example, examining supply chains for climate risks: is a factory increasingly exposed to flood risks, or a workforce to heat-related stress? Such approaches are well understood and often embedded within practice. However, with hazards changing rapidly, and exposure and vulnerability being difficult to predict (partly due to the socio-economic environment), predicting and managing specific hazards is increasingly too conservative to manage the range of risks. For example, a disruption in a country's food supply could come from a very large range of hazards affecting production (drought, floods, heat, fire, pests, etc.), transport (rail, road, or ship transport, including port infrastructure), health-related impacts on the workforce along the supply chain (exposure to heat, new pests, and diseases), changes in market conditions and local retail environments and so on. If there are a large number of potential hazards, even if each is low probability, the overall risk may be large: it may not be possible to predict a given disruption, but it is possible to predict one may be likely (Challinor et al. 2021). Such radical uncertainty calls for a strategic focus on resilience building. Resilience is broadly defined as the maintenance of function in the face of perturbation/shock and variously includes robustness and capacity to withstand or absorb shocks and the ability to recover rapidly from shocks.

Resilience arises from the following principal elements, especially economically:

- *Redundancy*. Redundancy in the name of better resilience (also called 'functional redundancy' as it is a deliberative strategy) refers to the opposite of 'lean-ness'. For example, maintaining buffer stocks of key commodities to avoid operations being interrupted by supply chain issues. Alternatively, it may involve reducing the lengths of supply chains through 'onshoring'.
- *Diversity*. Diversity refers to diversifying supply chains or operations (such as transport routes or suppliers) to avoid the case of having a sole supplier whose business gets interrupted, with knock-on consequences for operations.
- *Modularity*. Modularity is about distributing functions across operations rather than centralising them (e.g. a central deport or processing plant, or a single large-scale producer).
- *Substitutability*. Substitutability is about designing the ability to substitute components, particularly if one's supplies get disrupted. This can occur both in terms of a particular product, but also perhaps in incentivising

consumers to accept alternatives. For example, a food retailer unable to supply imported fruit, being able to promote home-grown fruit as an alternative.

* *Flexibility.* Flexibility requires being able to change business operations to track a crisis. For example, food producers for hospitality outlets typically suffered disproportionally during COVID-19 as they struggled to find alternative markets for their products. Flexibility is also termed 'strategic agility' (Elali 2021).

While many of these facets of resilience help to sustain processes and functions despite an external shock, agile management is relevant for dealing with those shocks at all scales, whether it involves dealing with a supply chain or creating mechanisms to resolve tensions between communities.

There are conceptual frameworks (e.g. Carter et al. 2021) that can be used to walk through the dynamics of hazard-impact pathways to identify more concretely where to build resilience to mitigate cascading risks (e.g. to help identify crucial bottlenecks that may be subject to failure and to identify the need to diversity or decentralise).

## Conclusions

Risks that cascade across borders and sectors by causing ripple effects on the flow of goods, people, information, finance, and conflict are likely to increase. This is in part because of the increasing nature and severity of climate and other environmental hazards. It is also from changing exposure – as institutions and states become more deeply embedded in complex, global networks, and vulnerability – as inequality exposes a greater proportion of people to impacts caused by shocks. Such cascades can, in extremis, lead to the systemic failure of the smooth functioning of our societies, as illustrated by COVID-19.

Risk mitigation necessarily (but insufficiently) requires a transition to a net-zero world to halt further climate change, but even then environmental hazards are likely to expose us to significantly more frequent and severe, disruptive impacts. Adaptation can be made to identifiable direct risks (predictable impacts on a place), but the way risks can propagate across the world, suggests that adaptation to build more strategic resilience is necessary. Such adaptation includes building functional redundancy, diversity, and modularity, but this also comes at a cost: these properties have largely been removed for efficiency gains.

Cascading risks are likely to grow in the future, and without hazard mitigation and risk adaptation, they are likely to grow significantly enough to become periodically systemic. These future costs must be balanced against the transition costs of building resilience.

## References

Ba, B. and S. M. Cold-Ravnkilde 2021. When jihadists broker peace. *DIIS Policy Brief.* https://www.diis.dk/en/research/when-jihadists-broker-peace (accessed January 2023).

Barriopedro, D., E. M. Fischer, J. Luterbacher, R. M. Trigo, and R. García-Herrera 2011. The hot summer of 2010: redrawing the temperature record map of Europe. *Science* 332(6026): 220–224.

Battiston, S., D. Delli Gatti, M. Gallegati, B. Greenwald, and J. Stiglitz, 2012. Liaisons dangereuses: increasing connectivity, risk sharing, and systemic risk. *Journal of Economic Dynamics and Control* 36(8): 1121–1141.

Bebber, D.P., M. A. T. Ramotowski, and S. J. Gurr 2013. Crop pests and pathogens move polewards in a warming world. *Nature Climate Change* 3(11): 985–988.

Benton, T. G. 2017. Food security, trade and its impacts. *resourcetrade.earth.* https://resourcetrade.earth/publications/food-security-trade-and-its-impacts (accessed 1 May 2023).

Benton, T. G., A. Froggatt, L. Wellesley, O. Grafham, R. King, N. Morisetti, J. Nixey, and P. Schroder 2022. *The Ukraine War and Threats to Food and Energy Security: Cascading Risks from Rising Prices and Supply Disruptions.* London: Chatham House.

Berman, N., M. Couttenier, and A. Leblois 2021. Crop prices and deforestation in the tropics. 10.21203/rs.3.rs-658539/v1 (accessed 1 May 2023).

Bhardwaj, P. 2018. Fiber optic wires, servers, and more than 550,000 miles of underwater cables: here's what the internet actually looks like. *Insider.* https://www.businessinsider.com/how-internet-works-infrastructure-photos-2018-5?r=US&IR=T (accessed 1 May 2023).

Bisson, L., 2020. Decentralisation and inclusive governance in fragile settings. *Clingendael.* https://www.clingendael.org/publication/decentralisation-and-inclusive-governance-fragile-settings (accessed 1 May 2023).

Brooks, D. R., E. P. Hoberg, and W. A. Boeger 2019. *The Stockholm Paradigm: Climate Change and Emerging Disease.* Princeton, NJ: Princeton University Press.

Carroll, C. R., I. Hobbs, N. Williams, and J. Manthorpw 2020. London matters 2020: a global analysis of the key trends in the insurance market. *Kennedys.* https://kennedyslaw.com/thought-leadership/reports/london-matters-2020-a-global-analysis-of-the-key-trends-in-the-insurance-market/ (accessed 1 May 2023).

Carter, T. R., M. Benzie, E. Campiglio. H. Carlsen, S. Fronzek, M. Hilden. C. Reyer, and C. West 2021. A conceptual framework for cross-border impacts of climate change. *Global Environmental Change* 69: 102307.

Cepero, O. P., S. Desmidt, A. Detges, F. Tondel, P. van Ackern, A. Foong, and J. Volkholz 2021. Climate change, development and security in the Central Sahel. https://www.cascades.eu/wp-content/uploads/2021/06/Climate-Change-Development-and-Security-in-the-Central-Sahel.pdf (accessed 1 May 2023).

Challinor, A. and T. G. Benton 2021. Third UK Climate Change Risk Assessment Technical Report UK Climate Risk Independent Assessment (CCRA3: International Dimensions. https://www.ukclimaterisk.org/independent-assessment-ccra3/technical-report/ (accessed 1 May 2023).

Court, C. D., J-P. Ferreira, G. J. Hewings, and M. Lahr 2021. Accounting for global value chains: rising global inequality in the wake of COVID-19? *International Review of Applied Economics* 35(6): 813–831.

Desmidt, S., O. Puig, A. Detges, P. van Ackern, and F. Tondel 2021. Climate change and resilience in the Central Sahel. https://www.cascades.eu/wp-content/uploads/2021/06/Climate-Change-Resilience-Central-Sah-el-CASCADES-Policy-Paper-June-2021-1.pdf (accessed 1 May 2023).

Elali, W. 2021. The importance of strategic agility to business survival during Corona crisis and beyond. *International Journal of Business Ethics and Governance* 4(2): 1–8.

Fujita, M. and N. Hamaguchi 2011. Japan and economic integration in East Asia: post-disaster scenario. *The Annals of Regional Science 2011* 48(2): 485–500.

Future Power Technology 2020. The world's longest power transmission lines. *Future Power Technology*. https://www.power-technology.com/features/featurethe-worlds-longest-power-transmission-lines-4167964/ (accessed 1 May 2023).

Gifford, E. and R. Gifford 2016. The largely unacknowledged impact of climate change on mental health. *Bulletin of the Atomic Scientists* 72(5): 292–297.

Goda, T. and A. Torres García 2017. The rising tide of absolute global income inequality during 1850–2010: is it driven by inequality within or between countries? *Social Indicators Research* 130(3): 1051–1072.

Haldane, A. G. and R. M. May 2011. Systemic risk in banking ecosystems. *Nature* 469(7330): 351–355.

Hoag, H. 2014. Russian summer tops 'universal' heatwave index. *Nature*. 10.1038/nature.2014.16250 (accessed 1 May 2023).

Hooijer, A. and R. Vernimmen 2021. Global LiDAR land elevation data reveal greatest sea-level rise vulnerability in the tropics. *Nature Communications* 12(1): 1–7.

Hossain, N. and D. Green 2011. Living on a spike: how is the 2011 food price crisis affecting poor people? *Oxfam Policy and Practice: Agriculture, Food and Land* 11(5): 9–56.

Human Development Report 1994. *United Nations Development Programme 2022*. https://www.undp.org/library/human-development-report-1994 (accessed January 2023).

IDMC 2021. *Global Report on Internal Displacement*. Geneva: IDMC.

Ijaz, M., M. Kashif Yar, I. Hussain Badar, S. Ali, M. S. Islam, M. H. Jaspal, Z. Hayat, A. Sardar, S. Ullah, and D. Guevara-Ruiz 2021. Meat production and supply chain under COVID-19 scenario: current trends and future prospects. *Frontiers in Veterinary Science* 8: 432. 10.3389/fvets.2021.660736 (accessed 1 May 2023).

Jones, A. and B. Hiller 2017. Exploring the dynamics of responses to food production shocks. *Sustainability* 9(6): 960.

Kay, J.A. and M. A. King 2020. *Radical Uncertainty: Decision-Making beyond the Numbers*. London: W. W. Norton & Company.

Kornhuber, K., S. Osprey, D. Coumou. S. Petri, V. Petoukhov, S. Rahmstorf, and L. Gray 2019. Extreme weather events in early summer 2018 connected by a recurrent hemispheric wave-7 pattern. *Environmental Research Letters* 14(5): 054002.

Kornhuber, K., D. Coumou, E. Vogel, C. Lesk, J. F. Donges. J. Lehmann, and R. M. Horton 2020. Amplified Rossby waves enhance risk of concurrent heatwaves in major breadbasket regions. *Nature Climate Change* 10: 48–53.

Livingstone, H. 2021. Huge two-day underwater avalanche sent mud 1,000km into ocean. *The Guardian*. https://www.theguardian.com/world/2021/jun/08/

underwater-avalanche-africa-sent-mud-1000km-nigeria-south-africa (accessed 1 May 2023).

Lu, J., X. Mao, M. Wang, Z. Liu, and P. Song 2020. Global and national environmental impacts of the US-China Trade War. *Environmental Science and Technology* 54(24): 16108–16118.

Mann, M. E., S. Rahmstorf, K. Kornhuber, B. A. Steinman, S. K. Miller, and D. Coumou 2017. Influence of anthropogenic climate change on planetary wave resonance and extreme weather events. *Nature: Scientific Reports* 7: 45242.

Mosello, B., C. König, E. Wright, and G. Price 2021. Rethinking human mobility in the face of global changes A focus on Bangladesh and Central Asia. *Adelphi.* https://www.adelphi.de/en/publication/rethinking-human-mobility-face-global-changes (accessed 1 May 2023).

Natalini, D., G. Bravo, and A. W. Jones 2017. Global food security and food riots: an agent-based modelling approach. *International Society for Plant Pathology* 11(5): 1153–1173.

Panu, P. 2020. Anxiety and the ecological crisis: an analysis of eco-anxiety and climate anxiety. *Sustainability* 12(19): 7836.

Papadopoulou, T.C. and M. Özbayrak 2005. Leanness: experiences from the journey to date. *Journal of Manufacturing Technology Management* 16(7): 784–807.

Philippe, M., J. A. Carr, J. Dell'Angelo, M. Fader, J. A. Gephart, M. Kummu, N. R. Magliocca, M. Porkka. M. J. Puma, Z. Ratajczak, M. C. Rulli, D. A. Seekell, S. Suweis, A. Tavoni, and P. D'Odorico 2016. Reserves and trade jointly determine exposure to food supply shocks. *Environmental Research Letters*, 11 (9): 95009.

Puma, M. J., S. Bose, S. Y. Chon, and B. I. Cook 2015. Assessing the evolving fragility of the global food system. *Environmental Research Letters* 10(2): 24007.

Rees, S. and J. Fisher 2020. COVID-19 and the mental health of people from refugee backgrounds. *International Journal of Health Services* 50(4): 415 417.

Scheelbeek, P. F., C. Moss, T. Kastner, C. Alae-Carew, S. Jarmul, R. Green, A. Taylor, A. Haines, and A. D. Dangour 2020. UK's fruit and vegetable supply is increasingly dependent on imports from climate-vulnerable producing countries. *Nature Food* 1(11): 705–712.

Spratt, S., 2013. Food price volatility and financial speculation. *Future Agricultures Working Paper* 47. https://www.future-agricultures.org/publications/working-papers-document/food-price-volatility-and-financial-speculation/ (accessed 1 May 2023).

UN 2016. *Human security handbook.* https://www.un.org/humansecurity/wp-content/uploads/2017/10/h2.pdf (accessed 1 May 2023).

US Chamber of Commerce 2017. International Index of Energy Security Risk 2016. *Institute for 21st Century Energy.* https://www.globalenergyinstitute.org/sites/default/files/energyrisk_intl_2016.pdf (accessed 1 May 2023).

Watanabe, M., H. Shiogama, Y. Imada, M. Mori, M. Ishii, and M. Kimoto 2013. Event attribution of the August 2010 Russian heat wave. *SOLA* 9(0): 65–68.

Wegren, S. K. 2011. Food security and Russia's 2010 drought. *Eurasian Geography and Economics* 52(1): 140–156.

Welton, G. 2011. The impact of Russia's 2010 grain export ban. *Oxfam Policy and Practice: Agriculture, Food and Land* 11(5): 76–107.

# 2

# GEOPOLITICS AND SECURITY IN THE CHANGING ARCTIC

*Kimberly Marten*

## Introduction

The Arctic region is warming four times faster than the rest of the planet on average (Rantanen et al. 2022). Climate change there is proceeding at a pace and scope that has surprised even specialists, and this environmental shock is reshaping a wide range of security and trade interests in the region. The immediate change with the widest geopolitical impact is occurring along Russia's Northern Sea Route (NSR). Linking Asia to Europe across Russia's north, and dotted with huge oil, gas, and mineral enterprises, the NSR is becoming increasingly ice-free. This has potentially enormous implications for global trade and security, especially because China added the NSR to its global Belt and Road Initiative (BRI) in 2018.

Russia's key role in the Arctic means that a different shock reverberated in the region on 24 February 2022, when Russia invaded Ukraine. Russia's unprovoked war against its neighbour precipitated crushing, Western-led (and, so far, mostly globally observed) economic and political sanctions against Moscow. Even though the war has not directly involved the Arctic region, it had immediate repercussions there that interact with the impact of climate change.

These two independent but intertwined shocks in Russia are interacting with three other political factors shared across the Arctic region: (1) the desire by each of the Arctic coastal states (the so-called A5) to uphold sovereign control over nearby waters, giving them an incentive to maintain existing international law in the region even as rivalries and conflicts grow; (2) uncertainty about how much future development of the Arctic's rich oil and natural gas reserves will be politically possible, as global climate activism may turn petroleum into a stranded asset with no net return; and (3) the rapidly

DOI: 10.4324/9781003377641-4

changing power dynamics for indigenous peoples across the region as new pressures and opportunities arise.

The goal of this chapter is to describe and analyse the effects of these interacting factors on the changing and complex geopolitics of the Arctic.

## Climate change in the Arctic

Climate change is causing three self-reinforcing environmental shifts in the Arctic region. First and most fundamental, the increasing loss of thick, older Arctic Ocean ice and glaciers and the resultant opening of darker open ocean and land is causing a decline in Earth's ability to reflect (rather than absorb) sunlight and heat. This natural reflection of solar heat is called the albedo effect, and its loss accelerates warming at both the local and global levels. Sea ice loss is also driving rapid acidification of Western Arctic Ocean waters, at a level three to four times as high as elsewhere (Qi et al. 2022).

Second, Arctic permafrost – soil or rock mixed with ice that has been frozen for at least two years and in some cases up to 740,000 years (Stone 2020) – is permanently thawing as its constituent ice particles melt. Up to half of Earth's organic carbon may be found in dead matter that is currently locked in permafrost, much of which is located in Arctic or sub-Arctic areas (Bykova 2020). Thawing exposes this preserved matter to new decay, releasing large amounts of carbon into the atmosphere and thereby intensifying the global greenhouse effect. Permafrost thaw also releases methane, an even more potent (if shorter-lived) greenhouse gas, both from decaying organic matter and from subsurface natural gas. Some natural gas is released through sudden blowout craters, but deep channels in newly exposed Arctic limestone deposits are potentially releasing even more (Froitzheima et al. 2021). Permafrost thaw on Arctic coastlines causes erosion that in turn intensifies albedo loss, and the melted ice from thawed permafrost raises ocean levels and further aggravates erosion.

Third, climate change is causing northern (boreal) forests to experience temperature rises without an associated increase in precipitation. While wildfires have always been a significant part of Arctic ecology, this warmer, drier weather is associated with a greater number of much larger forest and even tundra fires, changing boreal forests from a carbon sink to a carbon source (Gauthier et al. 2015). Carbon is released not only by the burning of trees and other living plants, but also even more significantly from burning of the rich and deep organic matter that makes up much Arctic soil (Viñas 2019). These huge fires cause yet more permafrost to thaw.

## Arctic implications of Russia's invasion of Ukraine

Russia's industrial and commercial developments in the Arctic depend on international investment and trade. Russia has the Arctic's largest landmass and coastline. It is also the state with the largest population in the Arctic, concentrated in often dilapidated cities designed in Soviet times to support mining and other facilities originally built with GULAG prison camp labour. Russian President Vladimir Putin sought to take advantage of the increasingly ice-free NSR to revitalise and champion the country's Arctic region, now home to major energy developments as well. But sanctions have thrown these grandiose plans into doubt.

Russia's Ukraine War also led to a Western pause in Arctic Council diplomacy. The Arctic Council was founded in 1996 by the eight states with Arctic territory (Canada, Denmark/Greenland, Finland, Iceland, Norway, Russia, Sweden, and the United States) as a forum to encourage cooperation between Russia and the West on Arctic environmental, scientific, marine safety, and indigenous well-being issues. It had long been hailed for its successes even as tensions escalated under Putin, including three legally binding agreements between 2011 and 2017 (on search and rescue operations, pollution control, and scientific access). But Russia's decision to launch a war at a time when it headed the Council's rotating chair led the other seven member states to halt temporarily the Council's meetings and other activities, including scientific cooperation. The seven non-Russian members announced in June 2022 a 'limited resumption' without Russian participation of 'projects … contained in the workplan approved by all eight Arctic States'. Yet the suspension of Russian participation has created uncertainties in a region long known as 'high North, low tension'.

The implications of the Council's Russia hiatus extend beyond its member states. Unique among international organisations, the Council includes representatives of indigenous peoples in all of its discussions, giving them significant consultative influence even though they lack voting rights. The Council also includes 13 non-Arctic silent observer states from Europe and Asia, and a number of observer international organisations, including the European Union (EU). While these observers do not participate directly in discussions, they have used Council meetings to engage in other Arctic-related diplomatic activity on the sidelines. China in particular considers its observer role on the Council to be a key marker of its status in the Arctic region, while (as will be discussed below) observer status rules have simultaneously served as a brake on China's assertiveness in the Arctic. What, if any, effect the Arctic Council's partial hiatus has on China's influence and activities in the region remains to be seen.

The Barents cooperation framework, established in 1993 on Norway's initiative, was also suspended in March 2022. Barents cooperation occurs at an intergovernmental level through the Barents Euro-Atlantic Council (where the five Scandinavian states, the European Commission, and Russia are members) and at an interregional level through the Barents Regional Council (involving 13 counties or their equivalents in Norway, Finland, Sweden, and

Russia, and centring on non-governmental organisations). The original Barents' goals were to normalise relations between Russia and the West after the collapse of the Soviet Union, and it has fostered business and investment, cultural and educational, and health and environmental cooperation of various kinds, with a special role for indigenous peoples (Leland and Hoel 2008).

Compounding these uncertainties, Putin's invasion of a sovereign neighbouring state, and especially the atrocities Russian troops have committed in Ukraine, led Russia's previously nonaligned Arctic neighbours Finland and Sweden to seek membership in the North Atlantic Treaty Organisation (NATO) in May 2022. This means that the Arctic Council, whose mandate is designed to avoid military security issues, will likely soon be made up of seven NATO member states plus Russia. It also means that Russia's increasing remilitarisation of the Arctic under Putin, which had already raised tensions in the region, may now become even more entrenched as NATO's presence extends further along Russia's Arctic borders.

### Russia's Northern Sea Route

Canada's Arctic Northwest Passage is also undergoing sea ice melt and attracting new international trade and investment, including from China. But the location and relative ease of navigating the NSR and its well-established infrastructure makes it (or at least did so, before Putin provoked sanctions) much more commercially significant in the near term (Lackenbauer et al. 2018). The Russian empire began transporting timber to the United Kingdom along its Arctic coastline in the 1800s, at a time when coastal ice always underwent some summer melting (Bond 1983). Starting with Soviet leader Joseph Stalin's 1930s GULAG forced labour camps, the Soviet Union then built huge mining and processing enterprises and associated cities across its Far North, leaving the Russian Arctic much more heavily populated (and polluted) than its Western counterparts (Hill and Gaddy 2003). The NSR began to be used to connect nickel mining and refining enterprises in the western and central parts of the Soviet Arctic in 1939, and summer traffic along the route increased markedly in the late 1970s with improvements in Soviet ice-breaker technology (Bond 1983).

In the early 1990s, advanced satellite imaging revealed that NSR summertime sea ice was retreating year by year, with scientists hypothesising this was due to human activity. By the early 2000s, some climate change models began to predict an ice-free summer Arctic by 2050 (Luedtke 2015). But it was not until well into Putin's regime, in 2007–2008, that Arctic politics went through what Jonathan N. Markowitz calls an 'exogenous shock' because of melting sea ice (Markowitz 2020). Record-breaking ice melt several years in a row, including 'extreme summer minimum ice coverage' in 2007 and 2008, led mainstream modellers to predict that the entire Arctic Ocean would be nearly ice-free in the month of September (its traditional annual low-ice point) by 2037 (Wang and Overland 2009). In 2020, updated models began to foresee completely ice-free Arctic Ocean summers by 2035 (Guarino et al. 2020).

The Kremlin today worries about the increasing damage to infrastructure and dwellings caused by thawing permafrost and forest fires, and for the first time, in 2021, Russia's National Security Strategy listed climate change as a threat (Trenin 2021; Duclos 2021). Yet President Putin has by and large championed climate change as a win for Russia. In addition to the benefits seen from expanded northern agriculture and reduced heating costs, the Kremlin has focused on the commercial possibilities of an increasingly ice-free NSR (Gustafson 2021; Staalesen 2021d; Conley et al. 2021). The Putin regime has embarked on gargantuan plans for new ports, rail lines, and even cities tied to the region's oil, natural gas, coal, and mineral reserves while touting the NSR as a global shipping route to rival the Suez Canal.

Putin also oversaw the remilitarisation of Russia's Arctic coastline, particularly in the western province of Murmansk. Western analysts have debated whether Moscow's motives there are defensive, to protect a newly accessible (and hence vulnerable) national border, or offensive, to use for new power projection attempts beyond the Arctic region against Europe and North America (Boulègue 2019, 2021; Conley et al. 2021). The huge naval port at Severomorsk, on the Kola Peninsula just 45 kilometres from the Norwegian border, is the home of Russia's Northern Fleet, which, in 2014, was given command over an emerging new military district that includes two Arctic motorised infantry brigades and significant air defence forces, beyond its longstanding naval assets. Approximately two-thirds of Russia's strategic nuclear deterrent arsenal is based in the Arctic, and the Arctic is where Putin has also chosen to test some of the new nuclear-capable 'super weapons' he unveiled in 2018 (Connolly 2021).

Beijing meanwhile released a new Arctic strategy in 2018, with the NSR as a key component. China began referring to itself as a 'near-Arctic state', and in cooperation with Russia added the NSR to its global commercial BRI as the Polar (or Ice) Silk Road (Nakano and Li 2018). Beijing's new-found interest in the Arctic, and especially in the region's energy and mineral resources, is in turn alarming some in the West about its intentions there. One important question is whether Russia and China will forge a new commercial and/or military alliance in the Arctic – or instead struggle over access rights as Russia asserts control over NSR passage (Foxall 2021; Doshi et al. 2021).

Predictions about the NSR's economic future were already quite contested by 2021. Broken ice and sudden freezes are likely to bedevil all but high summer passage there for quite some time, requiring the use of scarce and expensive icebreakers or ice-class ships in dangerous conditions (Humpert 2021). Russia's own Arctic commercial enterprises and military forces have developed well-established means for using the route, but foreign interest was always likely to suffer as a result. Russian sources noted that the state would have to subsidise international shipping there, at least initially, to compete with Suez Canal costs (Staalesen 2021f). There is also great scientific uncertainty about the ultimate effects of climate change phenomena along the NSR. Permafrost thaw and coastal erosion could make port facilities and the transportation infrastructure connected to them unstable and unreliable, even as ice melt makes the waterway more navigable (Conley et al. 2021).

Now the future of the NSR has become even more difficult to predict, given the stiff and still evolving investment, trade, and banking sanctions imposed against Moscow by the United States, the EU, and their allies. Most Russian industry depends on imported Western technology, and as existing stocks are depleted this will further hamper opportunities for growth. Putin insisted that his Artic development plans would continue anyway, using 'countries and alliances from outside the region' (Staalesen 2022). One country to watch in this regard is the United Arab Emirates (UAE). UAE President Mohamed bin Zayed Al Nahyan visited Putin in St Petersburg in October 2022 and spawned talk of Emirati investment in Novatek, a publicly traded Russian liquid natural gas (LNG) firm based in the Arctic, even though the necessary liquefaction technology is under EU sanctions (Humpert 2022).

Yet China, for example, has much greater trade and investment interests in North America and Europe than it does in Russia, and even state-owned Chinese banks and firms (except for its autarkic sovereign wealth fund) largely abided by the initial Western sanctions against Russia following Russia's seizure of Crimea in 2014 (Gabuev 2016). That Chinese pattern of observing the letter of Western sanctions, even while increasing trade with Russia, appeared to continue in 2022. The power of the US Treasury Department and the US banking system to punish sanctions-busting countries and businesses through secondary sanctions means that other potential partners will likely think twice about long-term investment and trade deals with Russia now, even if they publicly disavow the sanctions or attempt to remain neutral on the war in Ukraine.

The fact of Russia's dependence on external technology and investment likely means that the commercial viability of the NSR will have to wait for a post-Putin future. How this might affect conflict and cooperation in the Arctic, and how soon that future will become clear, may depend on who Putin's eventual successor is, and how that person chooses to approach Ukraine, Europe, and China.

## The importance of sovereignty in the Arctic

It is possible that an enraged Putin (or perhaps an extreme nationalist successor in the Kremlin) could exit the global institutional order entirely if

Russia feels sufficiently excluded. The United Nations (UN) General Assembly voted overwhelmingly in March 2022 to recognise Ukraine's territorial sovereignty and demand Russia's immediate military withdrawal, and by a smaller margin, in April, to suspend Russia from the UN Human Rights Council. Then in May 2022, UN Secretary-General António Guterres called Russia's invasion a violation of the UN Charter in a speech before the UN Security Council (UNSC), where Russia is a permanent member. There is even a lively expert debate about whether Russia could be removed from UNSC membership (Maurer 2022).

Assuming that Russia continues at least minimal UN participation, the UN norm of sovereignty will likely limit the degree of conflict that occurs in a warming Arctic. Putin may not have recognised Ukraine's right to exist as a sovereign nation (Putin 2022), but he will certainly continue to insist that Russia has the right to control its own sovereign territory and to exercise its international legal rights in surrounding waters.

In particular, the United Nations Convention on the Law of the Sea (UNCLOS) benefits Russia and the other four states with Arctic Ocean coastlines (Canada, Denmark/ Greenland, Norway, and the United States). Even though the US Senate never ratified UNCLOS, Washington recognises UNCLOS as customary international law and abides by its rules (NOAA 2022). UNCLOS extends the sovereign territory of all coastal states 12 nautical miles beyond their shores and also gives those states the right to declare a 200-mile exclusive economic zone (EEZ) to govern the use and management of marine natural resources (such as fisheries, wind energy, and seabed mineral nodules, for example). While the central Arctic Ocean over the North Pole does not fall under any EEZs, UNCLOS makes clear that most of the Arctic is not ungoverned territory.

This matters because some Chinese statements have implied that the Arctic is a 'strategic frontier' whose resources should be divided equally among humankind based on state population (Doshi et al. 2021). Although China is a UNCLOS signatory, it does not recognise UNCLOS norms in the South China Sea; in 2016, Beijing rejected the UNCLOS-based ruling of an international tribunal in a territorial dispute there that favoured the Philippines. Nevertheless, a condition for Arctic Council observer status (which can be revoked at any time by any of the eight member states) is recognition of the existing legal framework in the Arctic region, including UNCLOS. China has thus far complied. This gives all Arctic states an incentive to maintain both the Arctic Council and UNCLOS jurisdiction.

The A5 emphasised this in their 2008 Ilulissat Declaration, asserting that their sovereign coastal status gave them, alongside 'other users of this ocean', a 'solid foundation for responsible management' of the entire Arctic Ocean with 'no need to develop a new comprehensive international legal regime to govern' it (A5 2008; Landriault et al. 2019). In other words, the Arctic is not like Antarctica, an uninhabited territory that needs to be governed by a global treaty regime.

The A5 then demonstrated their ability to lead Arctic Ocean management issues (alongside other users) with the Central Arctic Ocean Fisheries Agreement. Under the auspices of a new 'A5+5' format (where the plus-5 includes the fishing-intensive states of China, Iceland, Japan, and South Korea, alongside the EU), this agreement created a new regional fisheries management system in anticipation of the time when the central Arctic Ocean will be ice-free. The parties agreed not to permit commercial fishing by their citizens in the high seas of the central Arctic for the duration of the agreement (16 years) and to create a joint scientific research and monitoring programme to better understand the central Arctic ecosystem. The hope is that by the time the current agreement expires, they will have accumulated sufficient joint expertise to allow the creation of an agreed commercial fishery regime for the newly open ocean (Balton 2021). This US-initiated agreement was signed in 2018 and came into force in 2021 – in other words, more evidence of high North cooperation at a time of global high tension between the West and both Russia and China.

Beyond approving the fisheries agreement, Russia has also accepted the jurisdiction of the UN Commission on the Limits of the Continental Shelf (CLCS) (Hossain 2021). According to the UNCLOS regime, a state's continental shelf, like its EEZ, is assumed to extend 200 nautical miles, and states have exclusive rights to the resources lying on and below the seabed on their shelves (including oil and natural gas reserves, and once again, mineral nodules, but unlike the EEZ, not fisheries or other use of the water itself) (Overfield 2021). The CLCS allows a state to propose that its underwater geographical shelf extends beyond its EEZ. States must submit many rounds of scientific and mapping data to support these claims, and then the CLCS issues a recommendation (not a ruling) in response. Russia was the first state to file such a proposal in 2001 and has made several ever-larger amendments in ensuing years. It was followed by Denmark and Canada, and the shelf claims of these three Arctic states overlap. (The United States cannot file a proposal at present since it has not ratified UNCLOS.) At least thus far, Moscow has not acted in defiance of CLCS recommendations and seems content to negotiate with Denmark and Canada about their overlapping shelf claims using CLCS guidelines.

A submarine scientific expedition crew raised hackles in August 2007 when it planted a Russian flag under the sea ice at the bottom of the Arctic Ocean in international waters at the North Pole, in a stunt lauded by Russian state media. Yet in the following years, there has been no attempt by Russia to seize the North Pole or any other Arctic territory by force. In spite of all of the conflict between Russia and the West, Russia has acted largely in accordance with well-established international legal norms in the Arctic.

There are legal grey areas contested by the United States against both Russia and Canada. Article 234 of UNCLOS gives states special rights to control environmental pollution in EEZ locations where ice is a navigational hazard for most of the year. Both Russia and Canada have interpreted this to mean that they can oversee commercial shipping along their Arctic coastlines by excluding ships that flout their national regulations (Fletcher School of Law and Diplomacy 2022). In 2018, Russia enacted a law limiting commercial transit between Russian ports, as well as ice-breaker service and the international export of oil, natural gas, and coal from NSR ports, to ships that are Russian flagged. While the Russian government can make exceptions, and there are procedures for switching flags, this is a clear assertion of legal sovereignty over the NSR (Moe 2020).

Ottawa and Moscow also both claim that their Arctic coastal waterways are internal, given the presence of historically inhabited islands off their coasts, and hence subject to domestic control. The United States claims in contrast that they are straits used for international navigation and instead subject to the free naval transit rules of the high seas (Auerswald 2019). After a French naval vessel made a surprise 2019 voyage along the NSR from Norway to Alaska, Russia changed its laws to force foreign navies to seek permission for transit 45 days in advance and to take a Russian pilot on board for the trip (Staalesen 2019). While US naval officers have talked about carrying out a Freedom of Navigation Operation (FONOP) in either the NSR or Northwest Passage to assert the right of passage, Washington has not yet taken that potentially escalatory step.

Moscow's assertion of legal control over NSR transit has implications for its relationship with Beijing (Foxall 2021). Twice (in 2003 and 2012) Russia has blocked Chinese vessels from using the NSR, and in 2013 it even prevented Chinese researchers from renting Russian ships for a scientific expedition, fearing that Beijing's goals were actually security related. This pattern changed in 2014, following Russia's seizure of Crimea from Ukraine, when the Kremlin began viewing China as an alternative to the West after the latter began imposing sanctions on Moscow. But the message remains clear: Russia holds the cards in the NSR, and China must play by Moscow's rules. Once again this gives Russia an incentive to uphold norms of sovereignty in the Arctic, so that its own assertions of legal rights there continue to be respected.

### The problem of stranded assets

Russia has portrayed the opening of the NSR as a bonanza for its petroleum industry, but the truth is more complicated. At least in the short term, global petroleum investments and development plans have indeed boomed as oil prices skyrocketed in early 2022 – in part because of Russia's war in Ukraine.

Yet oil, gas, and coal companies across the globe have all become increasingly attuned to the likelihood that at some future point, global climate activism and the political push for net-zero carbon emissions will leave their products without a market, and their existing physical assets without buyers (Livsey 2020). Some analysts believe that this spurning of carbon, now more gradual and backsliding than many would prefer, will come to resemble a J-curve with a sudden acceleration, perhaps as early as the mid-2020s (Dalman et al. 2022). If modellers are correct that petroleum assets will become abruptly 'stranded' or unusable, the impact will hit not only the states and international companies that own or trade in petroleum, but also their financers, insurers, and ordinary people with pension funds or other broadly diversified investments. The result could be a global depression, and fears of this have motivated US President Biden's climate action policies (US White House 2021).

Stranded asset concerns are especially acute in the Arctic, where a number of major international banks and insurance firms have recently decided not to support additional petroleum projects (Mikhailova 2022). The A5 states all have significant known petroleum deposits in the Arctic but with very different political backdrops shaping their current policy choices.

The government of Greenland, an island country with an 89% indigenous population and self-rule within the sovereign Kingdom of Denmark, decided in 2021 to suspend all oil exploration on its territory even though it has potential reserves of over 30bn barrels. Several oil majors had prospected there since the 1970s but came up mostly dry (Skydsgaard 2021). The ruling Inuit Ataqatigiit Party said its decision was based on both environmental factors and economic considerations that amount to a recognition of the stranded assets problem. The new policy is interesting (with its longevity called into doubt by the popular Danish television drama series 'Borgen') because Nuuk needs to boost its economic self-sufficiency if it is to achieve independence from Copenhagen. Denmark currently subsidises more than half of Greenland's budget. As melting ice makes its resources more accessible, Greenland is increasingly the site of geopolitical competition between China, which seeks access to its mineral wealth in return for economic support, and the United States, which has long operated the Thule Air Force base there and is now being pushed by Nuuk to expand its commercial investments (Walt 2021).

Canada has not gone quite as far as Greenland but seems at the moment to have also tipped the balance against Arctic petroleum. The most important reserves in the Canadian Arctic are located offshore. Although shrinking Beaufort Sea ice is making those huge reserves more accessible, and helps to explain the Canadian filing of a CLCS extended continental shelf claim, commercial interest waned after the 1980s amid both regulatory uncertainty and global price instability. Prime Minister Trudeau placed a moratorium on new offshore Arctic oil and gas drilling in 2016, subject to a five-year scientific evidence review, and in 2019 the government suspended all existing offshore drilling as well. In 2021, the ban was extended through 2022, pending additional scientific research (Petroleum and Mineral Resources Management Directorate 2022). These decisions were controversial, opposed by both the Northwest Territories regional government and the Inuvialuit Regional Corporation (IRC) of local indigenous people, who in past decades had relied on oil industry jobs. In 2023 the moratorium was extended, under a new agreement with the IRC to ensure their full participation in policy going forward.

US decision making on Arctic petroleum is even more politically fraught. It is caught in a battle between industry-supporting Republicans and environmentalist Democrats at the national level, with local Alaskan indigenous communities split over whether to prioritise environmental or employment interests, and with federal court rulings adding to uncertainties. The state of Alaska has long depended on the petroleum industry, largely based in two Arctic regions on the North Slope of the Brooks Mountain Range: the National Petroleum Reserve to the west (owned by the US federal government since the early 20th century, with limited areas leased to private companies), which contains the environmentally fragile Teshekpuk Lake area; and the Prudhoe Bay field in the centre (owned by the state of Alaska and leased to a consortium of private oil majors). In 2019, two-thirds of Alaska's state budget stemmed from oil revenues, and all state residents receive an annual dividend from the Alaska Permanent Fund, an investment vehicle based on the state's petroleum and mineral leases and royalties. Yet oil production in the state peaked in 1988, and revenues have been in decline ever since (Marohl 2021). While new reserves continue to be discovered, older fields are in decline. The trans-Alaska pipeline system, with the largest line running south from Prudhoe Bay, now carries only a quarter of its 1980s volume (Harball 2017). Oil major BP sold its huge Alaska holdings to Hilcorp, a local company, and left the state, in 2019, because of the political and commercial uncertainties it encountered, but ExxonMobil and ConocoPhillips continue significant Alaskan operations.

New offshore drilling was banned by the administration of Democratic President Barack Obama in 2016, in a joint act with Trudeau (but without Canada's five-year review provision). One preexisting near-coast Arctic development was unaffected by the ban, the BP/ Hilcorp Liberty Project, but its authorisation was later revoked by a federal court order in a suit brought by environmentalists in 2020.

The question of whether to open the protected Arctic National Wildlife Reserve (ANWR, a huge tract of federally owned land east of Prudhoe Bay) to petroleum drilling has been contested for over forty years. A well was dug there in the 1980s by a Chevron-led consortium but with reportedly poor results (Bourne 2015). A Republican-led US Congress voted to open a vast new region of the ANWR to drilling in late 2017, with support from the Arctic Slope Regional Corporation (ASRC), a private company owned by the area's Inupiat indigenous residents that controls subsurface rights (Hardin 2018). Republican President Trump put 22 parcels of ANWR land up for lease as one of his last acts in office in January 2021, but no major oil corporations were interested because of the political and commercial uncertainties involved (Fountain 2021). Democratic President Biden then suspended all drilling in the ANWR in June 2021, pending scientific review of its environmental impact. Two existing leaseholders, Chevron and Hilcorp, later paid the ASRC US$10 m to terminate their existing contracts there (Herz 2022).

Meanwhile, ConocoPhillips received federal approval for a major new Alaskan development, the Willow Project in the National Petroleum Reserve. Biden had reversed much of Trump's Arctic development plans by closing off new land to oil leases in much of the National Petroleum Reserve in 2022, after Trump had opened it despite the objection of environmentalists. The ultimate fate of Alaskan oil remains unclear, but stranded assets concerns are already having a great deal of impact there.

In contrast to these trend lines in Greenland, Canada, and the United States, the Norwegian government is substantially increasing its Arctic offshore drilling, adding 31 exploration blocks in the Barents Sea in 2022 and planning to lift national oil output by 9% by 2024 (Solsvik 2022). Norway now justifies its oil exports at least in part as an alternative to Russian oil for Europe. The Norwegian state-controlled Equinor oil major (formerly Statoil) announced that it was exiting all cooperative ventures with its Russian counterparts, in February 2022, after having been involved with joint ventures in the Russian Arctic since the 1990s with assets worth US$1.2bn (Nilsen 2022).

But Oslo's approach to Arctic petroleum is 'paradoxical', and it has begun to face agonising decisions around the stranded assets question (Lahn 2021). Beyond the income it gains from its 67% stake in Equinor, Oslo earns additional income from petroleum licences, pipelines, and refining. In the words of a government brochure, 'The industry plays a vital role in the Norwegian economy and the financing of the Norwegian welfare state', providing 28% of Norwegian GDP, 42% of state revenues, and 58% of export earnings (Norskpetroleum 2022). Yet the government also wants to be known as a leader on climate issues. Most of the country's electricity consumption comes from renewable hydropower, and electric vehicles (EVs) made up two-thirds of its automobile market by 2021. Most domestic Norwegian oil use goes only into powering offshore rigs.

Like the United States, Norway has also faced court filings by environmentalists. In 2020, a suit was brought to the Norwegian Supreme Court, arguing that permits issued in 2016 for Arctic offshore oil drilling contributed to climate change and hence violated the human right to a clean environment guaranteed by the constitution and the European Convention on Human Rights (Libell and Taylor 2020). When activists lost that case, they tried again at the European Court of Human Rights, where the case is still pending as of 2022. One argument the state has raised against the activists is that if Norway does not supply the oil, someone else with less stringent drilling and transit environmental practices will, as long as there is global demand.

In comparison with the complex politics of its Western counterparts, Russia's Arctic petroleum development seems rather simple: petroleum enterprises, including the state-controlled Rosneft oil and Gazprom natural gas companies and the publicly traded Novatek LNG firm, have begun major new development and drilling projects that appear to undercut any professed Russian concern about climate change (Hurst and Khrennikova 2021; Staalesen 2021a, 2021b). Rosneft CEO Igor Sechin, a longtime friend of Putin, claims that his firm's new Arctic developments will produce 'green barrels' of oil, even while building 15 new company towns and 3 new airports (Staalesen 2021e). A 2021 state programme for the expansion of LNG exports, focused on new Arctic fields, seems almost Soviet: the Energy Ministry is assigned to coordinate the plan and report back to the Prime Minister annually on how well it is being fulfilled (Mishushtin 2021).

Activism of all kinds has been progressively shut down under Putin, and environmentalism in Russia has always tended to focus on local ecological concerns, not global climate change (Newell and Henry 2016; Davydova 2021). This gives Moscow great leeway in its development choices. The distinction between state and private interests in Russia is blurred, especially in the case of huge Arctic petroleum enterprises. For example, Leonid Mikhel'son, the CEO and Chairman of Novatek (based entirely in the Arctic), saw both his business opportunities and his personal fortune skyrocket when he sold shares to Putin crony Gennady Timchenko in 2008 (Kaz'mina and Todorova 2016). In turn, Russia's kleptocratic system results in vastly overpriced state contracts and shady offshore deals that hide graft and corruption (Dawisha 2014; Buckley 2018; Åslund 2019).

Russia's economy depends on oil and natural gas exports, and the Putin regime has consistently skewed its discussions about Arctic petroleum development to emphasise rewards and downplay risks (Sidortsov 2019). The Putin regime may not care much about the long-term stranded assets risk, since its leaders have already made a great deal of money by approving new petroleum contracts even if those businesses ultimately fail. Yet Russia's recent hope has been that its Arctic oil and gas will be increasingly demanded by Asian customers, especially China, even if it progressively loses its European market (Spivak and Gabuev 2021; Eiterjord 2022; Greenwood and Luo 2022). China currently imports around 10% of its imported natural gas from Russia, including through Gazprom's Power of Siberia pipeline (first opened in 2019); a Power of Siberia 2 is on the books and would increase exports significantly. China is also a partial financer of, and investor in, Novatek's Yamal LNG project and is a smaller partner in the Arctic LNG 2 project now underway. Beijing has purchased LNG from Novatek since 2013, and its purchases have increased with time. In 2020, China also took a shipment of Russian Arctic oil from Gazprom-Neft, the oil production arm of Gazprom. (China imports a great deal of oil from Russia, but most comes from Siberia and the Russian Far East below the Arctic Circle.) Additionally, China is involved in some shipbuilding and port development projects related to the NSR.

While China has increased its overall trade with Russia since Western sanctions began in 2014, and again after the Russian invasion of Ukraine in 2022, it has nonetheless been very careful (at least through to the end of 2022) not to violate the Western sanctions regime. India also began buying up discounted Russian oil in 2022, but from the Urals, not the Arctic; the grades of the two oils are not easily substitutable by end users. It thus remains unclear whether Russia can meet the expectations that Putin has for the NSR and its associated energy companies, given their dependence on international trade, financing, and technologies. Moscow may find itself with near-term stranded assets that stem from sanctions, not climate change activism.

### The political influence of indigenous peoples in the Arctic

Climate change has already had a noticeably negative impact on the security of indigenous Arctic peoples who choose to maintain their traditional lifestyles. Permafrost thaw and coastal and river erosion have threatened indigenous homes, infrastructure, and cultural artefacts. The Yup'ik village of Newtok, Alaska, has become famous for its 20-year process to permanently relocate residents ten miles away to a safer village still under construction (Welch 2019). That ongoing move, along with more general transportation difficulties because of thawing permafrost, also imperilled the accuracy of the 2020 US federal census. Indigenous peoples across the United States are already undercounted by census workers, and this exacerbated fears that Alaskan indigenous communities would receive fewer government resources than they deserved, including relocation support (Wang 2020).

Climate change is also altering the animal populations that traditional indigenous communities rely on for their livelihoods. The effects may be mixed. For example, fish stocks in the Arctic may be increasing overall as waters warm, but the stocks moving in (like pink salmon and common cod) are not part of traditional diets, and large and aggressive fish may drive the smaller and more vulnerable native Arctic fish species into extinction (Kennedy 2015; NOAA 2020). Reindeer herds are being affected, too. In the Yamal-Nenets Autonomous Okrug (YNAO) of Russia an anthrax outbreak that killed thousands of reindeer and infected dozens of humans in 2016 may have been caused by permafrost thaw, since it was associated with a temperature rise (Liskova et al. 2021; Ezhova et al. 2021).

Meanwhile, growing commercial interest in Arctic extraction has interacted with these physical and biological processes, leading to differing results for indigenous peoples' political power in different locations. In Russia's YNAO, the expansion of natural gas exploration and development by Gazprom and Novatek has significantly reduced traditional reindeer grazing grounds for the Nenets people. Perhaps in anticipation of this, the Putin regime cracked down particularly hard on indigenous activism as part of its general imposition of control over politics and the media (Magomedov 2020). Some Russian indigenous activists were forced to emigrate after being labelled foreign agents, and some advocacy groups have been shuttered; the largest Russian indigenous non-governmental organisation, the Russian Association of Indigenous Peoples of the North (RAIPON) was taken over by pro-Putin representatives under threat of elimination (Scollon 2020).

Yet in another Russian Arctic region, the Taimyr Peninsula, increasing global attention to the minerals used in sustainable production became a political lever for indigenous activists. Nornickel, a gigantic private Russian firm partially based in Taimyr, is the world's largest producer of high-grade nickel, a component of rechargeable batteries used in EVs. Nornickel has a long history of being a terrible polluter, and in 2020 one of its Taymyr facilities created a notorious oil spill that dumped 21,000 tons of diesel fuel in 20 minutes when a poorly constructed holding tank collapsed from permafrost thaw. (The thaw *per se* was not related to climate change; some permafrost thaw happens every year, and Arctic construction is supposed to plan for it.) This leak turned surrounding rivers bright red and may have reached the Arctic Ocean. It damaged the surrounding lands where indigenous peoples live, hunt, and fish. Despite Putin's crackdown against activism, the Association of the Indigenous Peoples of the Taimyr Peninsula was able to wring concessions out of Nornickel, including a US$25 m assistance plan to support indigenous culture and the reindeer herding economy (Nornickel 2021). What contributed to their ability to do this was that an indigenous activist living in exile, Pavel Sulyandziga, was able to mobilise indigenous groups internationally to write open letters lambasting Nornickel's actions. A well-publicised one was addressed to Elon Musk, the CEO of Tesla, asking him not to buy nickel from Nornickel (Meduza 2020). Probably even more effective were those addressed to BASF, whose nickel refining joint venture with Nornickel in Finland is designed to supply EV batteries to the European market, and to Nornickel's major Western lenders, UBS Switzerland and Credit Suisse (Nilsen 2021). Sulyandziga remained dissatisfied with Nornickel's actions in response, but it nonetheless seems that indigenous climate activism can make a difference even in Russia.

Where Arctic indigenous voices are having the most sway is Greenland, known in the local Inuit language as Kalaallit Nunaat. As noted above, Greenland has recently banned oil exploration, and an additional limit was placed in November 2021 on some types of mining: parliament passed a law banning exploration of deposits where there is even a small amount of uranium present (Reuters 2021). This effectively quashed the development of a rare earth minerals mine, the Kvanefjeld project, which had been on the books since 2007. Although led by Australian firm, Greenland Minerals, what gave this project particular geopolitical significance is that the largest shareholder of Greenland Minerals is actually a Chinese firm, Shenghe Resources (Treadgold 2021). China already has the largest global holdings of rare earth minerals, which are used to construct computers, EVs, and wind turbines in addition to advanced weapons, although its global share slipped from 80% to 60% in recent years as Western countries upped their own production in response (Chang 2022). Greenland's ruling Inuit Ataqatigiit Party made clear that it was not against all mines – which may be a ticket out of dependence on Denmark – just this one.

## Conclusion

There are so many factors, both physical and political, that interact in such complex ways across the Arctic that it is impossible to make any definitive predictions about future geopolitics.

Yet despite rising concerns that the melting Arctic will become some kind of dangerous new 'great game' as states scramble for resources there, a variety of factors are likely to limit both their actions and the potential for conflict in the region. The desire by the coastal Arctic A5 states to uphold their sovereignty, and hence the power of international law, will limit the scope for territorial contention and competition over ocean resources. The commercial reality that new petroleum developments in the Arctic face is the likelihood of becoming stranded assets at some future point – alongside the impact of sanctions against Russia – is increasingly serving as a limit on Arctic development. These trends may eventually mean that the full potential of Arctic carbon-based resources remains locked in the earth, as renewables supplant petroleum. Meanwhile, the political actions of indigenous peoples, in resource-rich Greenland and even in authoritarian Russia, are demonstrating that great powers lack the sway over local communities they held in an earlier era.

The West should continue to be wary about Russia's and China's assertiveness in the Arctic, but avoid buying the hype of autocratic leaders who are overselling their own Arctic ambitions. Russia's NSR will likely be plagued by both physical and commercial limits, even if sanctions are eventually lifted and Putin's plans are able to go forward. China may want to see itself as a near-Arctic state, but Nuuk has pushed back against its influence attempts, and Moscow may too if Beijing takes the NSR too much for granted.

Indeed, the most useful geopolitical action the world might take in the Arctic is to cooperate more with each other, to mitigate the climate change that is truly harming the region. Unfortunately, such cooperation does not seem to be making much headway at the moment.

## References

A5 2008. *The Ilulissat Declaration*. https://arcticportal.org/images/stories/pdf/Ilulissat-declaration.pdf (accessed 1 May 2023).

Åslund, A. 2019. *Russia's Crony Capitalism: The Path from Market Economy to Kleptocracy*. New Haven, CT: Yale University Press.

Auerswald, D. 2019. Now Is Not the Time for a FONOP in the Arctic. *War on the Rocks*. https://warontherocks.com/2019/10/now-is-not-the-time-for-a-fonop-in-the-arctic/ (accessed 1 May 2023).

Balton, D. 2021. No. 9: The Arctic Fisheries Agreement Enters into force. *Wilson Center (Polar Institute)*. https://www.wilsoncenter.org/blog-post/no-9-arctic-fisheries-agreement-enters-force (accessed 1 May 2023).

Bond, A. R. 1983. *Noril'sk: Profile of a Soviet Arctic Development Project*. Ph.D. dissertation. Milwaukee, WI: University of Wisconsin.

Boulègue, M. 2019. Russia's military posture in the Arctic: managing hard power in a 'low tension' environment. *Chatham House*. https://www.chathamhouse.org/sites/default/files/2019-06-28-Russia-Military-Arctic_0.pdf (accessed 1 May 2023).

Boulègue, M., 2021. Mitigating Russia's military posture in the European Arctic: towards a High North hard security architecture. (In) D. Depledge and P. W. Lacenbauer (eds) *On Thin Ice: Perspectives on Arctic Security*. Peterborough, ON: Trent University Press, pp. 71–77.

Bourne, J. K. 2015. What Obama's drilling bans mean for Alaska and the Arctic. *NationalGeographic*. https://www.nationalgeographic.com/science/article/150205-obama-alaska-oil-anwr-arctic-offshore-drilling (accessed 1 May 2023).

Buckley, N., 2018. Corruption and power in Russia. *Foreign Policy Research Institute*. https://www.fpri.org/article/2018/04/corruption-and-power-in-russia/ (accessed 1 May 2023).

Bykova, A. 2020. Permafrost thaw in a warming world. *The Arctic Institute*. https://www.thearcticinstitute.org/permafrost-thaw-warming-world-arctic-institute-permafrost-series-fall-winter-2020/ (accessed 1 May 2023).

Chang, F. K. 2022. China's rare earth metals consolidation and market power. *Foreign Policy Research Institute*. https://www.fpri.org/article/2022/03/chinas-rare-earth-metals-consolidation-and-market-power/ (accessed 1 May 2023).

Conley, H. A., C. N. Newlin, C. W. Wall, and A. Lohsen 2021. *Russia's Climate Gamble: The Pursuit and Contradiction of its Arctic Ambitions*. Washington, DC: Center for Strategic and International Studies.

Connolly, R. 2021. *Putin's 'Super Weapons'*. London: Chatham House.

Dalman, A., M. Coffin, and M. Fulton 2022. Managing peak oil: why rising oil prices could create a stranded asset trap as the energy transition accelerates. *Carbon Tracker Initiative*. https://carbontracker.org/reports/managing-peak-oil/ (accessed 1 May 2023).

Davydova, A. 2021. Environmental activism in Russia: strategies and prospects. *Center for Strategic & International Studies.* https://www.csis.org/analysis/environmental-activism-russia-strategies-and-prospects (accessed 1 May 2023).

Dawisha, K. 2014. *Putin's Kleptocracy: Who Owns Russia?* New York, NY: Simon & Schuster.

Doshi, R., A. Dale-Huang, and G. Zhang 2021. *Northern Expedition: China's Arctic Activities and Ambitions.* Washington, DC: Brookings Institution.

Duclos, M. 2021. Russia's National Security Strategy 2021: the Era of 'information confrontation'. *Institut Montaigne.* https://www.institutmontaigne.org/en/blog/russias-national-security-strategy-2021-era-information-confrontation (accessed 1 May 2023).

Eiterjord, T. 2022. What does Russia's invasion of Ukraine mean for China in the Arctic? *The Diplomat.* https://thediplomat.com/2022/03/what-does-russias-invasion-of-ukraine-mean-for-china-in-the-arctic/ (accessed 1 May 2023).

Ezhova, E., D. Orlov, E. Suhonen, D. Kaverin, A. Mahura, V. Gennadinik, I. Kukkonen. D. Drozdov, H. K. Lappalainen, V. Melnikov, T. Petaja, V-M. Kerminen, S. Zilitinkevich. S. M. Malkhazova, T. R. Christensen, and M. Kulmala 2021. Climatic factors influencing the Anthrax outbreak of 2016 in Siberia, Russia. *Ecohealth* 18(2): 217–228.

Fletcher School of Law and Diplomacy 2022. The Arctic and the LOSC. (In) J. Burgess, L. Foulkes, P. Jones, M. Merighi, S. Murray, and J. Whitcare (eds) *Law of the Sea: A Policy Primer.* Medford, MA: Tufts University. https://sites.tufts.edu/lawofthesea/files/2017/07/LawoftheSeaPrimer.pdf (accessed 1 May 2023).

Fountain, H. 2021. Sale of drilling leases in Arctic refuge fails to yield a windfall. *New York Times.* https://www.nytimes.com/2021/01/06/climate/arctic-refuge-drilling-lease-sales.html (accessed 1 May 2023).

Foxall, A. 2021. The Sino-Russian partnership in the Arctic. (In) D. Depledge and P. W. Lackenbauer (eds) *On Thin Ice? Perspectives on Arctic Security.* Peterborough, ON: Trent University, pp. 82–90.

Froitzheima, N., J. Majkab, and D. Zastrozhnov 2021. Methane release from carbonate rock formations in the Siberian permafrost area during and after the 2020 heat wave. *Proceedings of the National Academy of Sciences of the United States of America* 118(32). https://www.pnas.org/doi/full/10.1073/pnas.2107632118

Gabuev, A. 2016. Did Western sanctions affect Sino-Russian economic ties?. *Carnegie Endowment for International Peace.* https://carnegieendowment.org/2016/04/26/did-western-sanctions-affect-sino-russian-economic-ties-pub-63461 (accessed 1 May 2023).

Gauthier, S., P. Bernier, T. Kuuluvainen, A. Shvidenko, and D. G. Schepashenko 2015. Boreal forest health and global change. *Science* 349(6250): 819–822.

Greenwood, J. and S. Luo 2022. Could the Arctic be a wedge between Russia and China?. *War on the Rocks.* https://warontherocks.com/2022/04/could-the-arctic-be-a-wedge-between-russia-and-china/ (accessed 1 May 2023).

Guarino, M., L. Sime, D. Schröeder, I. Malmierca-Vallet, E. Rosenblum, M. Ringer, J. Ridley, D. Feltham, C. Bitz, E. J. Steig, E. Wolff, J. Stroeve, and A. Sellar 2020. Sea-ice-free Arctic during the Last Interglacial supports fast future loss. *Nature Climate Change* 10(10): 928–932.

Gustafson, T. 2021. *Klimat: Russia in the Age of Climate Change.* Cambridge, MA: Harvard University Press.

Harball, E. 2017. Alaska's 40 years of oil riches almost never was. *NPR Weekend Edition Sunday*, 24 June.

Hardin, S. 2018. *The most powerful Arctic oil lobby group you've never heard of. CAP.* https://www.americanprogress.org/article/powerful-arctic-oil-lobby-group-youve-never-heard/ (accessed 1 May 2023).

Herz, N. 2022. 2 oil companies quietly spent $10 million to exit Arctic Refuge leases. *Anchorage Daily News*. https://www.adn.com/business-economy/energy/2022/05/27/two-oil-companies-quietly-spent-10-million-to-exit-arctic-refuge-leases/ (accessed 1 May 2023).

Hill, F. H. and C. G. Gaddy 2003. *The Siberian Curse: How Communist Planners Left Russia Out in the Cold*. Washington, DC: Brookings Institution Press.

Hossain, K. 2021. Russia's proposed extended continental shelf in the Arctic Ocean: science setting the stage for law. *American Society of International Law Insights* 25(8). https://www.asil.org/insights/volume/25/issue/8 (accessed 1 May 2023).

Humpert, M. 2021. Early Winter freeze traps ships in Arctic ice, highlighting weak safety regime. *High North News*. https://www.highnorthnews.com/en/early-winter-freeze-traps-ships-arctic-ice-highlighting-weak-safety-regime (accessed 1 May 2023).

Humpert, M. 2022. Russia's Novatek to use closer ties with UAE to secure key technology for Arctic LNG project. *High North News*. https://www.highnorthnews.com/en/russias-novatek-use-closer-ties-uae-secure-key-technology-arctic-lng-project (accessed 1 May 2023).

Hurst, L. and D. Khrennikova 2021. Rosneft's green pledge with BP challenged by its Artic plans. *Bloomberg*. https://www.bloomberg.com/news/articles/2021-02-04/rosneft-s-green-pledge-with-bp-challenged-by-its-arctic-plans (accessed 1 May 2023).

Kaz'mina, I. and M. Todorova 2016. Poleznye resursy: kto pomog Leonidu Mikhel'sonu vpervye vozglavit' spisok Forbes [Useful resources: who helped Leonid Mikhel'son first head the Forbes list (of Russian billionaires)]. *Forbes.ru*, 13 April.

Kennedy, C. 2015. Warming waters shift fish communities northward in the Arctic. *Climate.gov*. https://www.climate.gov/news-features/featured-images/warming-waters-shift-fish-communities-northward-arctic (accessed 1 May 2023).

Lackenbauer, P. W., A. Lajeunesse, J. Manicom, and F. Lasserre 2018. *China's Arctic Ambitions and What They Mean for Canada*. Calgary: University of Calgary Press.

Lahn, B. 2021. Norway wants to lead on climate change. But first it must face its legacy of oil and gas. *Vox*. https://www.vox.com/22227063/norway-oil-gas-climate-change (accessed 1 May 2023).

Landriault, M., A. Chater, E. W. Rowe, and P. W. Lackenbauer 2019. *Governing Complexity in the Arctic Region*. London: Routledge.

Leland, S. R. and A. H. Hoel 2008. Learning by doing: the Barents cooperation and development of regional cooperation in the north. (In) P. Aalto and H. Blakkisrud (eds) *The New Northern Dimension of the European Neighborhood*. Brussels: Centre for European Policy Studies, pp. 36–53.

Libell, H. P. L. and D. B. Taylor 2020. Norway's Supreme Court makes way for more Arctic drilling. *New York Times*. https://www.nytimes.com/2020/12/22/world/europe/norway-supreme-court-oil-climate-change.html (accessed 1 May 2023).

Liskova, E. A., I. Egorova, Y. O. Selyaninov. I. V. Razheva, N. A. Gladkova, N. N. Toropova, O. I. Zakarova. O. A. Burova, G. V. Surkova. S. M. Malkhazova, F. I. Korennoy, I. V. Iashin, and A. A. Blokhin 2021. Reindeer Anthrax in the Russian

Arctic, 2016: climatic determinants of the outbreak and vaccination effectiveness. *Frontiers in Veterinary Science* 8: 668420. 10.3389/fvets.2021.668420.

Livsey, A. 2020. Lex in depth: the $900 bn cost of 'stranded energy assets'. *Financial Times*. https://www.ft.com/content/95efca74-4299-11ea-a43a-c4b328d9061c (accessed 1 May 2023).

Luedtke, B. 2015. An ice-free Arctic Ocean: history, science, and scepticism. *Polar Record* 51(257): 130–139.

Magomedov, A. 2020. The Russian State and the Arctic indigenous peoples: is politics coming back? *Demokratizatsiya* 28(4): 541–564.

Markowitz, J. N. 2020. *Perils of Plenty: Arctic Resource Competition and the Return of the Great Game*. Oxford: Oxford University Press.

Marohl, B. 2021. Oil production in Alaska reaches lowest level in more than 40 years. *US Energy Information Administration*. https://www.eia.gov/todayinenergy/detail.php?id=47696 (accessed 1 May 2023).

Maurer, D. 2022. A UN Security Council Permanent Member's de facto immunity from Article 6 expulsion: Russia's Fact or Fiction?. *Lawfare*. https://www.lawfareblog.com/un-security-council-permanent-members-de-facto-immunity-article-6-expulsion-russias-fact-or-fiction (accessed 1 May 2023).

Mikhailova, E., 2022. The Arctic this week take five: week of February 28, 2022. *The Arctic Institute*. https://www.thearcticinstitute.org/arctic-week-take-five-week-28-november-2022/ (accessed 1 May 2023).

Meduza 2020. Indigenous groups in Northern Russia ask Elon Musk to boycott Russian mining company 'Nonickel'. *Meduza*, 7 August.

Mishushtin, M. 2021. *Cover Letter*. Government of the Russian Federation.

Moe, A. 2020. A new Russian policy for the Northern sea route? State interests, key stakeholders and economic opportunities in changing times. *Polar Journal* 10(2): 209 227.

Nakano, J. and W. Li 2018. *China Launches the Polar Silk Road*. Washington, DC: Center for Strategic and International Studies.

Newell, J. P. and L. A. Henry 2016. The state of environmental protection in the Russian Federation: a review of the post-Soviet era. *Eurasian Geography and Economics* 57(6): 779–801.

Nilsen, T., 2021. Indigenous peoples call on Nornickel's global partners to demand environmental action. *Barents Observer*. https://thebarentsobserver.com/en/indigenous-peoples/2021/03/russian-indigenous-people-lose-out-electromobility-industry-hunts-metals (accessed 1 May 2023).

Nilsen, T. 2022. Equinor exits Russia. *Barents Observer*. https://thebarentsobserver.com/en/2022/02/equinor-exits-russia (accessed 1 May 2023).

NOAA, 2020. Pink salmon may benefit as Pacific Arctic warms. *NOAA Fisheries*. https://www.fisheries.noaa.gov/feature-story/pink-salmon-may-benefit-pacific-arctic-warms (accessed 1 May 2023).

NOAA 2022. *Law of the Sea Convention*. https://www.gc.noaa.gov/gcil_los.html (accessed 1 May 2023).

Nornickel 2021. Nornickel boosts support for Taimyr's indigenous communities. *Nornickel*. https://www.nornickel.com/news-and-media/press-releases-and-news/nornickel-boosts-support-for-taimyr-s-indigenous-communities (accessed 1 May 2023).

Norskpetroleum 2022. The government's revenues. *Norskpetroleum*. https://www.norskpetroleum.no/en/economy/governments-revenues/ (accessed 1 May 2023).

Overfield, C. 2021. An off-the-shelf guide to extended continental shelves and the Arctic. *Lawfare*. https://www.lawfareblog.com/shelf-guide-extended-continental-shelves-and-arctic (accessed 1 May 2023).

Putin, V., 2022. Transcript: Vladimir Putin's televised address on Ukraine. *Bloomberg News*. https://www.bloomberg.com/news/articles/2022-02-24/full-transcript-vladimir-putin-s-televised-address-to-russia-on-ukraine-feb-24 (accessed 1 May 2023).

Petroleum and Mineral Resources Management Directorate 2022. *Northern Oil and Gas Annual Report 2021*. Government of Canada. https://www.rcaanc-cirnac.gc.ca/eng/1651254877470/1651254907033 (accessed 1 May 2023).

Qi, D., Z. Ouyang, L. Chen, Y. Wu, R. Lei, B. Chen. R. Feely, L. G. Anderson, W. Zhong, H. Lin, A. Polukhin. Y. Zhang, Y. Zhang, H. Bi, X. Lin, Y. Luo, Y. Zhuang, J. He, J. Chen, and W-J. Cai 2022. Climate change drives rapid decadal acidification in the Arctic Ocean from 1994 to 2020. *Science* 377(6614): 1544–1550.

Rantanen, M., A. Karpechko, A. Lipponen, K. Nordling, O. Hyvarinen, K. Ruosteenoja, T. Vihma, and A. Laaksonen 2022. The Arctic has warmed nearly four times faster than the globe since 1979. *Communications: Earth and Environment* 3: 168. 10.1038/s43247-022-00498-3

Reuters 2021. Greenland bans uranium mining, halting rare earths project. *Reuters*. https://www.reuters.com/world/americas/greenland-bans-uranium-mining-halting-rare-earths-project-2021-11-10/ (accessed 1 May 2023).

Scollon, M. 2020. At risk: Russia's indigenous peoples sound alarm on loss of Arctic, traditional way of life. *Radio Free Europe/Radio Liberty*. https://www.rferl.org/a/russia-arctic-indigenous-peoples-losing-traditional-way-life-climate-change/30973726.html (accessed 1 May 2023).

Sidortsov, R. 2019. Benefits over risks: a case study of government support of energy development in the Russian North. *Energy Policy* 129: 132–138.

Skydsgaard, N. 2021. Greenland ends unsuccessful 50-year bid to produce oil. *Reuters*. https://www.reuters.com/business/energy/greenland-puts-an-end-unsuccessful-oil-adventure-2021-07-16/ (accessed 1 May 2023).

Solsvik, T. 2022. Norway plans to expand Arctic oil and gas drilling in new licensing round. *Reuters*. https://www.reuters.com/business/energy/norway-plans-expand-arctic-oil-gas-drilling-new-licensing-round-2022-03-17/ (accessed 1 May 2023).

Spivak, V. and A. Gabuev 2021. The Ice Age: Russia and China's energy cooperation in the Arctic. *Carnegie Endowment for International Peace*. https://carnegiemoscow.org/commentary/86100 (accessed 1 May 2023).

Staalesen, A. 2019. Russia sets out stringent new rules for foreign ships on the Northern Sea Route. *Arctic Today*. https://www.arctictoday.com/russia-sets-out-stringent-new-rules-for-foreign-ships-on-the-northern-sea-route/ (accessed 1 May 2023).

Staalesen, A. 2021a. Igor Sechin: New Arctic oil discoveries are world's biggest. *Barents Observer*. https://thebarentsobserver.com/en/industry-and-energy/2021/02/igor-sechin-new-arctic-oil-discoveries-are-worlds-biggest (accessed 1 May 2023).

Staalesen, A. 2021b. Gazprom scrapped Shtokman, but does not want to abandon projected terminal site on coast. *Barents Observer*. https://thebarentsobserver.com/en/arctic-lng/2021/03/gazprom-scrapped-shtokman-does-not-want-abandon-projected-terminal-site-barents (accessed 1 May 2023).

Staalesen, A. 2021c. Green light for huge oil terminal on Taymyr coast. *Barents Observer*. https://thebarentsobserver.com/en/industry-and-energy/2021/03/green-light-huge-oil-terminal-taymyr-coast (accessed 1 May 2023).

Staalesen, A. 2021d. North Russian regions want extension of Arctic shipping route. *Barents Observer*. https://thebarentsobserver.com/en/arctic/2021/04/north-russian-regions-want-extension-arctic-shipping-route (accessed 1 May 2023).

Staalesen, A. 2021e. Under mounting pressure from renewables, Russia's top oilman says new Arctic giga-project will produce 'green barrels'. *Barents Observer*. https://thebarentsobserver.com/en/industry-and-energy/2021/06/under-mounting-pressure-renewables-russias-top-oilman-says-new-arctic (accessed 1 May 2023).

Staalesen, A. 2021f. Moscow mulls subsidies for shippers sailing Northern Sea Route. *Barents Observer*. https://thebarentsobserver.com/en/arctic/2021/09/moscow-mulls-subsidies-shippers-sailing-northern-sea-route (accessed 1 May 2023).

Staalesen, A., 2022. Isolated Russia says it will invite 'non-Arctic' states to develop its North. *Barents Observer*. https://thebarentsobserver.com/en/arctic/2022/04/isolated-russia-says-it-will-invite-non-arctic-states-develop-its-north (accessed 1 May 2023).

Stone, R. 2020. Siberia's 'gateway to the underworld' grows as record heat wave thaws permafrost. *Science Insider*. https://www.science.org/content/article/siberia-s-gateway-underworld-grows-record-heat-wave-thaws-permafrost (accessed 1 May 2023).

Treadgold, T. 2021. Greenland said no to Trump and now says no to Australia and China. *Forbes*. https://www.forbes.com/sites/timtreadgold/2021/04/09/greenland-said-no-to-trump-and-now-says-no-to-australia-and-china/?sh=53388ef34448 (accessed 1 May 2023).

Trenin, D. 2021. Russia's National Security Strategy: a manifesto for a New Era. *Carnegie*. https://carnegie.ru/commentary/84893 (accessed 1 May 2023).

US White House 2021. Fact sheet: Biden administration roadmap to build an economy resilient to climate change impacts. *US White House*. https://www.whitehouse.gov/briefing-room/statements-releases/2021/10/15/fact-sheet-biden-administration-roadmap-to-build-an-economy-resilient-to-climate-change-impacts/ (accessed 1 May 2023).

Viñas, M.-J. 2019. NASA studies how Arctic wildfires change the world. *NASA*. https://www.nasa.gov/feature/goddard/2019/nasa-studies-how-arctic-wildfires-change-the-world (accessed 1 May 2023).

Walt, V. 2021. The US has had a military presence in Greenland since 1941: a young new leader wants much more. *Time*. https://time.com/6049749/greenland-mute-egede-u-s-blinken/ (accessed 1 May 2023).

Wang, H. L. 2020. Climate change complicates counting some Alaska native villages for census. *NPR*. https://www.npr.org/2020/02/10/802218309/climate-change-complicates-counting-some-alaska-native-villages-for-census (accessed 1 May 2023).

Wang, M. and J. Overland 2009. A sea ice free summer Arctic within 30 years? *Geophysical Research Letters* 36(7). 10.1029/2009GL037820

Welch, C. 2019. Climate change has finally caught up to this Alaska village. *National Geographic*. https://www.nationalgeographic.com/science/article/climate-change-finally-caught-up-to-this-alaska-village (accessed 1 May 2023).

# 3

# GEOPOLITICS AND SECURITY IN THE CHANGING ANTARCTIC

*Samuel Jardine and Timothy Clack*

## Introduction

The Antarctic is an internationalised continent, the governing Antarctic Treaty System (ATS) having successfully contained the pre-1959 escalating military competition between claimants and potential claimants. It channelled this geopolitical competition into other arenas, particularly scientific activity. However, in a rush to quickly prevent the spread of the Cold War across the continent, the ATS purposefully avoided settling contentious issues. Claims were left frozen, rather than resolved. This has left the Antarctic simultaneously a global common and the largest global area affected by 'unrecognized borders' (Dodds 2021, p. 155).

This chapter argues that the potential for an escalation of the ATS' narrowed parameters for competition over claims and influence is highly possible if circumstances on the continent alter to regalvanise national interests and weaken the ATS. The environmental impact of climate change does this in two interlinked ways. Firstly, through galvanising 'suppositional imperialism'. This is a new term proposed by this chapter that captures the primary interest that all Antarctic states, stretching back to 1908, share. This being the hope or fear that the Antarctic has hidden economic and strategic opportunities yet to be realised. States have and will build and maintain a sovereign presence to ensure they are well-placed to capitalise on or influence these potential unknown opportunities. Climate change's environmental impact galvanises this by easing accessibility and teasing cost-effective conditions for economic exploitation. Secondly, exacerbating the battle over the spirit of the ATS in the run-up to the 2048 renegotiation of its Environmental Protocol by increasing the opportunities and frequency in which Antarctic Consultative

DOI: 10.4324/9781003377641-5

States must adapt or create new governance frameworks for the ATS to retain its legitimacy and relevancy.

As the ATS requires decisions to have the unanimous consensus of all Consultative States, updates are slow, with a significant chance that issues become highly politicised. This allows room for 'ATS-Changers', a loose group of states who wish for a more permissive ATS like China and Russia, to hold the ATS hostage, forcing 'ATS-Maintainers' to compromise for their agreement. This will steadily shift the nature of the ATS towards exploitation. If compromise is not forthcoming the threat looms of deadlocks and, due to the comparatively rapid pace of climate change (Fox 2019; Specktor 2020), governance gaps and grey areas where states can engage in competition and exploitation outside the scope of the ATS.

The narrowness of the ATS-enforced competition parameters which channel efforts into providing effective occupation through scientific activity and governance, while failing to address claims, means that an increasing number of states (will) engage in this area in the hope, or fear, that the ATS alters or collapses in 2048, or in its runup, themselves contributing to this outcome. In this context, it can be concluded that reform to the ATS in the face of this climate change exacerbated geopolitical competition is highly unlikely. The need to thus prepare for a return to pre-ATS competition is significant for all Antarctic-interested states.

## Physical impact of climate change on the Antarctic

The impact of climate change on the Antarctic is one of sweeping, if paradoxical, change with the region containing places that are some of the world's most rapidly warming, and some the least. The Antarctic mainland has largely 'not warmed in the last seven decades', despite the consistent rise of greenhouse gases in the atmosphere, due to the scale of its ice sheet (Singh and Polvani 2020). However, the Antarctic peninsula has been found to be warming faster 'than any other terrestrial environment in the Southern Hemisphere' (Siegert et al. 2019). This is due to its closer proximity to the warming Southern Ocean (Sato et al. 2021), whose own heating threatens its role as a key carbon sink (Bates 2021).

The warming of the Southern Ocean is also contributing to the Antarctic's steady loss of sea and glacier ice (Wang et al. 2022), the latter's rate of melting is unprecedented in the past 5,500 years (Braddock et al. 2022). This has significant repercussions. Globally, as the Antarctic ice contains roughly 60% of the earth's total freshwater, the release of just some of this would, and indeed has been, fundamentally affecting sea level, undersea ecosystems, and the ocean currents which regulate climate (British Antarctic Survey 2015; Gilbert 2021; NOAA 2022; Hancock 2022). It also will see sea level rise between 3.3 m and 6 m, which will be 'catastrophic' for the millions inhabiting low-lying regions around the world (Mancilla and Roberts 2022).

Over the longer term in the Antarctic, there is the real risk that it becomes irrevocably ice-free (Garbe et al. 2020; United Nations 2022). In the shorter term, one of the world's most inhospitable places will start to see conditions ease, particularly in warming areas of the peninsula and western ice sheet (Schleussner et al. 2016).[1] In 1989, the US Office of Technology in a congressional report highlighted that ice sheet thickness was one of the key practical obstacles for economically viable mining but that the Western Antarctic and coastal areas were then already viable sites for mining due to their thinner ice cover (US Congress Office of Technology Assessment 1989, pp. 176–180). The melting ice, then, will facilitate more cost-efficient human activity, specifically potential exploitation.

### Suppositional imperialism: climate change as exacerbator

Antarctic-interested states hold a myriad of self-defined, often-similar factors to justify their interest at different times. Argentina and Chile conceive their Antarctic claims as core national territory defined by the 1494 Treaty of Tordesillas (Scott 2011) and geographic proximity (Howkins 2017, p. 11). The UK emphasises discovery and stewardship (International Court of Justice 1956). The US and USSR's justification for reserving the right to make claims was partly based on security concerns (Beck 2009; KNA 1955a). However, a common thread for all actors across space and time has been 'suppositional imperialism'. This is where the reason for maintaining an economically and politically costly presence in the Antarctic is based not on a tangibly current or clearly time-scaled economic or strategic opportunity But that at an unknown point, something of this sort may appear. States thus hope to capitalise on this unknown potential, or fear others will if not checked, and so seek to build and maintain Antarctic presence and influence.

'Imperialism' is the term used commonly for this unifying interest because policies and activities to manufacture 'ownership' by Antarctic-interested states fit comfortably into its model. As defined by the Encyclopaedia Britannica (2019), imperialism is 'the state policy, practice, or advocacy of extending power and dominion' through direct territorial acquisition or by establishing political and economic control. Suppositional imperialism can be epitomised, for example, by the UK's Falkland Islands Dependencies Survey (FIDS) 1950 long-term justification for its Antarctic presence that, '[i]t can be expected that the Antarctic as a whole will be developed in the future in ways which cannot be precisely seen' (KNA 1950a).

### Suppositional imperialism in practice

Suppositional imperialism is found embedded throughout the history of Antarctic competition. In 1920, Britain's Under-Secretary of State for the

Colonies committed to pursuing the 'definite and consistent' extending of control, aiming that 'the whole of the Antarctic should ultimately be included within the British Empire', an endeavour that would not be without significant cost, sparked by concerns that France should be prevented from expanding its own claims (BL 1926). This was despite most existing claims remaining unexplored with no known tangible value (Dykes 2020).

In 1940, Chile's formalisation of an Antarctic claim was significantly influenced by US Admiral Byrd's testimony to Congress regarding 147 mineral samples he had taken from the region, including iron, coal, and petroleum. His expedition had included Chilean and Argentine observers (Howkins 2017 p. 70). Byrd did not actually mention the quantities of these samples, and indeed contemporaries, like the British Colonial Office, doubted their veracity (KNA 1944). Regardless they all, including the Colonial Office, continued to increase their competitive Antarctic presence in case the continent was found to be an economic treasure trove (BL 1948).

This suppositional imperialism has continued into the ATS era from 1961, whose Article IV saw claims 'frozen' though not resolved or renounced (Secretariat of the Antarctic Treaty 1959, hereafter, SAT). While the continent was not fully mapped until 1983 (BAS 2015) and the Antarctic's mineral wealth was still unknown and disputed (US Geological Survey 1974), the potential that the ATS might be changed to facilitate mining saw the number of ATS signatories expand from 25 to 38 as a sleuth of new states joined (BAS 2015), fearing being left without influence over any potential finds. Tangibility plays second fiddle to hopes and fears.

To utilise a recent example which forms one element of the Antarctic's current geopolitical competition, in February 2022, Australia announced an extra AUS$800 million to 'shore up' its Antarctic claim amid increasing strategic competition with China (Belot 2022). According to China's Antarctic Institute's internal newspaper, China's new bases in Australia's 'area' are tasked with carrying out 'resources exploration' (Brady 2017a). This is in keeping with President Xi Jinping's statement, in 2014, that China stands ready to better 'exploit the continent' (Brady 2017b). Both Australia and China – like many other states – are keen to deploy significant resources in the hope that economic wealth can be found, or the fear that it might be utilised by a strategic competitor first.

Climate change is – and will further – exacerbate the suppositional imperialism fuelling Antarctic-interested actors by increasing the prospect, if not reality, of tangible returns. This further pushes states to compete for influence to protect their interests, while climate change also potentially provides a legitimising cover for these actions. For instance, China left out the 'resource exploration' element of its new bases, in favour of telling other ATS members it was solely for 'climatic studies' (Brady 2017b). Likewise, Australia justifies its recent funding package primarily through the 'dual-use' lens of climate change (Belot 2022). The same holds for the UK's significant increase in Antarctic investment (UK Research and Innovation 2022), despite these activities also providing the advertised aim of the increase in sovereign presence (Tossini 2020).

## ATS fragility amidst ongoing competition

The ATS is often framed as one of the most successful international agreements of all time (Triggs 2011; Beck 2009). It prevented the Antarctic from becoming a zone of militarised competition and, through its vagueness, allowed vastly different interests and perceptions to be maintained, if not accommodated under its auspices (Dodds 2012). Simultaneously, it promoted three cornerstones for all Antarctic-interested states: peace, scientific research, and environmental protection (Norwegian Polar Institute 2022). However, it is now facing significant pressure from a changing geopolitical and climate environment (Yermakova 2021).

The Antarctic is now far more accessible due to the impact of climate change and technological development than when it was drawn up in the 1950s to stem potential Cold War militarisation (KNA 1957a). It has also expanded from 12 original signatories, who are all Consultative members with a say in the ATS' consensus-based executive to now 54 signatories, 29 of which now have Consultative status (SAT 2014). This growth presents two interlinked problems. Its consensus-based approach means that decisions to prepare the Antarctic and ATS for climate change lack the agility and rapidity needed, particularly, on pressing and politically contentious matters such as tourism (McGee and Liu 2019). Every issue has the potential to become a long-running affair, requiring a significant investment of time and political capital. This makes it slow to adapt and leaves gaps in its governance that can be exploited (Garrick 2021). The increase in Consultative States at different periods of time has widened the number and diversity of national interests within the ATS. As of 2015, the original 12 Consultative signatories, who created the ATS to resolve a shared Cold War problem, are now outnumbered by newer Consultative States. These usually joined due to the significant ATS drawn-out controversies, such as the 1983–1991 mineral and environmental debates (Jakob 2004). This has seen a greater diversity of thought regarding the future shape of the ATS, particularly how exploitation orientated it should be, compared to founding parties.

When taken together these two issues, added to suppositional imperialism will see the consensus-based umbrella of the ATS stretched, as its necessary 'collective hegemony' is undermined (Yermakova 2021). Climate change's environmental impact opens new potential opportunities – tangible or otherwise – mandating more frequent discussion to regulate and update the ATS providing more opportunity for states to steadily shift, through forcing compromise, the nature of the ATS towards their interests lest it risks irrelevance through long-term governance gaps (McGee et al. 2020). This will escalate as states seek to position themselves for the 2048 renegotiations. For instance, a Chinese diplomat in 2010 urged Chinese tour operations to become active in the Antarctic to corner market share, and so a political stake in the issue, before the ATS attempted to introduce restrictions on tourism (Brady 2017b).

## Claims retaining their power

The issues are exacerbated by the ATS' failure to resolve claims. Making the success of its core stipulation to remove the Antarctic as a 'scene or object of international discord' questionable (SAT 1959). This failure was partly due to the Cold War context of its conception with participants favouring a rapid, non-controversial agreement for fear of negotiations breaking down or states refusing to sign (KNA 1957b). Frozen and unresolved claims continue to be a key consideration for all Antarctic-interested states as no renunciation of sovereign rights has been made by claimants. The competition, for instance between Britain, Argentina, and Chile, over their overlapping claims remains unresolved. In this context, no Antarctic-interested state can risk ignoring that one day the ATS may fail, and claims go live again, and so they seek to position themselves for such an eventuality. In an extreme example, despite the ATS forbidding new claims, in 1967, Ecuador made one, putting it into a frozen dispute with Norway (Ecuadorenlaantartida 2017).

States who still reserve the right to make their own claims such as the United States and Russia are now joined by others like China (Brady 2017a). Antarctic claims in a *de-facto* sense then have retained their relevance and importance and most activity on the continent has a dual purpose in also contributing to the maintenance or potential creation of claims. This particularly means that, in the lead-up to the 2048 renegotiation, states are positioning themselves competitively to both have the best leverage for either shifting the ATS framework or as a contingency measure to support their claims – current or potential – if the ATS collapses at any point. This is clearly seen in the increasing presence of China, whose scientific and physical position can constitute the basis for a *de-facto* physical claim (Brady 2017a), particularly around a perceived 'resource rich' triangular strip in Eastern Antarctica (Brady 2017b). Here China's concentration of infrastructure weakens Australia's frozen claim to the area, while strengthening China's own courtesy of superior effective occupation and activity. This is comparable to measures taken by Argentina and Chile against Britain in their competition between 1942 and 1959 (KNA 1955c, 1957d).

Likewise, the United States maintains a significant Antarctic presence that can potentially justify claims to most of the continent (Scott 2011), something which, in the past, had concerned British policymakers (KNA 1957d). Meanwhile, claimant states like Britain and Australia openly acknowledge that their largely scientific activity is to also provide a sovereign presence, maintaining and securing their claims. Britain's Antarctic Survey is noted by its director as having two roles: 'world class' scientific research and 'to be the UK permanent presence in Antarctica' (UK Parliament 2020). Australia's Minister for the Environment directly highlighted that Australia's increased Antarctic science funding reflects, 'Australia's commitment to our sovereignty in the Australian Antarctic Territory and its leading voice in the region' (Dingwall 2022).

### Science, base racing, and sovereignty

The link between Antarctic scientific activity, encouraged by the ATS, and legitimising a sovereign presence is not a new conception, nor one that came with the ATS. Due to the continent's lack of an indigenous people whose administration by a colonial power could constitute the legitimisation for a sovereign claim (Howkins 2017, p. 7), simply maintaining a 'physical' sovereign presence became key. This could take the form of having commercial enterprises, such as whaling companies, which accept your regulation and jurisdiction as, for example, Britain did to Norway and Chile, with the latter's companies being a significant boon due to overlapping claims (International Court of Justice 1956).

More commonly to ensure a consistent presence, states turned to funding specialist scientific institutions to maintain Antarctic bases. For Britain, this was the FIDS, which became the British Antarctic Survey (BAS) in 1962. This was a civilian-manned scientific organisation which competed with its United States, Argentine, and Chilean counterparts. The scientific research conducted by FIDS was less important than it simply having bases and people persistently deployed in the region. As the Foreign Office in 1950 noted FIDS bases are, 'established and manned for reasons not primarily concerned with scientific research and exploration'. The fact scientific work was undertaken was considered a way to get some 'positive return for the effort and expense' (KNA 1950a). This is not to say that the scientific research which took place did not also have a role in securing claims. The publication of research, particularly meteorological data or cartographic knowledge, was considered by all claimant states as a valid way to legitimise sovereign claims (Dodds 2002, p. 29; Dodds 2009, p. 157).

The predominant method of competition from 1942 to 1959, particularly between Britain, Chile, and Argentina regarding their overlapping claims was the 'base race'. The physical, persistent presence epitomised by permanent or semi-permanent infrastructure on the harsh continent provided 'effective occupation' (Dodds 2022). The competition between these states intensified as each funnelled more funding into the construction of bases to cover more ground. Britain's Foreign Office provided the Cabinet with a running commentary of comparative base numbers (KNA 1955b), and Argentina's pulling ahead – nine to Britain's six by 1955 – was considered a 'substantial deterioration in our [sovereign] position' (KNA 1955c).

The ATS took these methods of recognised geopolitical competition and sanctioned them within its articles. This meant that despite its best intentions and success in demilitarising the competition, the Antarctic persisted to be a scene of discord, i.e. geopolitical competition over claims remained. Current scientific activity, like its antecedent endeavours, is inexorably linked to the production and maintenance of political influence and claims. Because of this, since the ATS' inception, all claimant powers have continued to largely conduct research in 'their' areas, with only a limited internationalisation of scientific infrastructure and multilateral cooperation. Out of the 110 main facilities, only 2 are joint stations (Hemmings 2011). Similarly, 'joint proposals' made by Consultative States through the ATS are very low, with concrete cooperation 'still a long way to go' (Portella Sampaio 2019).

Assertions in this context that, in the face of climate change, Consultative States should, or would, be able to reform the ATS (Yermakova 2021; McGee and Haward 2019) to increase its resilience and secure the Antarctic as a conservationist-centric global commons are highly unrealistic. This is because national interests are at best, not allayed by the ATS, and at worst encouraged with the means of competition.

### The 2048 competition: maintainers and changers

The picture is not helped through climate change exacerbations, including suppositional imperialism (above) and also, increasingly, the opportunities and frequency in which Consultative States must spar politically for the future of the ATS. Climate change adds to the slow-moving agenda a comparatively rapid set of issues requiring new or updated regulations. Two broad battle lines over the future spirit of the ATS are emerging from this. Expanding on the conception highlighted by Dodds (2021, p. 159) regarding two emerging attitudes to the Antarctic, these can be dubbed the ATS-maintainers and ATS-changers. The two sides are not coherent alliances, but simply groupings whose members' interests align on most, though not all, Antarctic issues.

### *ATS-Maintainers*

Maintainers wish to safeguard the ATS' current status quo built on scientific diplomacy and conservation. This is largely, of course, for their own advantage (Yao 2021) as the status quo delays a new frontier of competition; allowing time to further consolidate positions while limiting escalation and keeping the costs of competition comparatively low. Maintaining the current situation as far as possible would heavily favour the United States, Australia, and the United Kingdom (Scott 2011; Young 2021). Unsurprisingly, these states tend to fall into this camp. Argentina and Chile are also maintainers as they see the Antarctic as extensions of their national territory and so are keen to limit potential inroads into their claims.

### *ATS-Changers*

Changers, headed by China and Russia, aim to shift the ATS towards a state permissive to economic exploitation by steadily 'chipping away' at its conservation focus, potentially if needed to prepare for making claims in 2048 (Hoare 2020; Dodds and Boulegue 2022). As noted above, exploitation of the Antarctic's wealth has largely been prohibited by the ATS' 1991 Environmental Protocol (SAT 1991). However, this opens for renegotiation in 2048, with the potential to also revitalise the issue of claims (Dodds 2018). Prior to this, with climate change providing the need for new and updated regulations, ATS-changers have found increasing opportunities to push their agenda. This threatens to deadlock the ATS process, which would create increasing governance gaps and grey areas for changers to exploit regarding their Antarctic activities, while also raising the international profile of the Antarctic question and attracting more states to join the ATS, making consensus even harder to achieve.

In the face of this significant pressure, maintainer states are likely to compromise with changers to preserve the ATS' relevance and functionality. This steadily pushes the ATS towards the wider permissive basis which the changers seek. Between now and 2048, the further the ATS shifts from its conservation cornerstone, the more likely it is that renegotiation around the Environmental Protocol will go in their favour as the baseline is closer to what these states aspire towards.

### Conservation as a weapon

Both maintainers and changers utilise the language and tools of conservation and climate change integrated into the ATS to justify their Antarctic interests. This is partly because it is the ATS-recognised way to gain political capital and legitimisation and provides cover for 'dual-use' undertakings. For instance, China frequently uses climate change to justify and shield its Antarctic interest and increasing presence from criticism; it justified its latest controversial base, deemed 'unnecessary' by other ATS members on the grounds of climate research. Bases of course constituting 'effective occupation' and internal documents highlighting the outpost's 'resource exploration' focus alongside the publicly proffered reason (Brady 2017b).

Supporting conservationist policies is a key means of limiting the potential influence and claim-building capacity of other states. The United Kingdom proudly denotes it sponsored the first high-seas Antarctic Marine Protected Area (MPA) in 2009 (BAS 2015), Australia is pushing for the East Antarctic Marine Protection Area (EAMPA) (Australian Antarctic Program 2021), and Russia, too, an ATS-changer, fashions itself as a champion of MPAs (Boulegue 2022). The geopolitical reason is that they disrupt a state's ability to up-scale activity in these areas, making claim case construction more difficult. Also, as highlighted by Adrian Howkins, the pre-ATS conception of environmental authority and stewardship as accepted ways to justify a claim is still relevant (Howkins 2017, p. 209).

The creation of Antarctic Specially Managed Areas (ASMA) fulfils a similar geopolitical role inland. These are areas managed directly by the proposing states to protect the environment from activity. Again, they provide an ATS-approved tool which can be utilised to showcase environmental authority and limit the potential activity of other actors in an area. China courted controversy by pushing for one around its Kunluw base on grounds of environmental protection (Liu 2019). Maintainer states blocked this due to its break with precedent that ASMA's are established when two states operate in the same area, unlike at Kunluw (Hataya, 2020). This stoked fears that Beijing was laying the ground for a potential future claim (Brady 2017a).

Incidents of competition and political manoeuvres in the lead-up to 2048 are likely to increase significantly, exacerbated by the impact of climate change in three key policy arenas: minerals, bioresources, and tourism.

### Minerals

The mythologisation of Antarctic minerals has, as highlighted above, contributed significantly since 1939 to competition driven by suppositional imperialism. Admiral Byrd's 147 samples from 1939 included iron, coal, and petroleum (Howkins 2017, p. 70) and were a significant contributor to driving pre-ATS competition, even among sceptical states who feared simply being left out (KNA 1944). The difficulties of the 1980s mineral controversy, for both the ATS and Consultative states, would see the mineral-exploitation can simply 'kicked down the road', banning mineral extraction for non-scientific purposes and setting 2048 as a date to revisit (SAT 1991). This 'ticking clock' caused the current competition towards 2048, which imperils the ATS.

The potential for the extraction of Antarctic minerals remains a key battleground for ATS-maintainers and ATS-changers. Relevant here is the geopolitically driven global competition – most intensively observed between China and the United States – surrounding strategic mineral and hydrocarbon supply chains, with expected shortages on the horizon as early as the 2030s due to their politicisation, increasing global populations, and the demands of the 'green revolution' to meet Paris Agreement targets (London Politica 2022 p.13). In this context, as a melting Antarctic eases conditions for more cost-efficient exploration, its untapped mineral potential, real or imagined, will potentially increase in significance well before 2048.

China joined and became a Consultative member during the 1980s mineral controversy (SAT 2014) and, as evidenced previously, promoted a rebalancing of the ATS between 'protection' and 'utilisation' at the highest levels (Gaoli 2017). Indeed, regarding the ban on mining, a diplomat in charge of China's Antarctic policy at Canberra put forward that it simply, 'won preparation time' for China to position itself to capitalise in 2048 (Yilin 2009). To this end, Beijing has utilised the ATS' ambiguity and lack of enforcement mechanisms (Yermakova 2021) to prospect for potential mineral areas despite the ban, under the guise of ATS-permitted scientific geological exploration (Young 2021). Russia has also sent prospecting missions into the Antarctic as part of Moscow's 2030 strategy to explore its value, with a key emphasis on minerals. However, these measures are opposed directly by the United States, the United Kingdom, and Australia who have publicly committed to upholding the extraction ban (Final Report of the Thirty-second Antarctic Treaty Consultative Meeting 2009).

This lays the groundwork for a significant political clash as China and Russia continue to ramp-up their efforts as 2048 approaches. Indeed, a clash would suit Beijing and Moscow as an opportunity to pressure ATS-maintainers to compromise nudging the ATS towards being exploitation-permissive or risk the implication or threat of non-ATS adherence, diminishing the ATS' relevance and legitimacy. Indeed, if maintainers refuse, it gives Beijing a platform to reiterate its criticism of the ATS being a 'rich man's club' (Brady 2012). This narrative is aimed at other ATS-changers and the 85% of the world who currently are not ATS members (Howkins 2017, p. 209) and would struggle for Antarctic influence due to the high ATS-mandated costs to become a Consultative State (Hemmings 2022). In the context of publicising potential wealth, this could generate a significant global backlash against the current ATS, in turn, putting pressure on maintainers.

### Bioresources

Antarctic bioresources, particularly from its ocean, represent a growing commercial industry. Historically, fishing and whaling were the key reasons for early Antarctic competition (Howkins 2017, pp. 45–46), and its regulation was a factor in showcasing the legitimacy of claims (BL 1926). Informed by this historic context, the regulation of Antarctic krill, for example, presents a growing issue for ATS states. Indeed, the ATS' Commission for the Conservation of Antarctic Marine Living Resources (CCAMLR) has become increasingly important in safeguarding sustainable krill fishing and the species' future.

Krill are key to the Antarctic's fragile ecosystem and play a significant role in carbon capture (National Geographic 2010; Cavan et al. 2019). Climate change, however, is seeing krill concentrated increasingly at the poles due to warming waters (Atkinson et al. 2019; Sylvester et al. 2021). Krill oil is a huge growth market expected to grow from US$740.3 m in 2021 to US$2.25bn by 2031 (Future Market Insights 2022). As such, in 2020, the quota limits for some Antarctic areas were reached after only 69 days compared to an average of 130 days over the previous five years (CCAMLR 2020). The key drivers for this have been China and Russia. The former has reported its intent to dramatically increase its krill share (Hongqiao 2016) and has significantly invested in the required infrastructure and specialist vessels (Godfrey 2019). Russia has made krill exploitation a key part of its national strategy to increase domestic seafood production (Stupachenko 2020). Both states have poor track records regarding illegal fishing (IUU Fishing Index 2021) which presents a significant risk to the functioning of the consensus-based ATS. Russia has previously vetoed punishments for its own vessels caught illegally fishing in the Antarctic, leaving the perpetrators able to continue (Allen 2020). Such a breakdown in ATS governance does not bode well for the increased fishing activity, particularly of Chinese vessels, which Beijing often utilise as a grey zone tool of geopolitical competition (Sinclair 2021).

Partly in response to this, though also in preparation for the impact of climate change on the Southern Ocean's ecosystem, MPAs have been set up. The latest, in 2016, though was 'hard-won' (Knoss 2019); having been proposed in 2011. China and Russia are vocal critics due to both the potential economic loss, and that, historically, the regulation of fishing is an act of environmental authority. This is particularly the case as the boundaries of MPAs proposed by claimant states tend to correlate with their claimed areas (Dodds and Brooks 2018).

It is unlikely that a new MPA, such as the EAMPA, will be established anytime soon. China and Russia, for example, vetoed three further MPA proposals in 2020 (Harvey 2020). This leaves the question of how to manage the Antarctic's resources in an increasingly fragile position. Passing MPAs requires compromise, which gives further 'nudging' opportunities to exploitation-minded states. This is also exacerbated by the fact that climate change will shift the geographical dispersion of various species, rendering hard-fought MPAs ineffective and needing replacement (Duong 2021).

### *Tourism*

The Antarctic is growing as a tourist destination. In 2018–2019, over 56,000 tourists visit the continent. From October 2019 to April 2020, this increased, despite the Covid-19 pandemic, to nearly 74,000 people (Kokyay 2022). The majority of tourists are from the United States and then China. Chinese visitation may exceed US visitation in the 2020s (Dodds 2021, p. 128). Beijing seeks actively to make China, in its own words 'a major tourism nation in Antarctica' (Brady 2017a) and is expanding its Antarctic cruise-ship fleet by nine ships, deliverable by the end of the decade (The Maritime Executive 2019).

In the face of the increasing focus on tourism from ATS-changer states, the ATS conservationist framework is struggling to keep up. A recent study highlighted that already, even the current tourist activity has had a significant warming effect on the Antarctic, with 'black soot' increasing the rapidity of ice melt (Hager 2022). Likewise, the ATS regulation surrounding tourism is vague, relying on recommendations to national governments and leaving much up to individual states to enforce. The 2022 resolution regarding 'permanent facilities for tourism and other non-governmental activities in Antarctica', for example, recognises that such facilities on the continent are likely to be planned, yet simply recommends that national governments 'make every effort' to prevent their citizens from doing it (SAT 2022).

This leaves a significant loophole as climate change makes the continent increasingly accessible. Allowing a state, such as China, which promotes the expansion of Antarctic tourism and which aims already to have significant numbers of China-based tour operators and travel agents working in the Antarctic (see Brady 2017a) to unilaterally allow the construction of such facilities. The current ATS-affiliated body that regulates tourism, the International Association of Antarctica Tour Operators (IAATO), despite having higher standards for its members than the ATS recommends (Howkins 2017, p. 206), is self-regulating and, moreover, membership is voluntary. Limiting oversight, it is estimated that up to 50% of tourist enterprises are operating independently of it (Bray 2016).

In geopolitical terms, this is a looming problem. The construction of such facilities on the mainland would immediately raise questions concerning taxation, regulation, and sovereignty (Howkins 2017, p. 196). This type of non-scientific, effective occupation is not within the scope of the narrowed ATS competition, and so, unless addressed in the treaty, risks being an escalatory factor regarding claims and influence. Moves to try and shift the ATS to undertaking a firmer regulatory stance on the topic to prevent this will require unanimous consensus and so can become a new opportunity for ATS-changers to shift the ATS.

### Risks of pre-ATS geopolitical competition

The weakening of the ATS – as it struggles with the interests of changers and an increasing array of issues which challenge its narrow parameters of competition – could see the steady re-introduction, ATS collapse or not, of pre-1959 tools of competition. This is particularly the case as claims and potential claims increase in political estimation as the 'safe' way to secure influence over the Antarctic's future. This is made possible by state actors, such as the Chinese government and armed forces, already envisaging the Antarctic in national security frameworks, and not simply as political arena (Garrick 2021). Such a development carries major risks regarding geopolitical competition, spill-over, and militarisation – warnings about which are evident in the history of pre-ATS competition.

Lessons on the perils of a weakened ATS are clear from history. The Falklands/ Malvinas War (in 1982) had a significant Antarctic dimension (Dodds 2008; Clack and Pollard 2022); essentially being a continuation of the regions 1942–1959 competition in an area that the ATS did not cover. This Falklands-related competition is still ongoing, though currently through other means (Jardine 2021). The Antarctic that is no longer specifically exclusionary to armed conflict could see it re-emerge due to the Falklands' key role as the UK's 'gateway' to the region, acting now, as it did historically, as a strategic hub (Jardine 2021).

Revitalised Antarctic competition in the modern era would see more points of dispute regarding claims (potential or otherwise). China and Australia now significantly share space, for example. Moreover, in our multipolar world, the Antarctic world attracts the attention of other states who, much as in previous bouts of competition, feel they should have a stake in the situation (e.g. India in 1956; KNA 1956a). The rapid upscaling of infrastructure and activity in the arena would increase. As noted above, base racing to provide 'effective occupation' is still the default for exerting sovereignty over the uninhabited continent. This would, in sum, be environmentally damaging for a continent already suffering the effects of (stringently regulated) human activity.

Linked to current and near-term geopolitical competition in the Antarctic, there are three key, interrelated areas of risk: grey zone activities, militarisation, and conflict.

### Grey zone

The construction of infrastructure was, and remains, an act of sovereignty. However, the 1942–1959 era of Antarctic competition also saw infrastructure as targets for destruction, particularly if left unattended (BL 1948). This could have severe consequences. When Britain destroyed the Chilean base on Deception Island in 1953, for example, the Chilean Foreign Minister pushed for retaliation, and the threat of war lingered for weeks (Howkins 2017, p. 123). The employment of such grey zone strategies – acts just below the threshold of war – offers the cover of deniability when required (KNA 1955e) although an issue is whether all participants share the same levels of tolerance (Mazarr 2015, p. 36).

With the increase in utilisation of grey zone activities in modern geo-political competition, epitomised by Russia's deployment of 'little green men' in Ukraine in 2014 and China's deniable 'maritime militia' (Morris et al. 2019), attribution can be difficult and misunderstandings, particularly with increased numbers of Antarctic-interested states, more likely, or even engineered. Indeed, the expansion of Antarctic competition could present, as was feared historically (KNA 1956c), more opportunities for rival states like Russia to sow discord among global partners (Karlsen 2019). While it might be assumed some states, like the United Kingdom and the United States, would, given their ties, be more communicative in the region, it should be noted that, historically, they were keen to compete actively over claims and through underhand methods, e.g. during the World War II and the Cold War despite their close partnership (KNA 1957c, 1948a, 1955f).

### *Militarisation*

A core part of the 1942–1959 era competition was the utilisation of military assets to protect infrastructure, exert sovereignty, and 'take forcible action against intruders' (KNA 1955g). An escalatory military build-up thus took place in the region. For example, in 1943, Argentina deployed one military vessel and Chile two to the Antarctic. By 1947, this had jumped to 28 and 3, respectively, with cruisers on the cards (KNA 1950b). Britain also frequently deployed warships, including cruisers (KNA 1948b). The increasing military presence of states in one area can, of course, lead to misunderstandings that trigger conflict.

In January 1949, Britain, Argentina, and Chile signed a tripartite naval agreement to limit the build-up of warships to only 'routine' visits (KNA 1950c) in order to somewhat mitigate the spiralling competition. Overall, though, this did little to prevent escalatory build-up. In 1952, Argentina sent a 50-strong military detachment to establish a base near the ruins of Britain's Hope Bay outpost tasked with preventing its reconstruction (Howkins 2017, p. 2). When the British civilian 10-strong civilian party arrived from the FIDS, they refused to leave and were surrounded. Shots were fired over their heads, and, at gunpoint, they were forced back to their boats being informed that further landings would be fired upon (KNA 1952b). Two Argentine naval ships were present to add pressure utilising weapons Britain deemed 'unlawful' under the tripartite agreement (KNA 1952a).

This incident could easily have sparked a conflict. The British Governor of the Falkland Islands argued that it was an act of war (KNA 1952a) and sent a frigate with full Admiralty approval to, 'act as the situation demands'. The FIDS party re-landed and worked on their base under the cover of the frigate's guns (KNA 1952b; Howkins 2017, p.3). Further incident was avoided due to this military pressure. As was the case for most of the 1942–1959 competitions incidents (KNA 1952c, 1956b), press coverage of the incident was downplayed purposefully by both sides to avoid public calls for escalation. If such incidents happened today, with the lack of government control over information dissemination through the modern media environment (Suciu 2022; Clack and Johnson 2021), policymakers would be more exposed to the pressure of public opinion to pursue potentially escalatory policies.

Only a year later, a second incident occurred at Deception Island. A large Argentine and Chilean military force had set up bases inside the boundaries of Britain's Port Foster (KNA 1953a). Despite fearing bloodshed (KNA 1953b), Britain sent local police, supported by Royal Marines and a frigate, with instructions to arrest the Latin Americans as 'illegal immigrants'. Escalation was anticipated across the region, but no plan was made for this, nor a contingency for what the party would do if challenged (KNA 1953b). It was only by chance that the British party arrived after most of the Latin American force had left and so a potentially escalatory confrontation was avoided. The two remaining soldiers were arrested without incident (KNA 1955d).

That the Antarctic could return to this state of unpredictable and highly tense competition is not far-fetched in today's multipolar world. The Arctic, previously dubbed 'High North, Low Tension', is now an arena of escalating geopolitical competition. This has seen an increasingly militarised presence courtesy of climate change and competition integrating it more directly into global affairs (Clack and Nugee 2022; Evans 2021). Moreover, the current spill-over from Russia's invasion of Ukraine saw the region's governing forum, the Arctic Council, suspended (Zinan 2022). Although Russia ceded the chair to Norway in May 2023, the future of the Arctic Council as a forum for international collaboration remains in doubt (Last 2023). A similar 'spill-over' is emerging in the Antarctic, too, with Russia rowing back on cooperation and engagement in favour of asserting strongly its national interests in the region, a position not criticised by China (Dodds and Boulegue 2022).

### Conflict

The Russian invasion of Ukraine also reconfirms the persistence of not merely escalatory competition, but of open conflict – declared or otherwise – as a tool of geopolitical competition. In 1946–1947, the United States demonstrated through Operation High Jump that even in harsh conditions, a significant military presence can be operated (Yusoff 2010). As climate change eases circumstances in the locale, the environment will become more permissive for operations. An Antarctic which suffers expanded escalatory competition and a limited civilian presence would potentially be vulnerable to hosting a geographically limited undeclared conflict.

## Conclusion

It is potentially too late to reform the ATS in order for it to be resilient to the impact of climate change on the exacerbation of geopolitical competition in the Antarctic. In no small part, this is due to the ATS' approach to channelling, not resolving, Antarctic competition. The result is that escalation has become somewhat inbuilt, with the ATS itself increasingly undermined. The consensus-based approach of the ATS leaves it vulnerable to an expanding set of Antarctic actors holding increasingly diverse interests in how the continent should be governed. These actors can compromise the conservation-focused ATS into irrelevancy or risk its collapse. As the lure of climate-exacerbated suppositional imperialism continues to drive this shift, governance gaps will continue to appear due to new, climate-induced opportunities. Even if the ATS survives, it will likely become more permissive. The trend will see a shift from a narrow-science, permissive focus to one of economic exploitation. This will, in turn, widen the parameters and tools of competition, and place further pressure on the treaty to tackle sovereign-based disputes over taxation and location (and so claims) for which there is simply no consensus. A return to pre-1959 competition and its escalatory potential is a real possibility for which states and militaries must prepare.

## Note

1 Longer-term, due to climate change, the prospect of self-sufficient human inhabitation exists in parts of the Antarctic landmass, particularly in coastal areas. Despite difficulties in growing certain crops due to seasonal darkness, if the temperature and precipitation were to increase – as is forecast over the next two centuries, other parts of the world demonstrate that conditions would permit grazing animals and forms of pastoralism and a legitimate growing season for certain hardier crops.

## References

Allen, L. 2020. Russian vessel suspected of illegal fishing in Antarctic waters will face no consequences. *Forbes*. https://www.forbes.com/sites/allenelizabeth/2020/10/31/russian-vessel-suspected-of-illegal-fishing-in-antarctic-waters-will-face-no-consequences/?sh=2118455e44a2 (accessed 1 May 2023).

Atkinson, A., S. L. Hill, E. A. Pakhomov, V. Siegel, C. S. Reiss. J. Loeb, K. Steinberg, K. Schmidt, G. A. Tarling, L. Gerrish, and S. F. Sa. (2019). Krill distribution contracts southward during rapid regional warming. *Nature Climate Change* 9(2): 142–147.

Australian Antarctic Program 2021. Proposals for new Marine Protected Areas. *Department of Climate Change, Energy, the Environment and Water*. https://www.antarctica.gov.au/about-antarctica/law-and-treaty/ccamlr/marine-protected-areas/eampa/ (accessed 1 May 2023).

Bates, S. 2021. Study confirms importance of Southern Ocean in absorbing carbon dioxide. *NASA*. https://climate.nasa.gov/news/3136/nasa-supported-study-confirms-importance-of-southern-ocean-in-absorbing-carbon-dioxide/ (accessed 1 May 2023).

Beck, P. J. 2009. Fifty years on: putting the Antarctic Treaty into the history books. *Polar Record* 46(1): 4–7.

Belot, H. 2022. Australian bid to shore up Antarctic claim. *ABC News.* https://www.abc.net.au/news/2022-02-22/australia-to-invest-800-million-shoring-up-antarctic-claim/100849734 (accessed 1 May 2023).

Boulegue, M. 2022. Antarctica, the Southern Ocean and the South Pole. *Chatham House.* https://www.chathamhouse.org/2022/06/militarization-russian-polar-politics/05-antarctica-southern-ocean-and-south-pole (accessed 1 May 2023).

Braddock, S., B. L. Hall, J. S. Johnson, G. Balco, M. Spoth, P. L. Whitehouse, S. Campbell, B. M. Goehring, D. H. Rood, and J. Woodward 2022. Relative sea-level data preclude major late Holocene ice-mass change in Pine Island Bay. *Nature Geoscience* 15(7): 568–572.

Brady, A-M. 2012. China's Antarctic interests. In A-M. Brady (ed.) *The Emerging Politics of Antarctica.* London: Routledge, pp. 31–50.

Brady, A-M. 2017a. China's expanding Antarctic interests: implications for Australia. *ASPI.* https://ad-aspi.s3.ap-southeast-2.amazonaws.com/2017-08/SR109%20Chinas%20expanding%20interests%20in%20Antarctica.pdf?VersionId=L_qDGafveA4ogNHB6K08cq86VoEzKQc (accessed 1 May 2023).

Brady, A-M. 2017b. China's expanding Antarctic interests: implications for New Zealand. *Canterbury.* https://www.canterbury.ac.nz/media/documents/research/China's-expanding-Antarctic-interests.pdf (accessed 1 May 2023).

Bray, D. 2016. The geopolitics of Antarctic governance: sovereignty and strategic denial in Australia's Antarctic policy. *Australian Journal of International Affairs* 70(3): 256–274.

BAS 2015. *British Antarctic Survey.* https://www.bas.ac.uk/ (accessed 1 May 2023).

BL 1926. IOR/L/E/8/585.

BL 1948. IOR/L/PS/12/1292.

Cavan, E. L, A. Belcher, A. Atkinson. S. L. Hill. S. Kawaguchi, S. McCormack, B. Meyer, S. Nicol, L. Ratnarajah, K. Schmidt, D.K. Steinberg, G. Tarling, and P. W. Boyd 2019. The importance of Antarctic krill in biogeochemical cycles. *Nature Communications* 10(1): 1–13.

CCAMLR 2020. Thirty-ninth meeting report. *CCAMLR.* https://www.ccamlr.org/en/system/files/e-cc-39-rep.pdf (accessed 1 May 2023).

Clack, T. and R. Johnson (eds) 2021. *The World Information War: Western Resilience, Campaigning, and Cognitive Effects.* London: Routledge.

Clack, T. and R. Nugee 2022. Climate change is creating security threats around the world – and militaries are responding. *The Conversation.* https://theconversation.com/climate-change-is-creating-security-threats-around-the-world-and-militaries-are-responding-173668 (accessed 1 May 2023).

Clack, T. and T. Pollard 2022. The Falklands war: background context. (In) T. Clack and T. Pollard (eds) *1982 Uncovered: The Falklands War Mapping Project.* Oxford: Archaeopress, pp. 2–10.

Dingwall, D. 2022. Funds to extend Australia's reach into inland Antarctica. *Canberra Times.* https://www.canberratimes.com.au/story/7629476/funds-to-extend-australias-reach-into-inland-antarctica/ (accessed 1 May 2023).

Dodds, K. 2002. *Pink Ice: Britain and the South Atlantic Empire.* New York: I.B. Tauris, pp. 1–206.

Dodds, K. 2008. The great game in Antarctica: Britain and the 1959 Antarctic Treaty. *Contemporary British History* 22(1): 43–66.

Dodds, K. 2009. Assault on the unknown: geopolitics, Antarctic science and the international geophysical year (1957-8). (In) S. Naylor and J. R. Ryan (eds) *New Spaces of Exploration: Geographies of Discovery in the Twentieth Century*. London: Bloomsbury Publishing, pp.148–172.

Dodds, K. 2012. *The Antarctic: A Very Short Introduction*. Oxford: Oxford University Press.

Dodds, K. 2018. In 30 years the Antarctic Treaty becomes modifiable. *The Conversation*. https://theconversation.com/in-30-years-the-antarctic-treaty-becomes-modifiable-and-the-fate-of-a-continent-could-hang-in-the-balance-98654 (accessed 1 May 2023).

Dodds, K. 2021. *Border Wars: The Conflicts That Will Define Our Future*. New York: Penguin.

Dodds, K. and M. Boulegue 2022. Ukraine: the impact on international collaboration in the Antarctic. *Council on Geostrategy*. https://www.geostrategy.org.uk/britains-world/ukraine-the-impact-on-russias-posture-and-international-collaboration-in-the-antarctic/ (accessed 1 May 2023).

Dodds, K. and C. Brooks 2018. Antarctic geopolitics and the Ross Sea Marine Protected Area. *E-International Relations*. https://www.e-ir.info/2018/02/20/antarctic-geopolitics-and-the-ross-sea-marine-protected-area/ (accessed 1 May 2023).

Duong, T. 2021. Marine Protected Areas are less effective than we thought. *EcoWatch*. https://www.ecowatch.com/marine-protected-areas-illegal-fishing-2654643519.html (accessed 1 May 2023).

Dykes, N. 2020. Antarctica: a brief history in maps, part 2. *British Library*. https://blogs.bl.uk/magnificentmaps/2020/05/antarctica-a-brief-history-in-maps-part-2.html (accessed 1 May 2023).

Ecuadorenlaantartida (2017). Historia. *Ecuadorenlaantartida*. http://ecuadorenlaantartida.mil.ec/?page_id=4017 (accessed 1 May 2023).

Evans, J. 2021. The history and future of Arctic state conflict. *Arctic Institute*. https://www.thearcticinstitute.org/the-history-and-future-of-arctic-state-conflict-the-arctic-institute-conflict-series/ (accessed 1 May 2023).

Fox, A. 2019. East Antarctica's ice is melting at an unexpectedly rapid clip. *Science*. https://www.science.org/content/article/east-antarctica-s-ice-melting-unexpectedly-rapid-clip-new-study-suggests (accessed 1 May 2023).

Future Market Insights 2022. Krill. *Future Market Insights*. https://www.futuremarketinsights.com/reports/krill-oil-market (accessed 1 May 2023).

Gaoli, Z. 2017. 中华人民共和国驻孟买总领事馆. *PRC Foreign Ministry*. https://www.fmprc.gov.cn/ce/cgmb/eng/zgyw/t1465158.htm (accessed 1 May 2023).

Garbe, J., T. Albrecht, A. Levermann, J. Donges, and R. Winkleman 2020. The hysteresis of the Antarctic ice sheet. *Nature* 585(7826): 538–544.

Garrick, J. 2021. The Antarctic Treaty System is on thin ice. *ASPI*. https://www.aspistrategist.org.au/the-antarctic-treaty-system-is-on-thin-ice-and-its-not-all-about-climate-change/ (accessed 1 May 2023).

Gilbert, E. 2021. Antarctica's 'doomsday' glacier. *The Maritime Executive*. https://maritime-executive.com/editorials/antarctica-s-doomsday-glacier-could-raise-sea-level-by-two-feet (accessed 1 May 2023).

Godfrey, M. 2019. China's demand for krill. *Seafood Source*. https://www. seafoodsource.com/news/supply-trade/chinas-demand-for-krill-may-result-in-changes-to-ccamlr-convention (accessed 1 May 2023).

Hager, J. 2022. Research and tourism leave soot deposits in Antarctica. *Polar Journal*. https://polarjournal.ch/en/2022/02/24/research-and-tourism-leave-soot-deposits-in-antarctica/ (accessed 1 May 2023).

Hancock, L. H. 2022. Why are glaciers and sea ice melting? *WWF*. https://www. worldwildlife.org/pages/why-are-glaciers-and-sea-ice-melting#:~:text=Rapid %20glacial%20melt%20in%20Antarctica (accessed 1 May 2023).

Harvey, C. 2020. New Antarctic reserves fail to win backing. *Science*. https://www. science.org/content/article/once-again-new-antarctic-reserves-fail-win-backing (accessed 1 May 2023).

Hataya, S. 2020. Legal implications of China's proposal for an Antarctic Specially Managed Area (ASMA) at Kunlun Station. *The Yearbook of Polar Law Online* 12(1): 75–86.

Hemmings, A. D. 2011. Why did we get an International Space Station before an International Antarctic Station? *The Polar Journal* 1(1): 5–16.

Hemmings, A. D. 2022. The functional exclusion of Least Developed Countries from the Antarctic regime. *The Polar Journal* 12(1): 88–107.

Hoare, C. 2020. China 'weaponising' pandemic for covert Antarctic mission sparking 'real concerns'. *The Express*. https://www.express.co.uk/news/world/1324506/china-weaponising-pandemic-covid19-antarctica-secret-mission-treaty-minerals-spt (accessed 1 May 2023).

Hongqiao, L. 2016. China aims high on Antarctic Krill. *Future Oceans*. https:// futureoceans.earthjournalism.net/china-aims-high-on-antarctic-krill/index.html (accessed 1 May 2023).

Howkins, A. 2017. *Frozen Empires. An Environmental History of the Antarctic Peninsula*. New York: Oxford University Press, pp.1–194.

Encyclopedia Britannica 2019. Imperialism. *Encyclopedia Britannica*. https://www. britannica.com/topic/imperialism (accessed 1 May 2023).

International Court of Justice 1956. Antarctica (United Kingdom v. Chile). *International Court of Justice.* https://www.icj-cij.org/en/case/27 (accessed 1 May 2023).

IUU Fishing Index 2021. https://iuufishingindex.net/? (accessed 1 May 2023).

Jakob, A. 2004. Mineral conflict in Antarctica during the 1980s. *University of Canterbury*. https://ir.canterbury.ac.nz/handle/10092/13993 (accessed 1 May 2023).

Jardine, S. 2021. Argentina's 'Cold War' strategy. *London Politica*. https:// londonpolitica.com/latamcaribb/argentinas-cold-war-strategy-to-take-the-falk-lands-islands/islas-malvinas-steadily-progresses (accessed 1 May 2023).

Karlsen, G. H. 2019. Divide and rule: ten lessons about Russian political influence activities in Europe. *Palgrave Communications* 5(1). doi:10.1057/s41599-019-022 7-8.

KNA (Kew National Archives) 1944. CO78/217/1, 15 May.

KNA (Kew National Archives) 1948a. CAB129/28/35, 20 July.

KNA (Kew National Archives) 1948b. DEFE4/13, 11 June.

KNA (Kew National Archives) 1950a. FO371/81131, 24 January.

KNA (Kew National Archives) 1950b. FO371/81131. 9 June.

KNA (Kew National Archives) 1950c. FO371/81131, 17 January.

KNA (Kew National Archives) 1952a. ADM1/23580, 1 February.
KNA (Kew National Archives) 1952b. ADM1/23580, 2 February.
KNA (Kew National Archives) 1952c. ADM1/23580, 4 February.
KNA (Kew National Archives) 1953a CAB129/58/32, 30 January.
KNA (Kew National Archives) 1953b. ADM1/25082, 1-11 February.
KNA (Kew National Archives) 1955a. FO371/113959, 19 March.
KNA (Kew National Archives) 1955b. FO371/113971, 26 January.
KNA (Kew National Archives) 1955c. FO371/113971, 13 January.
KNA (Kew National Archives) 1955d. FO371/113976, 20 October.
KNA (Kew National Archives) 1955e. FO371/113972, 3 March.
KNA (Kew National Archives) 1955f. FO371/113976, 28 October.
KNA (Kew National Archives) 1955g. FO371/113976, 26 August.
KNA (Kew National Archives) 1956a. FO371/119835, 17 February.
KNA (Kew National Archives) 1956b. FO371/119820, 28 August.
KNA (Kew National Archives) 1956c. FO371/119835, 21 February.
KNA (Kew National Archives) 1957a. FO371/126125, 31 January.
KNA (Kew National Archives) 1957b. FO371/126129, 4 November.
KNA (Kew National Archives) 1957c. FO371/126125, 11 February.
KNA (Kew National Archives) 1957d. FO371/126125, 11 January.
Knoss, T. 2019. Antarctic marine protection treaty offers lessons for global conservation. *Physics.* https://phys.org/news/2019-09-antarctic-marine-treaty-lessons-global.html (accessed 1 May 2023).
Kokyay, F. 2022. Impact of security dilemma on Antarctic militarization. *Polish Polar Research* 43(2): 165–185.
Last, J. 2023. Future of the Arctic Council in doubt after end of Russian chairmanship. *CBC News.* https://www.cbc.ca/amp/1.6836000 (accessed 14 August 2023).
Liu, N. 2019. The heights of China's ambition in Antarctica. *Lowy Institute.* https://www.lowyinstitute.org/the-interpreter/heights-china-s-ambition-antarctica (accessed 1 May 2023).
London Politica 2022. Overlooked political risks for 2022: the Politicization of 'everything'. *London Politica.* https://londonpolitica.com/forecast (accessed 1 May 2023).
Mancilla, A. and P. Roberts 2022. How do we solve the paradox of protection in Antarctica? *Aeon.* https://aeon.co/essays/how-do-we-solve-the-paradox-of-protection-in-antarctica (accessed 1 May 2023).
Mazarr, M. J. 2015. *Mastering the Gray Zone.* Carlisle Barracks: United States Army War College Press.
McGee, J. and M. Haward 2019. Antarctic governance in a climate changed world. *Australian Journal of Maritime & Ocean Affairs* 11(2): 78–93.
McGee, J. and N. Liu 2019. The challenges for Antarctic governance in the early twenty-first century. *Australian Journal of Maritime & Ocean Affairs* 11(2): 73–77.
McGee, J., B. Arpi, and A. Jackson 2020. 'Logrolling' in Antarctic governance: Limits and opportunities. *Polar Record* 56. 10.1017/S003224742000039X.
Morris, L. J., M. Mazarr, J. W. Hornung, S. Pezard, A. Binnenddijk, and M. Kepe 2019. Gaining competitive advantage in the gray zone. *RAND.* https://www.rand.org/pubs/research_reports/RR2942.html (accessed 1 May 2023).
National Geographic 2010. Krill. *National Geographic.* https://www.nationalgeographic.com/animals/invertebrates/facts/krill (accessed 1 May 2023).

NOAA 2022. How does the ocean affect climate and weather on land? *NOAA.* https:// oceanexplorer.noaa.gov/facts/climate.html#:~:text=Ocean%20currents%20act %20much%20like (accessed 1 May 2023).

Norwegian Polar Institute 2022. International cooperation in Antarctica. *Norwegian Polar Institute* https://www.npolar.no/en/themes/international-cooperation-in-antarctica/ (accessed 1 May 2023).

Portella Sampaio, D. 2019. The Antarctic exception: how science and environmental protection provided alternative authority deployment and territoriality in Antarctica. *Australian Journal of Maritime & Ocean Affairs* 11(2): 107–119.

Sato, K., J. Inoue, I. Simmonds, and I. Rudeva 2021. Antarctic Peninsula warm winters influenced by Tasman Sea temperatures. *Nature Communications* 12(1). 10.1038/s41467-021-21773-5.

Schleussner, C. F, J. Rogelj, M. Schaeffer, T. Lissner, R. Licker, E. M. Fischer, R. Knutti, A. Levermann, K. Frieler, and W. Hare 2016. Science and policy characteristics of the Paris Agreement temperature goal. *Nature Climate Change* 6(9): 827–835.

Scott, S. V. 2011. Ingenious and innocuous? Article IV of the Antarctic Treaty as imperialism. *The Polar Journal* 1(1): 51–62.

SAT 1959. The Antarctic Treaty. *Secretariat of the Antarctic Treaty.* https://www.ats. aq/e/antarctictreaty.html (accessed 1 May 2023).

SAT 1991. Environmental Protocol. *Secretariat of the Antarctic Treaty.* https://www. ats.aq/e/protocol.html (accessed 1 May 2023).

SAT 2014. Parties. *Secretariat of the Antarctic Treaty.* https://www.ats.aq/devAS/ Parties?lang=e (accessed 1 May 2023).

SAT 2022. ATS Resolution 5 ATCM XLIV. *Secretariat of the Antarctic Treaty.* https:// www.ats.aq/devAS/Meetings/Measure/778?s=1&from=1/1/1958&to=1/1/2158&cat= 14&top=0&type=0&stat=0&txt=&curr=0&page=1 (accessed 1 May 2023).

Siegert, M., A. Atkinson. A. Branwell, M. Brandon, P. Convey, B. Davies, R. Downie, T. Edwards, B. Hubbard, G. Marshall, J. Rogelj, J. Rumble. J. Stroeve, and D. Vaughan. 2019. The Antarctic Peninsula under a 1.5°C global warming scenario. *Frontiers in Environmental Science* 7. 10.3389/fenvs.2019.00102.

Sinclair, M. 2021. The national security imperative to tackle illegal, unreported, and unregulated fishing. *Brookings.* https://www.brookings.edu/blog/order-from-chaos/ 2021/01/25/the-national-security-imperative-to-tackle-illegal-unreported-and-unregulated-fishing/ (accessed 1 May 2023).

Singh, H. A. and L. M. Polvani 2020. Low Antarctic continental climate sensitivity due to high ice sheet orography. *NPJ Climate and Atmospheric Science* 3(1). doi:10.1038/ s41612-020-00143-w.

Specktor, B. 2020. Antarctica could melt 'irreversibly' due to climate change. *Live.* https://www.livescience.com/antarctica-ice-free-climate-change.html (accessed 1 May 2023).

Stupachenko, I. 2020. Russia exploring move into Antarctic krill fishery. *SeafoodSource.* https://www.seafoodsource.com/news/supply-trade/russia-exploring-move-into-antarctic-krill-fishery (accessed 1 May 2023).

Suciu, P. 2022. Social media is impacting military performance and changing the nature of war. *Forbes.* https://www.forbes.com/sites/petersuciu/2022/06/07/social-media-is-impacting-military-performance-and-changing-the-nature-of-war/?sh= 11f5de06e394 (accessed 1 May 2023).

Sylvester, Z. T., M. C. Long, and C. M. Brooks 2021. Detecting climate signals in Southern Ocean krill growth habitat. *Frontiers in Marine Science* 8. doi:10.3389/fmars.2021.669508.

The Maritime Executive 2019. SunStone orders sixth infinity cruise vessel. *The Maritime Executive.* https://maritime-executive.com/article/sunstone-orders-sixth-infinity-cruise-vessel (accessed 1 May 2023).

Tossini, J. V. 2020. British Antarctic Territory and the British Antarctic Survey. *UK Defence Journal.* https://ukdefencejournal.org.uk/britain-in-antarctica-british-antarctic-territory-and-the-british-antarctic-survey/ (accessed 1 May 2023).

Triggs, G. 2011. The Antarctic Treaty System: a model of legal creativity and cooperation. (In) P. A. Berkman, M. A. Lang, D. W. Walton and O. R. Young (eds) *Science Diplomacy Antarctica, Science, and the Governance of International Spaces.* Washington: Smithsonian Press, pp.39–49.

UK Parliament 2020. Foreign Affairs Committee written evidence. *UK Parliament.* https://committees.parliament.uk/work/114/environmental-diplomacy/publications/written-evidence/ (accessed 1 May 2023).

UK Research and Innovation 2022. UK invests to modernise polar science. *UK Research and Innovation.* https://www.ukri.org/news/uk-invests-to-modernise-polar-science/#:~:text=The%20UK%20is%20investing%20in (accessed 1 May 2023).

United Nations 2022. NDC synthesis report. *UN.* https://unfccc.int/news/climate-commitments-not-on-track-to-meet-paris-agreement-goals-as-ndc-synthesis-report-is-published (accessed 1 May 2023).

US Congress Office of Technology Assessment (1989). *Polar Prospects: A Minerals Treaty for Antarctica.* Washington: Congress Office of Technology, pp.176–180.

US Geological Survey 1974. Mineral Resources of Antarctica. *US Department of the Interior.* https://pubs.usgs.gov/circ/1974/0705/report.pdf (accessed 1 May 2023).

Wang, J., H. Luo, Q. Yang, J. Liu, L. Yu, Q. Shi, and B. Han 2022. An unprecedented record low Antarctic sea-ice extent during Austral Summer 2022. *Advances in Atmospheric Sciences* 39: 1591–1597.

Yao, J. 2021. An international hierarchy of science: conquest, cooperation, and the 1959 Antarctic Treaty System. *European Journal of International Relations* 27(4): 995–1019.

Yermakova, Y. 2021. Legitimacy of the Antarctic Treaty System: is it time for a reform? *The Polar Journal* 11(2): 1–18.

Yilin, W. 2009. 从南极条约体系演化看矿产资源问题-中国知网. *CNKI.* https://gb.global.cnki.net/kcms/detail/detail.aspx?filename=ZGHZ200905004&dbcode=CJFQ&dbname=CJFD2009&v= (accessed 1 May 2023).

Young, C. 2021. Eyes on the prize: Australia, China, and the Antarctic Treaty System. *Lowy Institute.* https://www.lowyinstitute.org/publications/eyes-on-prize-australia-china-and-antarctic-treaty-system (accessed 1 May 2023).

Yusoff, K. 2010. Configuring the field: photography in early twentieth-century Antarctic exploration. (In) S. Naylor (ed.) *New Spaces of Exploration: Geographies of Discovery in the Twentieth Century.* London: Bloomsbury, pp.52–77.

Zinan, C. 2022. Arctic Council at a Crossroads. *China-US Focus.* https://www.chinausfocus.com/peace-security/arctic-council-at-a-crossroads#:~:text=The%20Arctic%20Council%20has%20not (accessed 1 May 2023).

# 4

# SECURITY POLITICS OF CLIMATE CHANGE IN THE LEVANT

*Ana Maria Kumarasamy and Simon Mabon*

## Introduction

On the 26th of November 2021, hundreds of protestors gathered in Amman demonstrating the water-for-energy agreement between Jordan and Israel. Protestors rejected the agreement on several counts arguing that the project would normalise the relationship with Israel while it continues to occupy Palestinian territories. Moreover, it would create long-term dependency on Israel for water and hinder sustainable water management solutions in Jordan. The authorities in Jordan responded to these events by increasing security and arresting 16 of the protestors, a strategy increasingly used across the region (Davis 2021). The demonstrations highlighted the increasing water stress intensified by climate change and mismanagement, the complex geopolitical relationships in the Levant, and the role of the regime in regulating these challenges.

In recent years, the Levant has endured an array of challenges related to climate change, including intense weather events, such as droughts and flooding, forest fires, and dust storms, leading to increasing human security challenges that intersect with tense political and economic conditions. In the coming years, climate change is also expected to further negatively impact the neglected and mismanaged water and food systems across the region and create shortages. This complex set of challenges will impact the most vulnerable in the region – including millions of refugees and Internally Displaced Persons (IDPs) – placing untold pressure on states already on the brink of political and environmental catastrophe.

This chapter argues that the devastating repercussions of climate change and efforts to regulate such developments across the Levant have posed an

DOI: 10.4324/9781003377641-6

array of challenges to political life across the region, exacerbating existing social, political, cultural, religious, ethnic, and class-based tensions in the process. With a focus on Lebanon and Syria in particular, the chapter argues that these schisms emerge both directly and indirectly as a consequence of efforts to regulate both climate developments and the political lives of people across the region. The intersectional challenges and interactions posed by climate change have been addressed extensively, but here we argue that biopolitical processes designed to regulate life are also deployed in an effort to include environmental challenges within the scope of sovereign power as an additional tool in broader efforts to retain power. This plays out in a number of different ways from control of resource allocation and distribution, but also includes deliberate attempts to create unsustainable and inhospitable conditions for particular communities as punishment for political activities.

In pursuit of this line of inquiry, we engage with a number of different sets of disciplinary questions. The deliberate manipulation, exploitation, and abandonment of environmental issues necessitate an analysis of the mechanisms of sovereign power and the ways in which this plays out across the region, with a focus on the work of Giorgio Agamben. Building on this, we turn to an exploration of the ways in which climate change and political life intersect with sovereign power and impact the lives of people across the region. This focus on the political requires an exploration of the contingencies and complexities of environmental challenges that have emerged in the region in recent years. In order to unpack these complexities, this chapter starts by exploring the links between climate change, sovereign power, and political life. Following this, the chapter explores the role of the state, demographic changes, and the recent protests that have emerged across the Levant. We conclude by stressing how the interplay of environmental challenges and security challenges stemming from political instability results in an existentially precarious context that will have a devastating impact on the lives of people.

## Climate change, sovereign power, and political life

Emerging in the aftermath of the *Stern Report*, climate security, and its interlinkages with human security became a key turning point in environmental security research. Since then, scholars have emphasised the emerging security implications for state policies arising from issues such as rising sea levels, flooding, drought, forced migration, and the breakout of civil war to increasing securitisation by the state. As a focal point of this chapter, climate change and the state have been conceptualised in three dominant ways, including violent conflict, fragmentation, and regime survival.

Scholars have long debated the role of climate change in aggravating conflict and civil war. Emerging from the environmental-conflict link that developed in the aftermath of the Cold War, the climate-conflict thesis underscores the complexities around climate change and conflict with most scholars agreeing that climate change in itself does not lead to conflict, but that it can play a role in triggering conflict or act as a threat multiplier. As noted by many critics, the role of climate change in conflict is almost impossible to interpret and pinpoint. The extent to which climate change can amplify conflict has been contested among academics, as seen in the context of the Syrian civil war. Focusing on the Climate–Migration–Conflict nexus, Gleick (2014, p. 338) argued that the Syrian conflict 'had many roots' including how a 'multiyear drought beginning in the mid-2000s, combined with inefficient and often unmodernised irrigation systems and water abstractions … contributed to the displacement of large populations from rural to urban centers, food insecurity for more than a million people, and increased unemployment—with subsequent effects on political stability'. Similarly, Kelley et al. (2015, p. 3241) concluded that 'human influences on the climate system are implicated in the current Syrian conflict'. Criticising these claims, Selby et al. (2017, p. 233) argue that there is not enough evidence to support the claim that climate change and drought 'helped fuel' or 'sparked' or 'contributed to' or 'is implicated in' Syria's 'civil war', thus questioning the extent of migration and the role of migration in shaping conflict.

Other approaches seek to move beyond this debate. Hoffmann (2018, p. 9595) focuses on bringing in historical sociology and political ecology perspectives, arguing that the debate produces environmental orientalism and 'obscures political culpability for the ongoing violent oppression'. As such, what is at stake is not only conflicts as outcomes, but rather, social processes that can bring about inequitable forms of resource control and access in historically marginalised communities (Le Billon, 2018). By including discussions of political life within contexts of environmental challenges and climate stress, we consider interrelated regulatory processes by the state, which 'politicises' environmental problems and scarcities, and 'ecologises' existing contestations (Robbins 2012, p. 201). As such, questions are centred around control and regulation by various levels of power from the state to the community level, allowing for exploration of the impact of such developments on people.

Furthermore, scholars have debated the extent to which climate change can cause fragmentation in states. As noted by Crawford and Church (2021, p. 88):

[I]t would be wrong to assume that individuals and communities based in fragile states will automatically resort to violence when confronted with climate stress ... populations have, in fact, proven resilient to droughts, floods and extreme heat in the past and have honed their adaptive capacities and coping mechanisms over decades of living in severe climates.

Such questions – and the narrow perspective of 'state failure' and 'state fragility' more broadly – point to issues with Western-centric (and Weberian) notions of the state, and their limited ability to reflect on the impact of climate disasters and implications for resources and infrastructures. From a broader perspective, the result is not only state-led exploitation in which one group has greater access but also the segmentation of resources and services, as seen in the Levant, with a second-order consequence being the deepening of societal schisms. Scholarship on the region (and the Global South broadly) has sought to understand political ordering beyond the regime, factoring in the 'hybridity' and 'nestedness' of local agents and their ability to fill in the gaps, although unevenly, according to factors such as class, race, and identity politics. These networks of power are often discussed in the context of infrastructures as opposed to environmental security and climate change debates. In the Levant, scholars have typically sought to understand the political tensions of infrastructures within the context of sectarianism and class (Nucho 2016; Veridel 2018; Parreira 2021).

Despite the presence of serious questions about its relative power, the role of the state should by no means be neglected, especially when it comes to the ability of institutions and agents to regulate people and climate impacts through emergency politics. In the Global North, literature has sought to unpack 'the securitisation of the environment', including narrow and broader notions of security that include notions of human security (Floyd 2021, p. 235). Similarly, regimes in the Levant also regulate climate and human security challenges albeit amid corruption, political mediation, and patron–client relationships. Amid human security and climate change challenges, regimes attempt to regulate spaces through the establishment of exceptional measures. As explained by Colin H. Kahl (1998, p. 83), 'civil strife is initiated by the state elite who seek to capitalise on scarcities of natural resources and related social grievances to advance their patriarchal interests'. Hence acknowledging the role of infrastructures and climate challenges as a part of social and ecological processes that produce political exclusion. Following this line of thought, these processes are embedded within the slow and often silent nature of exclusionary politics. At the same time, the regulation of environmental processes can create inhospitable conditions for communities and even be used as a method of punishment for political activities, as seen with Saddam Hussein's measures to drain the marshlands in southern Iraq in the 1990s.

However, questions around the state, regulation, and security can be less clear-cut as it also intersects with violence which takes many forms in the Levant. Aside from direct forms of violent conflict between actors – rulers and ruled, or different communities – conflict and violence also take on different characteristics, reflecting the nature of the political and the contingencies and complexities of time and space. In particular, the regulation of life imposes forms of what Johan Galtung (1990) terms structural violence, an idea that impacts the ability of a person to reach their potential. Structural violence ranges from economic systems of oppression to political marginalisation, discrimination along identity-based lines, or the deliberate absence of regulatory processes which impact the health of people. Environmental factors feed into such forms of structural violence, once again in a range of different forms. From the deliberate practice of managing resources to decisions taken around waste management, the regulation of environmental factors and issues around climate security can have a devastating impact on the lives of people. Here, it is important to reflect on the ways in which sovereign power operates and the implications of such forms of power for climate change and people more broadly.

## Understanding sovereign power

Regulatory efforts by sovereign power are grounded in biopolitics, which concerns the organisation and control over human life. Fundamental to this, the state of exception is a space established by sovereign power (or by scales of sovereignties). Building on scholars like Hannah Arendt and Walter Benjamin, the Italian philosopher Giorgio Agamben (2005) argues that the state of exception is a paradigm of contemporary politics, allowing for analysis of the political in a range of forms. Across a volume of nine books, Agamben focuses upon the structures of sovereign power, engaging in a philological discussion of the ways in which sovereignty is actualised and operates, brought about by the declaration of a state of exception. Moving beyond the Weberian ideas of sovereignty, Agamben (1996, 1998, 2005) builds on the work of Carl Schmitt, Walter Benjamin, and Michel Foucault to argue that the legal and political order is defined by what is deemed exceptional to it. As such, only the sovereign can articulate when the law is suspended – and a state of emergency is declared – as a consequence of the sovereign's role as a lawgiver. Moreover, the sovereign is ultimately responsible for distinguishing between *bios* and *zoe,* creating a binary distinction between those people who are recognised legally as fully human as a consequence of their participation in political life and those people who remain outside, included in the *polis* by virtue of their exclusion into what he terms *bare life*. This idea of inclusion through exclusion locates bare life at the forefront of politics, bringing about a 'legal civil war that allows for the physical elimination not only of political adversaries but entire categories of citizen who for some reason cannot be integrated into the political system' (Agamben 2005, p. 2).

For Agamben (1998, p. 9), 'the realm of bare life – which is originally situated at the margins of the political order – gradually begins to coincide with the political realm, and exclusion and inclusion, outside and inside, bios and zoe, right and fact, enter into a zone of irreducible indistinction. At once excluding bare life from, and capturing it within, the political order, the state of exception actually constitutes, in its very separateness, the hidden foundation on which the entire political system rests'. Within the state of exception, zoe is reintroduced into politics, materialised in the form of camps. While clearly an inflammatory term, for Agamben this is a place where biological and political life meet, where private and public cannot be distinguished, existing outside of the normal juridical order. Agamben also draws upon the work of Michel Foucault, who, in *The History of Sexuality*, writes of biopolitics as an attempt to organise not only disobedient facets of society but also whole populations (Foucault 1979, 1992). For Agamben (1998, p. 131), 'one of the essential characteristics of modern biopolitics ... is its constant need to redefine the threshold in life that distinguishes and separates what is inside and what is outside'. When the inside becomes the outside, free from protection and stripped of human rights, the individual becomes included through its exclusion.

Ultimately, Agamben's thesis is that the camp is, 'the hidden paradigm of the political space of modernity' (1998, p. 23). Within the camp, which is seen as the site of exception, the individual can be reduced to the condition of bare life and – in this condition – is abandoned by the law, and left vulnerable to violations and non-sacrificial homicide. For Agamben, refugees should be seen as the ultimate 'biopolitical' subjects, individuals who can be regulated and governed at the level of population in a permanent 'state of exception', outside the normal legal framework, the camp. In camps, refugees are reduced to 'bare life': humans as animals in nature without political freedom and lacking agency. Through the emergence and ensuing normalisation of the state of exception, zones of indistinction are created. The distinction between zoe and bios blur with the emergence of a nation-state where 'human life is politicised only through an abandonment to an unconditional power of death' (1998, p. 56). The marginalisation of human life through forced displacement is one such means through which bare life is created. It is here where the concept of the nation-state returns, as a means of providing citizenship to those within its territorial borders. By its very nature, the nation-state is one, 'that makes nativity or birth (that is, naked human life) the foundation of its sovereignty' (Agamben 1996, p. 161).

The life caught in the sovereign ban is the life that is originarily sacred ... and, in this sense, the production of bare life is the originary activity of sovereignty. The sacredness of life, which is invoked today as an absolutely fundamental right in opposition to sovereign power, in fact originally expresses precisely both life's subjection to a power over death and life's irreparable exposure in the relation of abandonment.

Agamben (1998, p. 83)

The refugee is a prime example of the exclusionary politics that is inherent within notions of sovereignty and the state. As a result, 'by breaking the identity between the human and the citizen and that between nativity and nationality, [the refugee] brings the originary fiction of sovereignty to crisis' (Agamben 1996, p. 161). In operation, sovereignty excludes refugees from political and social action within a given territory, although, despite their exclusion, refugees remain subject to the law and included in the legal order through their 'constitutive exclusion' (Owens 2009). This idea of exclusion and distinction between citizen and non-citizen is integral to Agamben's thought:

> One of the few rules the Nazis faithfully observed in the course of the "final solution" was that only after the Jews and gypsies were completely denationalized (even of that second-class citizenship that belonged to them after the Nuremberg laws) could they be sent to the extermination camps. When the rights of man are no longer the rights of the citizen, then he is truly sacred, in the sense that this term had in archaic Roman law: destined to die.
>
> (1996, p. 162)

Exclusion thus means political irrelevancy but it also has much more severe repercussions: the sovereign makes decisions not just about the life and death of human beings but over who is a human being at all. As such, to paraphrase Agamben, within bare life, the *homo sacer* stands outside of the law while simultaneously being abandoned by it and relating to the law through his exclusion from it.

For our exploration, Agamben's ideas shed valuable light on the ways in which sovereign power operates and regulates life. In times of ongoing crisis, understanding the ways in which the sovereign regulates life is of paramount importance, particularly with the displacement and abandonment of people during environmental crises and climate change.

## The state and regulation

In the Levant, climate change and environmental challenges are interrelated with the state and human insecurity in various ways. States have manipulated environmental factors as a means of regulating group activity and potential, whilst also punishing political activities. One of the most extreme examples of environmental regulation includes Saddam Hussein's efforts to drain the marshlands located west of Basra city in the 1990s to punish his adversaries, having severe environmental and human security impacts, and displacing more than 200,000 people (IDMC 2007). The wetlands possess a unique ecosystem, situated at the intersection of the Euphrates and Tigris rivers. With the breaking of the dam in 2001, an estimated 90% of the marshlands had disappeared, prompting widespread displacement and a dramatic loss of biodiversity (ICRC 2021).

Regulating the lives of people through controlling water resources is not restricted to Iraq. Indeed, Israeli efforts to regulate the lives of Palestinians see the imposition of controls limiting access to water for Palestinian residents (Amnesty International 2017). In Hebron, a city home to a large Palestinian community and also Jewish settler communities, regulating access to water is a key feature of efforts to control the lives of indigenous communities. As an interview with an actvist in Hebron in 2013 confirmed, in some cases, this even involves putting holes in water tanks built to catch rainfall in an effort to prompt migration from the city and to transform its character into a Jewish site.

However, the regulation of infrastructures, such as water and electricity, can be less clear as nested agents act for the benefit of their communities whilst neglecting others – thus producing systematic political and socio-ecological inequalities. Elites controlling access to, and management of, infrastructure and resources have greater levels of political, social, and economic capital which embolden patronage structures and increase social inequalities. Researching public services and communal politics, Corstange (2016) finds that water services in Lebanon are worse in areas where one group dominates politically, meaning that areas with greater religious or sectarian diversity receive further services. In post-invasion Baghdad, similar issues were identified in the electricity sector. The emboldened Sadrist Movement 'was able to monopolize a scarce resource and divert it to areas in which it maintained its largest bases of support and operational headquarters' (Parreira 2021, p. 759). Through the control of infrastructures, political leaders are able to gain politically by maintaining systems of patronage which are often structurally violent and reproductive of intersectional difference.

Although not as lucrative as gas and oil, political leaders seek to gain from other infrastructures economically. Political elites in Lebanon are known to actively split the spoils of government, especially those connected with larger infrastructural projects and management resources (Leenders 2012). The distribution of resources between the political elite has led to overlapping responsibilities of services between ministries that work as sectarian bastions, which are followed by a lack of transparency and accountability, with Council for Development and Reconstruction (Sunni – Future Movement), the Council of the South (Shia – Amal), the Ministry of Displaced (Druze – Progressive Socialist Party), and Ministry of Energy and Water (Maronite – Free Patriotic

Movement) all having some stake in the rehabilitation and construction of water supply (Eid-Sabbagh 2015, p. 75; Salloukh 2019). As such, the most 'water rich' country in the MENA region – Lebanon – endured widespread water insecurity as 'water pollution is rampant, water conservation remains largely a slogan, chronic water shortages persist, access to safe and improved water resources remains low, unconventional water sources continue to be untapped, and institutions remain in need of financial and technical support' (MOE et al. 2021, p. 72). Much like many other aspects of Lebanese politics, water services, and other environment-related infrastructures are often treated as secondary to keeping the sectarian balance.

Environmental challenges are often shaped by corruption, mismanagement, and the lack of law enforcement, which have exacerbated impacts and scarcities. This is exemplified through the ongoing water management in Basra which led to the hospitalisation of 118,000 people, in 2018, due to the consumption of poor water quality (Al-Rubaie et al. 2021, p. 10). Like other places in the region, water in Basra has been shaped by connected issues, including decades of conflict, illegal extraction, maintenance neglect (of treatment plants and canals), and corruption. Political parties and militias have sought to hinder or gain control over water infrastructures (Mason 2022, p. 59). As such, infrastructures are extremely vulnerable to the range of political agents that seek to gain economically and politically in the region.

More broadly, the mismanagement of resources and infrastructures has a series of secondary effects on the environment. Across urban areas in Lebanon, flooding has frequently occurred due to old and faulty drain water systems and blockages related to the ongoing waste management crisis (UN-Habitat 2021, pp. 119–120). Another example includes the failure to put out the 2019 forest fires in Mount Lebanon due to the lack of maintenance of helicopters – which partly ignited the anti-governmental Thawra Movement as discussed below. In Syria, the pre-civil war agrarian crisis, which led to crop failures and livelihood loss, was caused by a series of factors, including extreme water stress, rural poverty, and public policies (De Chatel 2014). Although scholars disagree on the extent of the climate–migration–conflict nexus in the Syrian case (Selby 2019; Kelley et al. 2015; Gleick 2014), it is clear that the mismanagement of resources and infrastructures has left parts of the Levant extremely vulnerable. Adding to this are serious changes in the region's climate, with extreme heat and sandstorms becoming more common. From this, previously fertile land is disappearing, impacting jobs, food production, health, and prompting migration.

While easily reduced to a by-product of corruption and nepotism, such practices reveal a systematic biopolitical set of processes designed to ensure the survival of the sovereign at the expense of letting life live. In such conditions of environmental stress, individuals are routinely abandoned into conditions of bare life through neglect, exploitation, and regulation designed to ensure regime survival.

## Demographic changes

Anthropogenic change and the manipulation of resources and infrastructures have various implications for human activities in the Levant region, especially for livelihood, food, and water security. These implications are also nested – directly and indirectly – within various forms of demographic and socio-economic challenges that impact stability, security, and resource distributions. Migration patterns have played a significant role in shaping the region (Selby 2019). Urban areas have expanded significantly across the Levant due to urbanisation and displacement, with many refugees and IDPs settling in suburban areas with low-cost housing instead of formal and informal refugee camps.

The Levant is shaped by various scales of conflict-driven displacement that have impacted regional dynamics and (in)securities, including Palestinian, Syrian, and Iraqi IDPs and refugees, but also fewer numbers of refugees from outside the region, including from Yemen, Somalia, and South Sudan. Displacement originating in the Levant is mainly rooted in three main conflicts: the Palestinian *nakba* (catastrophe), post-invasion violence in Iraq, and the Syrian civil war (IDMC 2021; UNRWA 2022; UNHCR 2022). Moreover, rooted within the creation of the State of Israel and the Arab-Israeli war of 1948, Palestinian displacement continues to be shaped by a series of dynamics. In the aftermath of the nakba, Palestinian refugees fled to the neighbouring countries Jordan, Lebanon, and Syria – over 2 m, 0.47 m, and 5.6 m registered refugees in 2021, respectively – with many continuing to reside inside refugee camps but for the most part in suburban areas (UNRWA 2022). Internally, the number of IDPs is not only growing due to episodes of direct conflict, but also through demolitions, and forced evictions of Palestinians – an average of 940 Palestinians were evicted yearly between 2010 and 2019 (IDMC 2021, p. 44). In contrast, displacement in Iraq is shaped by various forms of violence including Saddam's dictatorship, the 2003 invasion, and the war against the Islamic State – with 1.2 m IDPs and over 270,000 registered refugees across the region in 2022 (UNHCR 2022, p. 1). In neighbouring Syria, the civil war has produced an unprecedented internal and cross-border displacement with over 6.7 m IDPs and 5.6 m registered Syrian refugees across the Middle East (UNHCR 2022, pp. 1–3). These conflicts have produced a sequence of humanitarian crises with displaced persons facing severe economic and socio-ecological insecurities. Moreover, these demographic changes have placed tremendous pressure on host communities and resource distributions.

Displaced persons across the Levant face a series of intersectional challenges related to politics, administration, and legal status. Jordan, Lebanon, Syria, and Iraq are not a party to the 1951 UN Convention on the Status of Refugees and thus lack substantive legislation to protect refugees. In Jordan, Lebanon, Iraq, and Syria, the experience with Palestinian refugees has shaped contemporary 'refugee regimes'. With the arrival of Palestinian refugees and increasing political mobilisation, the states – facing different dilemmas – settled on very different approaches. In Jordan, almost three-quarters of Palestinian refugees have received citizenship enabling them to access public goods and services, whilst the remaining non-naturalised refugees remain some of the most vulnerable people in the country often residing in one of the country's officials of unofficial camps (Amnesty International 2022).

The government in Lebanon, on the other hand, has supported strategies that avoided naturalisation – thus excluding Palestinian refugees from civil and social rights – fearing changes to the religious and sectarian balance in the country. To this day, the descendants of Palestinian refugees continue to be excluded from these rights, including children born by Lebanese mothers with foreign fathers. Legal restrictions also exist within more than 70 professions reserving them for Lebanese nationals (UNRWA 2016). In Syria and Iraq, Palestinian refugees were given equal rights but were not granted citizenship, having consequences for contemporary conflicts in the region (UNHCR 2017).

Although a relatively small number of Palestinian refugees settled in Iraq – about 30,000 – they were caught in the political and sectarian violence after 2003, where many were also targeted due to the 'preferential treatment – real or perceived – they received under Saddam' (Al-Khalidi et al. 2007, p. 14). As a result, Palestinian refugees in Iraq fled to Syria, with the government only starting to restrict the entry of Palestinians after 2006. The limited status of refugees in Syria and Iraq had severe implications for individuals and their continuous displacement during the Syrian civil war. Early on, the government of Jordan excluded a series of vulnerable groups: 'Palestinians living in Syria; all single men of military age; Iraqi refugees living in Syria; and any undocumented persons' (HRW 2021, p. 15). Similarly, Lebanon placed additional restrictions on the entry of Palestinian refugees from Syria (PRS) in May 2015 (UNHCR 2017). The protracted displacement of Palestinian refugees across the region has resulted in the diminishing willingness of states to protect this group of people.

At the start of the Syrian civil war, neighbouring Jordan and Lebanon initially welcomed Syrian refugees, albeit with different approaches. Pursuing a formalised approach, Jordan transferred Syrian refugees arriving at the border directly to formalised refugee camps where they could register with the UNHCR. In order to control the influx into urban areas, Jordanian authorities started to restrict border movement close to cities, thus forcing refugees to cross in remote areas (HRW 2021, p. 15). In contrast, Syrian refugees in Lebanon were initially received informally as guests, with the expectation that they would return to Syria at a later date, choosing not to have any formal refugee camps as a result of previous experiences with Palestinian refugee camps. However, with the prolonged and deteriorating situation in Syria, UNHCR registrations were banned in Lebanon after 2015. This was followed by the implementation of a sponsorship programme that led to the exploitation of refugee labour, and the increasing number of unregistered refugees that cannot for various financial and political reasons renew their legal status. A recent report found that only 16% of Syrian refugees in Lebanon aged 15 and older held legal residency (UNICEF et al. 2021, p. 13).

Refugees in Lebanon with illegal status face a series of intersectional challenges like exploitation, restrictive movement, and fear of arrest and detention. Recently, authorities in Jordan and Lebanon have pursued voluntary repatriation strategies for Syrian refugees. However, a recent report found that these refugees who have often returned to Syria have often done so under pressure and with limited information regarding the conditions within Syria – thus exposing them to considerable risks of violence and harm (HRW 2021). Again, Lebanese authorities have pursued these measures more aggressively than Jordan, implementing deportations for refugees who re-enter Lebanon. Despite these conditions and the compounding crisis in Lebanon, most Syrian refugees remain without protection from Lebanese authorities.

The Syrian refugee influx – together with forces like urbanisation, mismanagement, and corruption – has impacted services and infrastructures. In Jordan and Lebanon, the changing demographics have stretched authorities' capacity to provide vital services like healthcare, education, and waste management. As the crisis has extended, citizens in these states have started to increasingly blame refugees and governments for not securing services, the declining economy, and rising unemployment. However, refugees remain some of the most vulnerable people within these states struggling to cover their most basic needs. Moreover, resource scarcity has also increased, with water security being one of the most critical issues in Jordan (UNICEF Jordan and Economist Impact 2022). One of the most 'water poor' countries in the world, Jordan has been impacted by ageing water systems, insufficient planning, and population growth. Intersecting with food security, agricultural production is vulnerable to increasing costs, and import dependence. Food insecurity is an increasing challenge in Jordan, with 53% of Jordanians and 88% of the refugee population being insecure (UNICEF Jordan and Economist Impact 2022, p. 8). These interrelated resource scarcities, in turn, make the region particularly vulnerable to climate stress.

Refugees of other nationalities in Lebanon and Jordan – from countries such as Iraq, Sudan, Ethiopia, and Yemen – are faced with additional insecurities, with few receiving attention amidst the 'racial hierarchies' in humanitarian work (Janmyr 2022). In Lebanon, non-Syrian refugees without passports are unable to renew their residency and are often unable to access legal services from their embassy. The share of these refugees with legal residency in Lebanon has dropped significantly in recent years from 36% to 21% between 2019 and 2020 (UNHCR 2021). In Jordan, non-Syrian refugees are also regulated through the standard immigration rules unless they are registered with the UNHCR. A recent report found that these refugees registered 'to avoid harassment, arrest, and possible deportation by authorities', but that they 'have less access to services and often fewer legal rights' than Syrian refugees and vulnerable Jordanians due to bureaucratic obstacles (Johnston et al. 2019, p. 17). Due to the lack of targeted assistance, refugees in Jordan and Lebanon are more likely to spend more money on rent and utilities and less on food further contributing to their insecurity (UNHCR 2021, p. 52). Although numbers are relatively small, they also experience exclusionary practices by both humanitarian agents and authorities (Janmyr 2022).

IDPs face similar challenges with their legal status – some lacking documentation. In Iraq, IDPs with no identification are unable to exercise their full rights as Iraqi citizens making it difficult to move freely across checkpoints, cities, and regions, and unable to claim access to public services, such as healthcare and education (NRC 2019). Undocumented persons in Iraq face the risk of isolation and stigma – many being perceived as members of ISIL and/ or other extremist groups. Furthermore, children without birth certificates also face the risk of becoming stateless. A recent report found that 45,000 children born under ISIL have not been issued Iraqi-state documents depriving them of the most basic rights (NRC 2019, p. 3). The lack of civil documentation is, of course, a significant challenge in Iraq enforcing discrimination and political exclusion. IDPs in Iraq and Syria are extremely vulnerable lacking essential infrastructures and services. In northwest Syria, IDPs lack protection from political and gender-based violence, particularly since the Syrian authorities lost control of the area. With an estimated 2.8 m IDPs in this region alone, refugees struggle to access the most basic services across the 1,414 displacement camps in the region (Amnesty International 2022, p. 12). Inadequate tents in these camps make the situation particularly dire, with hazardous weather like the freezing winter months, sandstorms, and flooding severely impacting the quality of life of IDPs.

The parallels with Agamben's figure of *homo sacer* are quickly apparent amidst the abandonment of individuals into conditions of bare life. Much like the *homo sacer*, displaced people across the Levant have little hope of escaping their conditions given the absence of legal protection and the mechanisms of control in place to regulate their lives. These issues are then exacerbated by the actions of humanitarian hierarchies, adding to feelings of abandonment, marginalisation, and discrimination. The precarious situation facing many leaves them vulnerable to all manner of issues, from health and well-being, to more traditional questions of security and radicalisation.

## Protest

Amidst increasing environmental and intersectional challenges – and politicised responses to such issues – the landscape facing people across the Levant is precarious. In the aftermath of violent conflict, there is little wonder that the economies of Iraq, Lebanon, Palestine, and Syria have been devastated, facing heavy bills to repair infrastructures and urban spaces damaged during the war (Leenders 2012; Parreira 2021; World Bank 2022). Amid various stages of reconstruction, these economies are shaped by rapid demographic changes, and environmental challenges exacerbated by sectarian politics, mismanagement, and corruption. Specifically, in Lebanon, three decades after the Lebanese civil war, the economy is described by the World Bank (2022) as a 'deliberate depression' set out by the political leadership to stay in power in the wake of widespread anti-sectarian protests, COVID-19, and the 2020 Beirut Port explosion. Furthermore, economic challenges in Syria, already impacted severely by conflict and displacement, are exacerbated by the deepening crisis in Lebanon and Turkey, and also the escalation of US sanctions. In addition, Syria – like many other countries – is impacted by the soaring fuel and food prices that have emerged as a result of the war in Ukraine, together with food insecurity following 'record-low crop production in 2021' due to agricultural failures and drought, making Syria 'among the ten most food insecure countries globally' (World Bank 2022). The decreasing economic conditions in the Levant, including high unemployment and public debt, disproportionately impact vulnerable communities, deepening socio-ecological inequalities in the process.

The economic decline must be seen in the broader context of pre-existing political and socio-ecological dynamics. Protests that have emerged in the last decade underscore the role of socio-ecological systems within broader systems of political discontent. Moreover, we suggest that the interplay of socio-ecological degradation and other factors is more likely to lead to 'silent' suffering. While protest groups articulate a desire to move beyond communal cleavages, the counter-revolutionary strategies deployed by elites and, indeed, by the biopolitical machineries of sovereign power seek to exacerbate all modes of difference in pursuit of survival and self-interest (Mabon 2020). Once more, the creation of bare life and the inclusion of those abandoned by the state by virtue of their exclusion are mechanisms of control used by those in positions of power to ensure their survival.

Protests in the Levant have been nested – directly and indirectly – within policy and infrastructure failures. The agrarian crisis in Syria (in 2011), the waste management crisis in Mount Lebanon (in 2015), the water crisis in Basra (in 2018), and the water-for-energy protests in Amman (in 2021) display complex contexts facilitating the emergence of protest in the region. However, they must be understood within broader socio-ecological, economic, and political discontent. The protest in Amman described in the introduction, for example, was not just about the water deal but also the Israeli occupation and place of Palestinian refugees in the Jordanian state. The intersectionality of such issues – with climate change central – must not be overlooked.

The Arab uprisings were an outlet for the frustrations that people had with governance structures that had regulated life, underscoring growing anti-establishment, anti-sectarian, and pro-democracy sentiments across the region. As the years leading up to the Arab uprisings highlight, the agrarian crisis in Syria prompted mass migration from rural areas of the country to urban centres, altering the makeup of those spaces in the process (Selby 2019). The influx of farmers to urban centres, taken together with regular patterns of urbanisation, posed challenges to the ordering of life in major cities, resulting in large-scale unemployment and unrest, placing additional pressure on services that were already struggling. The agrarian crises display the interconnectedness with challenges, such as livelihood and migration, that deepened inequalities and led to the Arab uprisings.

Since the Arab uprising, increasing pressures on services and infrastructures have also led to involvement by activist movements. In Beirut and Mount Lebanon, the large-scale 2015 waste protests emerged as a result of the mismanagement of the waste sector. The streets piled up with waste as the political leadership failed to come to an agreement around the distribution of funds, and the location of a new landfill. Through #YouStink, protestors raised questions about corruption and awareness of the failures in power-sharing in Lebanon. Although the movement ultimately failed to bring about substantive changes within waste management and the political system, the protest was the first major anti-sectarian protest in Lebanon since the civil war. Similarly, in Basra in 2018, large-scale protests erupted as a result of the water epidemic described earlier in this chapter. Protestors demanded the improvement of water infrastructures and the poor health services received by those who had fallen ill.

Similar issues were apparent in the latest set of protests movement in the Levant and beyond. Between 2018 and 2022, protests materialised in varying degrees within compounding political, economic, and socio-ecological crises across the Levant. With the continuation of protests in Basra and across Iraq, the protests captured varying frustrations with government structures including corruption, unemployment, sectarianism, and inefficient infrastructures and public services (Al-Rubaie et al. 2021). In Lebanon, the Thawra movement (2019–2020) blamed the political leadership for the compounding political and socio-economic crisis. Data collected by *Lebanon Support* found that there were four main causes of collective action between 2019 and 2021: policy grievances, access to socio-ecological rights, corruption, and justice/ perceived justice (Civil Society Knowledge Centre 2022). The Thawra movement explored intersectional processes and demanded justice for all inhabitants beyond sect and class, including women, migrant workers, refugees, and LGBTQ+ people (Nagle 2022). Although the protests diminished over time, with the worsening of the economic crisis, COVID-19, and the Beirut Port explosion, the protests displayed a commitment to non-violent responses in the face of the power of sectarian elites backed by the machinery of sovereign power.

Responding to these protest movements, authorities have sought to undermine the activists. In Lebanon, authorities have utilised 'co-optation, counter-narratives, and repression' to minimise the impact of social movements (Geha 2019). In these cases, authorities responded with various forms of violence like rubber bullets and tear gas, in addition to arrest and detention. Clashes with authorities in Iraq killed at least 15 protestors; the victims being called 'water martyrs' (Al-Rubaie et al. 2021). Authorities in both countries also restricted public spaces by forcefully removing peaceful protestors. Moreover, armed militias clashed with protestors in both Lebanon and Iraq to divert and hijack the movements. Despite the widespread violence between armed groups and against activists in Lebanon, 70% of the 4,700 events between 2016 and 2020 remained peaceful (ACLED 2020). The willingness of movement activists to protest political, economic, and socio-ecological discontents within contexts of suppression and political violence provides grains of hope within a changing region. Nonetheless, increasing climate stress will further impact these mismanaged infrastructures and almost certainly further suffering and inequalities.

## Conclusion

This chapter has endeavoured to show the role of climate change and environmental regulation within the broader canvas of sovereign power and efforts to regulate life. The precarious political landscape of the Levant poses an array of challenges to rulers and ruled across the region, albeit conditioned by the complexities and contingencies of political life in each state. What remains constant, however, is the deliberate manipulation of environmental factors by the biopolitical machinery of sovereign power in pursuit of the accumulation of capital and to ensure survival. As Agamben's work on sovereign power highlights, political leaders have the ability to regulate access to resources and manufacture economic and socio-ecological inequalities.

Across the Levant, environmental challenges susceptible to climate change are shaped by corruption and mismanagement that have exacerbated qualities and quantities of infrastructures and public goods, such as water and food systems. Within these political systems, the environment is directly utilised in order to create unsustainable and inhospitable conditions for particular communities, such as the Palestinian community in Hebron. In other instances, the singular need for leaders and patronage structures takes precedence over the improvements of infrastructures, as seen in Lebanon and Iraq. While these structures are easily understood through arrangements of corruption and nepotism, they are designed to benefit and ensure the survival of, certain leaders, groups, and/ or militias whilst excluding others – who will become even more vulnerable with increasing climate stress.

Displaced persons are particularly vulnerable to the regulation of their political life and surrounding environments. Refugees and IDPs often lack access to ensure their most basic needs, with many lacking adequate shelter, healthcare, and food and water security inside and outside of encampments. For many displaced, political, social, and ecological rights are interrelated. Legal statuses, such as legal residency, UNRWA/UNHCR registrations, and civil documentation, not only extend social and political rights but also environmental rights through various degrees of humanitarian and governmental assistance – though often limited.

Amidst these compounding challenges, economic crises, unemployment, and grievances have multiplied in recent years. Recent activism has highlighted these frustrations around political systems displaying anti-establishment, anti-sectarian, and pro-democracy sentiments. In varying degrees across the Levant, these protests have reflected environmental insecurities and concerns about the future complications of climate change, including water-related protests in Basra and Amman. However, governments and armed groups have responded – with varying displays of force – to these protests across the Levant with repression and direct violence. The regulation of movement activists – seeking improved economic and socio-ecological conditions – displays the willingness of states to deploy additional tools to stay in power. In Lebanon and Iraq, the political deadlocks and extremely inhospitable conditions have reduced the number of collective actions. However, as long as movement activists are willing to protest there is still hope for sustainable and inclusive climate change adaption.

## References

ACLED 2020. Breaking the barriers: one year of demonstrations in Lebanon. *ACLED*. https://acleddata.com/2020/10/27/breaking-the-barriers-one-year-of-demonstrations-in-lebanon/ (accessed February 2023).

Agamben, G. 1996. Beyond human rights. (In) P. Virno and M. Hardt (eds) *Radical Thought in Italy: A Potential Politics*. Minneapolis, MN: University of Minnesota Press, pp. 159–165.

Agamben, G. 1998. *Homo Sacer: Sovereign Power and Bare Life*. Stanford, CA: Stanford University Press.

Agamben, G. 2005. *State of Exception*. London: The University of Chicago Press.

Amnesty International. 2017. The occupation of water. *Amnesty International*. https://www.amnesty.org/en/latest/campaigns/2017/11/the-occupation-of-water/ (accessed 1 May 2023).

Amnesty International 2022. Seventy+ years of suffocation. *Amnesty International*. https://nakba.amnesty.org/en/chapters/jordan/ (accessed 1 May 2023).

Al-Khalidi, A., S. Hoffmann, and V. Tanner 2007. *Iraqi Refugees in the Syrian Arab Republic: A Field-Based Snapshot*. Washington, DC: The Brookings Institution.

Al-Rubaie, A., M. Mason, and Z. Mehdi 2021. *Failing Flows: Water Management in Southern Iraq*. London: LSE Middle East Centre Paper Series.

Civil Society Knowledge Centre 2022. Interactive graphs. *Civil Society Knowledge Centre*. https://civilsociety-centre.org/cap/collective_action/charts (accessed 1 May 2023).

Corstange, D. 2016. *The Price of a Vote in the Middle East: Clientelism and Communal Politics in Lebanon and Yemen*. Cambridge: Cambridge University Press.

Crawford, A. and C. V. Church 2021. Climate change as a contributor to conflict. (In) R. A. Matthew, E. Nizkorodov and C. Murphy (eds) *Routledge Handbook of Environmental Security*. New York, NY: Routledge, pp. 82–91.

Davis, H. 2021. Hundreds protest in Jordan against water-energy deal with Israel. *Al Jazeera*. https://www.aljazeera.com/news/2021/11/26/hundreds-protest-in-amman-against-water-energy-deal-with-israel (accessed 1 May 2023).

De Chatel, F. 2014. The role of drought and climate change in the Syrian uprising. *Middle East Studies* 50(4): 521–535.

Eid-Sabbagh, K-P. 2015. A political economy of water in Lebanon: water resource management, infrastructure production, and the International Development Complex." Unpublished Ph.D. dissertation, SOAS, University of London.

Floyd, R. 2021. Securitizing the environment. (In) R. A. Matthew, E. Nizkorodov, and C. Murphy (eds) *Routledge Handbook of Environmental Security*. London: Routledge, pp. 227–240.

Foucault, M. 1979 [1976]. *The History of Sexuality: An Introduction*. London: Allen Lane.

Foucault, M. 1992 [1984]. *The History of Sexuality: The Use of Pleasure*. London: Penguin.

Galtung, J. 1990. Cultural violence. *Journal of Peace Research* 27(3): 291–305.

Geha, C. 2019. Co-optation, counter-narratives, and repression: protesting Lebanon's sectarian power-sharing regime. *Middle East Journal* 73: 9–28.

Gleick, P. H. 2014. Water, drought and climate change and conflict in Syria. *Weather, Climate and Society* 6(3): 331–340.

HRW 2021. 'Our lives are like death': Syrian refugees return from Lebanon and Jordan. *Human Rights Watch*. https://www.hrw.org/sites/default/files/media_2021/10/syria1021_web.pdf (accessed 1 May 2023)

Hoffmann, C. 2018. Environmental determinism as Orientalism: the geo-political ecology of crisis in the Middle East. *Journal of Historical Sociology* 31(1): 94–104.

IDMC 2007. Iraq: A displacement crisis. *International Displacement Monitoring Centre*. https://www.refworld.org/pdfid/4678d8872.pdf (accessed 1 May 2023).

IDMC 2021. A decade of displacement in the Middle East and North Africa 2010–2019. *International Displacement Monitoring Centre.* https://www.internal-displacement.org/publications/a-decade-of-displacement-in-the-middle-east-and-north-africa (accessed 1 May 2023).

International Committee of Red Cross 2021. Iraq's perfect storm - a climate and environmental crisis amid the scars of war. https://www.icrc.org/en/document/iraqs-perfect-storm-climate-and-environmental-crisis-amid-scars-war (accessed 1 May 2023).

Janmyr, M. 2022. Sudanese refugees and the 'Syrian refugee response' in Lebanon: racialised hierarchies, processes of invisibilisation and resistance. *Refugee Survey Quarterly* 41(1): 131–156.

Johnston, R., D. Baslan, and A. Kvittingen 2019. Realizing the rights of asylum seekers and refugees in Jordan from countries other than Syria with a focus on Yemenis and Sudanese. *NRC.* https://bit.ly/381qk57 (accessed 1 May 2023).

Kahl, C. H. 1998. Population growth, environmental degradation, and state sponsored violence: the case of Kenya, 1991–93. *International Security* 23(2): 80–119.

Kelley, C. P., S. Mohtadi, M. A. Cane, and Y. Kushnir 2015. Climate change in the Fertile Crescent and the implications of the recent Syrian drought. *Proceedings of the National Academy of Sciences* 112(11): 3241–3246.

Le Billon, P. 2018. Conflict ecologies: connecting political ecology and peace and conflict studies. *Journal of Political Ecology* 25: 239–260.

Leenders, R. 2012. *Spoils of Truce: Corruption and State Building in Postwar Lebanon.* New York, NY: Cornell University.

Mabon, S. 2020. *Houses Built on Sand: Violence, Sectarianism and Revolution in the Middle East.* Manchester: Manchester University Press.

Mason, M. 2022. Infrastructure under pressure: water management and state-making in southern Iraq. *Geoforum* 132: 52–61.

MOE, UNHCR, UNICEF and UNDP 2021. Lebanon State of the Environment and Future Outlook: Turning the Crises into Opportunities. *United Nations Development Programme.* https://www.undp.org/lebanon/publications/lebanon-state-environment-and-future-outlook-turning-crises-opportunities (accessed 1 May 2023).

Nagle, J. 2022. 'Where the state freaks out': gentrification, queerspaces and activism in postwar Beirut. *Urban Studies* 59(5): 956–973.

NRC 2019. Barriers form birth: undocumented children in Iraq sentenced to a life on the margins. *Norwegian Refugee Council.* https://www.nrc.no/resources/reports/barriers-from-birth/ (accessed 1 May 2023).

Nucho, J. R. 2016. *Everyday Sectarianism in Urban Lebanon: Infrastructures, Public Services and Power.* Princeton, NJ: Princeton University Press.

Owens, P. 2009. Reclaiming 'bare life'? Against Agamben on refugees. *International Relations* 23(4): 567–582.

Parreira, C. 2021. Power politics: armed non-state actors and the capture of public electricity in post-invasion Baghdad. *Journal of Peace Research* 58(4): 749–762.

Robbins, P. 2012. *Political Ecology: A Critical Introduction.* Chichester: John Wiley & Sons Ltd.

Salloukh, B. F. 2019. Taif and the Lebanese state: the political economy of a very sectarian public sector. *Nationalism and Ethnic Politics* 25(1): 43–60.

Selby, J. 2019. Climate change and the Syrian civil war, part II: the Jazira's agrarian crisis. *Geoforum* 101: 260–274.

Selby, J., O. S. Dahi, C. Frohlich, and M. Hulme 2017. Climate change and the Syrian civil war revisited. *Political Geography* 60: 232–244.

UN-Habitat 2021. *Beirut City Profile 2021*. Beirut: UN-Habitat.

UNHCR 2017. Return and readmission of Palestinian refugees from Syria (PRS) to Lebanon and Jordan. *UNHCR*. https://www.refworld.org/pdfid/5ab8cf9d4.pdf (accessed 1 May 2023).

UNHCR 2022. UNHCR Syria and Iraq Situations: 2022 response review. *UNHCR*. https://reliefweb.int/report/iraq/unhcr-syria-and-iraq-situations-2022-response-overview-february-2022?gclid=CjwKCAiAl9efBhAkEiwA4Torih_8A70i0Tm-TyUVhy09mQFks6FEYm1NL265CvGzFshrW5WHI9DPdBoCbJsQAvD_BwE (accessed 1 May 2023).

UNICEF, UNHCR, WFP, and VASYR 2021. Vulnerability Assessment of Syrian Refugees in Lebanon. 2021. *ReliefWeb*. https://reliefweb.int/report/lebanon/vasyr-2021-vulnerability-assessment-syrian-refugees-lebanon (accessed 1 May 2023).

UNICEF Jordan and Economist Impact 2022. Tapped out: the costs of water stress in Jordan. *UNICEF*. https://www.unicef.org/jordan/media/11356/file/water%20stress%20in%20Jordan%20report.pdf (accessed 1 May 2023).

UNRWA 2022 Where we work. *UNRWA*. https://www.unrwa.org/where-we-work/syria (accessed 1 May 2023).

UNRWA 2016. Employment of Palestine refugees in Lebanon: an overview. *UNRWA*. https://www.unrwa.org/resources/reports/employment-palestine-refugees-lebanon-overview (accessed 1 May 2023).

Veridel, E. 2018. Infrastructure crises in Beirut and the struggle to (not) reform the Lebanese State. *The Arab Studies Journal* 26(1): 84–112.

World Bank 2022. *Lebanon Economic Monitor, Fall 2020: The Deliberate Depression*. Washington, DC: World Bank Group.

# 5

# DECENTRING CLIMATE SECURITY

The research and policy implications of sudden-onset and slow-onset climate change

*Tom Deligiannis*

## Introduction

In recent years, broad agreement has developed among experts that climate change has an impact on violent conflict, with impacts most likely to affect internal, state-level violent conflicts, while the consequences on inter-state conflict are likely to be very limited, even in the case of water security (Mach et al. 2020). This consensus has not eliminated debates about climate change's role in causing violent conflict. In fact, leading voices argue that we need to better focus on the mechanisms at work between climate and conflict linkages, the conditions under which climate change's impacts operate, and the causal importance of climate change factors relative to other factors (von Uexkull and Buhaug 2021; Mach et al., 2020). There are significant, unsettled debates, for example, among scholars about whether and to what extent climate change has played a role in causing Syria's civil war or the ongoing insurgencies in parts of the Sahel (Charbonneau 2022; Ide 2018). Some researchers argue forcefully that other social, political, and economic factors are largely responsible, with little if any role for climate change (Daoudy 2020; Selby et al. 2017).

The conclusion that climate has some security role, but we are just not sure what kind of role it plays or how important it is in the end in causing ongoing and emerging violent conflicts, leads to obvious challenges for policymakers and practitioners who want to act. How can policymakers and researchers find a way forward to design climate security interventions in light of continued uncertainty about the role and impact of climate change on violent conflict? This chapter argues that significant advances have been made by scholars in untangling the linkages between climate change and insecurity;

DOI: 10.4324/9781003377641-7

however, progress is uneven and it hinges on distinguishing between two main types of climate change-security effects. In recent years, researchers have done a good job of outlining the climate security implications and causal processes involving sudden-onset climate change hazards. Research on the climate security implications of slow-onset climate change impacts, however, lags and faces significant challenges in disentangling slow-onset risks from wider processes and impacts of human-environment interactions. Thinking about the practical implications for policy interventions from sudden-onset events is similarly far ahead of work on policy interventions to reduce slow-onset climate security risks.

This chapter begins by reviewing the differences between slow-onset and sudden-onset climate change impacts and their security risks. Recent research about the linkages between sudden-onset climate change and violent conflict are then surveyed, and the policy implications of these findings are discussed. The underexplored security risks from slow-onset climate change are then examined, and an argument is outlined that the nature of these climate impacts complicates efforts to both understand the significance of these effects and what to do to reduce vulnerability to slow-onset risks. Climate change needs to be decentred from our analysis of human–environmental conflicts and contextualised within the range of processes and transformations people impose on nature. The chapter closes by outlining a research and policy approach to dealing with the slow-onset consequences of climate security risks through an adaptive learning approach.

## Distinguishing sudden-onset and slow-onset climate change

Disentangling the security impacts of climate change requires an understanding of the differences between two types of climate change impacts – impacts from sudden or intense onset natural hazards, like extreme rainfall or heat events or meteorological events like typhoons, and slow-onset climate change impacts, like gradually decreasing precipitation levels, increasing average temperature levels leading to increasing aridity, or gradual sea level rise in coastal zones. In the first case, the events are sudden or happen across a compressed timeline – sudden and intense downpours of a few hours or a few days, severe droughts that can strike one growing season or stretch across multiple seasons, or a hurricane that rages for a few days, with attendant wind and flooding damage that lingers long after the storm has passed. The social and human impacts are similarly intense and frequently unexpected when the hazard overwhelms local resilience, requiring various coping responses by households and individuals (Smith 2013). With slow-onset climate change, the impacts are gradual, incremental, and temporally varied – they mostly happen over years and usually lead to various adaptive responses by people, groups, and states. Gradual changes in local temperature or rainfall patterns, for example, may impact cultivator and herder production, leading agriculturalists to various local adaptive strategies to maintain agricultural livelihoods (Deligiannis 2012). When such responses are unavailable, agricultural livelihoods may decline and require more varied responses, like changes in economic activities or various forms of migration, in order to sustain livelihoods (McLeman et al. 2021).

To date, researchers have done a better job of appreciating the climate security risks of sudden-onset climate change and in thinking through the policy implications that follow from these risks, compared to the security and policy implications of slow-onset climate change risks. The complex ways that slow-onset climate changes interact with other human–natural processes may partly explain our analytical shortcomings.

## Security and policy implications of sudden-onset climate change

The past five years have seen a burst of research exploring the impact of sudden-onset climate hazards on societies, with several recent IPCC reports (Pörtner et al. 2022) and academic studies about the impacts of violent conflict and political protest. A recent review of this research noted that sudden-onset climate events can create societal shocks, particularly economic shocks, which impact the willingness and ability of those that seek to challenge the state by decreasing the opportunity costs of participating in violence, worsening group relations, and stunting socio-economic development (Mach et al. 2020). Joshua Busby's (2022) recent study and the work of Tobias Ide and his co-researchers (Ide et al. 2021, 2020) are representative of this recent wave of climate conflict research. While increasing the awareness of the conflict and security implications of sudden-onset climate change hazards, particularly for the conditions of conflict onset and conflict dynamics (Ide 2023a), this work is also highlighting how intermediate variables like, poverty, state capacity, and divisions within states mediate the conflict consequences of climate change.

Policy interventions that seek to head off the conflict implications of sudden-onset climate hazards focus both on reducing risks before hazards strike and more effective intervention in the immediate aftermath of a disaster. Carefully calibrated preparations can help build the capacity of states to withstand rapid-onset climate hazards, like Bangladesh's success with building typhoon shelters and Ethiopia's efforts to enhance its food security system (Busby 2022). To better design policy interventions, we can leverage our understanding of intermediate variables that help cause conflict, targeting underserved peripheral areas or marginalised groups, and including stakeholders outside of state structures, like traditional authorities and subnational representatives (Busby 2022). Ide's recent research shows that disasters impact power differentials between the state and challenger groups in ways that can either exacerbate ongoing violent conflict or create conditions for peace in the immediate aftermath of disasters (Ide 2023b). This finding suggests that international disaster response efforts must be sensitive to the impacts of interventions on power relations and tailor assistance to promote cooperation and peace, rather than exacerbate violent conflict. Such considerations can also create a safer environment for relief and reconstruction efforts (Ide 2023b). Given that resources for reducing risk and responding to disasters are always limited, focusing interventions on factors that also impact the peace and conflict consequences of climate change is a judicious use of resources.

## Climate security challenges and dilemmas of slow-onset climate change

The climate security implications of long-term, slow-onset climate change are less well studied (von Uexkull and Buhaug 2021), particularly for those who rely on natural resources for their livelihoods in developing countries – agricultural communities, herding communities, fishers, etc. In such cases, the climate change impacts are difficult to disentangle from other processes transforming human–environmental relationships and complicating any attempt to discern the causal role of climate change on conflict (Buhaug et al. 2023). A holistic analysis of human–environmental impacts over time is needed to contextualise and better isolate the unique slow-onset climate change impacts on violent conflict from other processes of environmental transformations driven by human actions that can impact violent conflict. For example, development projects (often led by abusive or exploitative local or national elites) have proven to transform traditional agricultural and natural environments in ways that marginalise many traditional resource users, leading to various adaptive livelihood responses that can worsen vulnerable local environments or push people into alternative livelihood activities that increase social tensions among groups and populations (Deligiannis 2012; Stonich 1993). Scholars are increasingly concerned about the negative implications of societal and household coping and adaption strategies on violent conflict (Deligiannis 2012), concerns that have amplified as attention has turned to climate change impacts. Calls to ensure 'conflict sensitive adaptation' and avoid maladaptation with climate change responses are part of a wider appreciation that human reactions to environmental change can lead to unfortunate outcomes (Dabelko et al., 2013, Jacobson et al., 2019, Buhaug et al., 2023, Barnett and O'Neill, 2010). As well, the environmental consequences of extractive projects have been shown to be particularly detrimental to the human security of groups impacted nearby, significantly increasing social tensions (Conde and Le Billon 2017; Gudynas 2020; Schilling et al. 2021) or providing fuel and opportunities for challenger groups (Le Billon 2012). Similarly, political and social exclusionary processes in some states disadvantage some resource user groups – like particular ethnic, religious, or class groups, leading to worsening pressures on remaining resources and exacerbating grievances and group tensions (Homer-Dixon 1999; Kahl 2006). Finally, broader economic and social changes of marketisation transform people's relationships with their local resources in ways that sometimes encourage unsustainable resource exploitation or aggravate divisions both within and between resource user groups (Baechler 1999; Deligiannis 2020). Slow-onset climate change impacts may worsen and aggravate many of these human–environmental processes in exceedingly complex ways. However, focusing policy interventions on the climate change dimension may produce only marginal benefits because the impacts and causes of other important processes may remain unaddressed.

Decentring the focus on climate change's role in causing violent conflict is the first step towards a better understanding of climate change's impact, particularly with slow-onset climate change. Scholars and practitioners need to contextualise climate change with other pressures on human–natural systems that may also contribute to violent conflict. Climate change is only one of many human-environment pressures, and failing to explore these impacts along with other processes and pressures can have unfortunate consequences. It can lead to distortions in policy-making that emphasise the wrong kind of interventions, focusing on climate-related interventions when other human–environmental processes may be equally or more deserving of attention. The first generation of research linking environmental change to violent conflict in the 1990s (Baechler 1999; Homer-Dixon 1999) did a better job of appreciating the conflict implications of a range of human pressures on the environment than much recent work on climate conflict. This first generation of research triggered important debates about the causal role of human–environmental pressures in causing violent conflict, debates that highlighted how difficult it is to discern efficacious policy interventions (Diehl and Gleditsch 2001; Floyd and Matthew 2013). These dilemmas have not disappeared with recent research on climate security, mirrored in recent calls for researchers to explore the relative contribution of climate factors among other conflict determinants or to solve the 'attribution' problem in climate security research (Buhaug et al. 2023; Mach et al. 2020).

Two brief examples highlight the importance of decentring climate change in any analysis of whether human pressure on the environment is causing violent conflict and about what to do about these pressures. Camilla Toulmin's multi-decade study of the village of Dlonguébougou, in central Mali, notes the impact that climate change has had on precipitation levels and precipitation frequency in the area, and the related challenges that this poses to cultivation and pastoral agriculture in the area (Toulmin 2020). Climate change is making rainfall patterns more intense and unpredictable in the area, even while the impacts of accelerated heating and drought periods remain uncertain. Although total rainfall levels remain stable and may even grow with future climate change, increasingly intense rains falling in a shorter period of time are causing havoc for cultivators and their livelihoods (Toulmin 2020). The livelihood security risks from these slow-onset climate changes seem self-evident. However, one could never understand the nature of local livelihood challenges, migration patterns, and settlement changes in Dlonguébougou – including the conflictual stresses and grievances that accompanied these changes – without also exploring processes of nearby land grabbing, population displacement, demographic growth, and economic transformation that have occurred in recent decades in the area. For example, sugarcane projects to the south of the village, developed by the Malian state and foreign interests, have displaced cultivators into the lands of Dlonguébougou and neighbouring villages, increasing competition for land with local cultivators and with itinerant pastoralists who have long used surrounding fields. While local leaders in Dlonguébougou have so far managed these competing stresses peacefully, research in other parts of Mali illustrates that the differential outcome of modernisation and development projects imposed on local communities can help stimulate violent conflicts among pastoralists and cultivators (Benjaminsen 2008; Benjaminsen and Ba 2019). Such development tensions are not unique to Mali and are affecting other nearby African nations, processes that may worsen as 'Green' development priorities lead national governments to push projects regardless of local impacts (Bergius et al. 2020). Slow-onset climate change adds another layer of challenges to local livelihoods, essentially creating additional structural risk factors that people have to manage (Busby 2022; Deligiannis 2020).

Similarly, research conducted on two neighbouring communities in highland Peru found that climate change impacts were evident and helped to alter cultivator and herding livelihoods in the latter decades of the 20th century. Long-standing land conflicts between the communities turned violent at times in the 1970s and 1980s, raising questions about whether climate change has aggravated competition and violent conflict between the communities over contested high-altitude grazing lands (Deligiannis 2020). Research on elevation-dependent warming shows that climate change usually causes mountain areas to warm faster than nearby lowland areas, though there is local variability (Pepin et al. 2022). Regional climate studies show significant warming trends in the Peruvian Andes in the second half of the 20th century, with warming in the study area happening at both high and low altitudes (Vuille et al. 2015). In this part of Peru, local slow-onset climate change has warmed higher altitudes and decreased cultivator and pastoral frost risks in high-altitude grazing zones of the communities, possibly helping to encourage increasing permanent settlement to these areas from other community settlements. However, these climate change impacts must be contextualised with other factors and processes that have also helped to encourage internal migration by members of both communities to contested areas.

Centuries-long processes of resource capture by Peruvian elites, economic pressures and opportunities introduced with marketization in the twentieth century, and pressures on resource endowments from changing demographic trends have also fueled household decisions to settle in contested high altitude zones.

If we conduct a counter-factual thought experiment in this case and remove the climate change pressures on the two communities from the causal story, the other processes and stressors would likely still have aggravated group conflict between the two communities. Long-term processes of pressure and change in land systems in these communities, combined with wider political, economic, and social trends in Peru, were likely more significant in aggravating group conflict between these two communities than the impact of climate change on productive opportunities in the villages (Deligiannis 2020).

### Dilemmas in dealing with the security implications of slow-onset climate change

Without appreciating the range of pressures on the human–natural systems in the cases discussed above in Mali and Peru, we have no way of contextualising the climate change impacts on livelihood risk and vulnerability relative to other processes and conditions. For effective policy interventions to slow-onset climate impacts, we need to de-emphasise climate security and refocus on the range of human and environmental security challenges that include climate change impacts. In the Peruvian case, climate change was seemingly enhancing the capacity of local farmers to expand their agricultural and livestock activities in high-altitude zones, leading to clashes between rival communities over resource control. Policymakers assessing the situation might reasonably conclude that policy interventions to clarify resource ownership between communities might end the conflict. In fact, this has been the dominant approach to settling land conflicts in the area. However, such a conclusion in the Peruvian case would ignore the other processes and livelihood stressors that were driving community members to settle in contested highland areas. Those stressors would remain unaddressed by solutions that merely clarify borders and resource ownership. What appears on the surface to be a case of disputes over resource control are, in fact, conflicts driven by migration and long-term resource constraints confronting residents. Left unaddressed, these pressures significantly worsened violence in the area in the 1980s, during the Shining Path's insurgency against the Peruvian state (Deligiannis 2020).

In the Malian case, policy interventions that focus on adapting to the slow-onset climate change challenges might include using faster ripening crops that mature in an agricultural calendar compressed by increasingly variable precipitation levels. Such adaptive responses are often mentioned by observers as a way to manage climate security risks for livelihood security (Puthalpet 2022). However, such interventions would do nothing to address increasing constraints on cultivated and grazing lands due to in-migration from farmers being displaced by national sugarcane development projects. The danger is that, in the rush to do something about climate change security, we will end up prioritising climate change interventions without an adequate under-standing of whether such interventions are actually addressing the range of key human–environmental processes in such situations that have transformed human–natural systems. Without a full diagnosis of the processes of human–environmental transformation, we might prescribe misguided or ineffective policy interventions, or even worsen conflictual relations with policy actions.

Successful adaptation and intervention to slow-onset climate change impacts pose a complex challenge because these impacts aggravate, amplify, and interact with other gradual processes of environmental change, like those accompanying economic development. Climate security interventions by international actors, development agencies, or national governments to address slow-onset impacts thus face challenges to identify how and where to act. Researchers striving to identify key climate security causal drivers for intervention – seeking to solve the attribution problem – may be misguided, given the complex interactive nature of such human–environmental systems (Beaumont and de Coning 2022; Schwartz et al. 2001). In a complex, inter-active human-environment system, linear causation is an illusion; there is no certainty that action on an identified cause will result in stopping a predicted outcome. One strategy to manage this complexity is to focus on identifying the unique set of factors in each case contributing to insecurity and risk (Ide 2023b), including the impacts from slow-onset climate change, and have international actors work with local actors to prioritise actions to build resilience and reduce vulnerabilities. Such an approach recognises that addressing slow-onset climate changes is a complex problem akin to state-building or peacebuilding (Busby 2022).

State development and peacebuilding are notoriously challenging to undertake successfully. Practical and conceptual challenges need to be appreciated and overcome. Institutional priorities of international actors, for example, may frustrate attempts to integrate local voices and local priorities for climate security action (Bremberg et al. 2022). As well, security actors operating in regions with both evident climate risks and violent conflict risks may prioritise climate security responses in ways that facilitate their missions, but which do little to reduce climate change risks (Charbonneau 2022). Military organisations are, in fact, poorly suited to respond to slow-onset climate risks. Such tasks are better left to development agencies and NGOs. The decades-long experience with international peacebuilding operations has also demonstrated that international actors, development agencies, and

NGOs have not been very successful in long-term peace and development efforts in fragile states (Chandler 2017). This record should provoke caution and humility in any actor planning slow-onset climate security interventions. Local actor-driven efforts to build climate security resilience should be supported by outside actors in iterative, adaptive capacity-building approaches, modelled on adaptive peacebuilding approaches and on locally managed disaster risk management practices (Collins 2009; De Coning et al. 2023). Crucially important, such efforts should strive to 'do no harm' and avoid aggravating conflicts or worsening human security (Dabelko et al. 2013; Lamain 2022).

## Conclusion

Researchers and practitioners have made considerable progress in diagnosing and treating climate change insecurities. However, that progress is uneven to date. We know much more about how to tackle the conflict challenges of rapid-onset climate change impacts than we do about slow-onset climate change impacts. This chapter has argued that the particular nature of slow-onset climate change complicates any attempt to both determine the significance of those climate impacts and the best way to handle the risks from slow-onset climate change, because they are difficult to disentangle from other processes of human-environmental change. It is not being argued that slow-onset climate impacts are irrelevant to the security consequences compared to other social, political, or economic factors. Instead, scholars and policymakers need to decentre the climate-security focus by approaching the problem with an environmental change-security focus to better appreciate the complexity of slow-onset climate change impacts in relation to various other human-environmental change processes. Determining how much climate change matters in provoking security risks is difficult in practice when dealing with complex, interactive systems. As well, addressing political or ethnic exclusion in regimes, building effective state capacity for fair development, and encouraging sustainable resource exploitation are exceedingly difficult tasks for both developing states and external development assistance programmes. Taking action to build climate resilience and reduce climate security risks may be ineffectual without a parallel commitment to also tackle these related challenges and sources of human pressure on environmental processes, conditions that are worsened over time by the impacts of slow-onset climate change.

## References

Baechler, G. 1999. *Violence through Environmental Discrimination: Causes, Rwanda Arena, and Conflict Model*. Dordrecht: Kluwer Academic Publishers.

Barnett, J. and S. O'Neill 2010. Maladaptation. *Global Environmental Change* 20: 211–213.

Beaumont, P. and C. de Coning 2022. Coping with complexity: toward epistemological pluralism in climate-conflict scholarship. *International Studies Review* 24(4). 10.1093/isr/viac055

Benjaminsen, T. A. 2008. Does supply-induced scarcity drive violent conflicts in the African Sahel? The case of the Tuareg rebellion in Northern Mali. *Journal of Peace Research* 45: 819–836.

Benjaminsen, T. A. and B. Ba 2019. Why do pastoralists in Mali join jihadist groups? A political ecological explanation. *The Journal of Peasant Studies* 46: 1–20.

Bergus, M., T. A. Benjaminsen, F. Maganga, and H. Buhaug 2020. Green economy, degradation narratives, and land-use conflicts in Tanzania. *World Development* 129: 104850.

Bremberg, N., M. Mobjörk, and F. Krampe 2022. Global responses to climate security: discourses, institutions and actions. *Journal of Peacebuilding & Development* 17: 341–356.

Buhaug, H., T. A. Benjaminsen, E. A. Gilmore, and C. S. Hendrix 2023. Climate-driven risks to peace over the 21st century. *Climate Risk Management* 39: 100471.

Busby, J. W. 2022. *States and Nature: The Effects of Climate Change on Security.* Cambridge: Cambridge University Press.

Chandler, D. 2017. *Peacebuilding: The Twenty Years' Crisis, 1997–2017.* Cham: Palgrave Macmillan.

Charbonneau, B. 2022. The climate of counterinsurgency and the future of security in the Sahel. *Environmental Science & Policy* 138: 97–104.

Collins, A. E. 2009. *Disaster and Development.* Abingdon: Routledge.

Conde, M. and P. Le Billon 2017. Why do some communities resist mining projects while others do not? *The Extractive Industries and Society* 4: 681–697.

Dabelko, G. D., L. Herzer, S. Null, M. Parker, and R. Sticklor (eds) 2013. *Backdraft: The Conflict Potential of Climate Change Adaptation and Mitigation.* Environmental Change and Security Program Report 14(2). Washington, DC: Woodrow Wilson International Center for Scholars.

Daoudy, M. 2020. *The Origins of the Syrian Conflict: Climate Change and Human Security.* Cambridge: Cambridge University Press.

De Coning, C., R. Saraiva, and A. Muto 2023. *Adaptive Peacebuilding: A New Approach to Sustaining Peace in the 21st Century.* Cham: Springer International Publishing.

Deligiannis, T. 2012. The evolution of environment-conflict research: toward a livelihood framework. *Global Environmental Politics* 12: 78–100.

Deligiannis, T. 2020. *Environmental and Demographic Change and Rural Violence in Peru: A Case Study of Cangallo, Ayacucho.* Unpublished PhD dissertation, University of Toronto.

Diehl, P. F. and N. P. Gleditsch 2001. *Environmental Conflict.* Boulder, CO: Westview Press.

Floyd, R. and R. A. Matthew 2013. *Environmental Security: Approaches and Issues.* London: Routledge.

Gudynas, E. 2020. *Extractivisms: Politics, Economy and Ecology.* Novia Scotia: Fernwood Publishing.

Homer-Dixon, T. 1999. *Environment, Scarcity, and Violence.* Princeton, NJ: Princeton University Press.

Ide, T. 2018. Climate war in the Middle East? Drought, the Syrian Civil War and the state of climate-conflict research. *Current Climate Change Reports* 4: 347–354.

Ide, T. 2023a. *Catastrophes, Confrontations, and Constraints: How Disasters Shape the Dynamics of Armed Conflicts*. Cambridge, MA: MIT Press.

Ide, T. 2023b. Rise or recede? How climate disasters affect armed conflict intensity. *International Security* 47: 50–78.

Ide, T., M. Brzoska, J. F. Donges, and C-F. Schleussner 2020. Multi-method evidence for when and how climate-related disasters contribute to armed conflict risk. *Global Environmental Change-Human and Policy Dimensions* 62. 10.1016/j.gloenvcha. 2020.102063

Ide, T., A. Kristensen, and H. Bartusevicius 2021. First comes the river, then comes the conflict? A qualitative comparative analysis of flood-related political unrest. *Journal of Peace Research* 58: 83–97.

Jacobson, C., S. Crevello, C. Chea, and B. Jarihani 2019. When is migration a maladaptive response to climate change? *Regional Environmental Change* 19: 101–112.

Kahl, C. H. 2006. *States, Scarcity, and Civil Strife in the Developing World*. Princeton, NJ: Princeton University Press.

Lamain, C. 2022. Conflicting securities: contributions to a critical research agenda on climate security. *Globalizations* 19: 1257–1272.

Le Billon, P. 2012. *Wars of Plunder: Conflict, Profits and the Politics of Resources*. London: Hurst & Company.

Mach, K. J., W. N. Adger, H. Buhaug, M. Burke, J. D. Fearon, C. B. Field, C. S. Hendrix, C. M. Kraan, J. F. Maystadt, J. O'Loughlin, P. Roessler, J. Scheffran, K. A. Schultz, and N. Uexkull 2020. Directions for research on climate and conflict. *Earth's Future* 8: e2020EF001532-n/a.

McLeman, R., D. Wrathall, E. Gilmore, P. Thornton, H. Adams, and F. Gemenne 2021. Conceptual framing to link climate risk assessments and climate-migration scholarship. *Climatic Change* 165. 10.1007/s10584-021-03056-6

Pepin, N. C., E. Arnone, A. Gobiet, K. Haslinger, S. Kotlarski, C. Notarnicola, E. Palazzi, P. Seibert, S. Serafin, W. Schöner, S. Terzago, J. M. Thornton, M. Vuille, and C. Adler 2022. Climate changes and their elevational patterns in the mountains of the world. *Reviews of Geophysics* 60: e2020RG000730.

Puthalpet, J. R. 2022. *The Daunting Climate Change: Science, Impacts, Adaptation and Mitigation Strategies, Policy Responses*. London: CRC Press.

Pörtner, H-O., D. C. Roberts, H. Adams, C. Adler, P. Aldunce, E. Ali, R. A. Begum, R. Betts, R. B. Kerr, and R. Biesbroek 2022. *Climate Change 2022: Impacts, Adaptation and Vulnerability*. Geneva: IPCC.

Schilling, J., A. Schilling-Vacaflor, R. Flemmer, and R. Froese 2021. A political ecology perspective on resource extraction and human security in Kenya, Bolivia and Peru. *The Extractive Industries and Society* 8: 100826.

Schwartz, D., T. Deligiannis, and T. Homer-Dixon 2001. The environment and violent conflict. (In) P. Diehl and N. P. Gleditsch (eds.) *Environmental Conflict*. Boulder, CO: Westview Press, pp. 273–294.

Selby, J., O. S., Dahi, C., Fröhlich, and M., Hulme 2017. Climate change and the Syrian civil war revisited. *Political Geography* 60: 232–244.

Smith, K. 2013. *Environmental Hazards: Assessing Risk and Reducing Disaster*. London: Routledge.

Stonich, S. C. 1993. *I am Destroying the Land! The Political Ecology of Poverty and Environmental Destruction in Honduras*. Boulder, CO: Westview Press.

Toulmin, C. 2020. *Land, Investment and Migration: Thirty-Five Years of Village Life in Mali*. Oxford: Oxford University Press.

Von Uexkull, N. and H. Buhaug 2021. Security implications of climate change: a decade of scientific progress. *Journal of Peace Research* 58: 3–17.

Vuille, M., E. Franquist, R. Garreaud, W. S. Lavado Casimiro, and B. Cáceres 2015. Impact of the global warming hiatus on Andean temperature. *Journal of Geophysical Research*. 120: 3745–3757.

# 6

# A NEW FRAMEWORK FOR UNDERSTANDING RISK

## A Closer Look at the Climate-Violence-Migration Connection in northern Central America

*Lauren Risi and Cynthia Brady*

### Introduction

Climate change is upending what we think we know about risk – and what to do about it. The impacts of climate change are compounding vulnerabilities across temporal and spatial scales in unprecedented ways. Long-established institutional frameworks for addressing threats to national security and global stability are no longer fit for purpose in our changing world. Taking early action to manage or reduce climate-related disruptions requires a re-wiring of our established systems and institutions to more effectively understand risk. We need a better understanding of the compound risks posed by climate change that is grounded in robust and timely quantitative and qualitative assessments.

This chapter will lay out a new framework for improving predictive capabilities developed by the Wilson Center in partnership with the National Oceanic and Atmospheric Administration and University Corporation for Atmospheric Research that provides a common conception of the relationship between security and weather and climate-related disruptions. Through an exploration of regional and country-specific cases, the project team has identified four core dynamics that, in their interactions with climate impacts, can lead to magnified, compounded, and cascading risks. Understanding how climate interacts with these vulnerabilities – whether structural, physical, or other – and how the dynamics interact with one another is key to unpacking the complex ways that climate change drives risk, and where there are entry points for more effective responses.

DOI: 10.4324/9781003377641-8

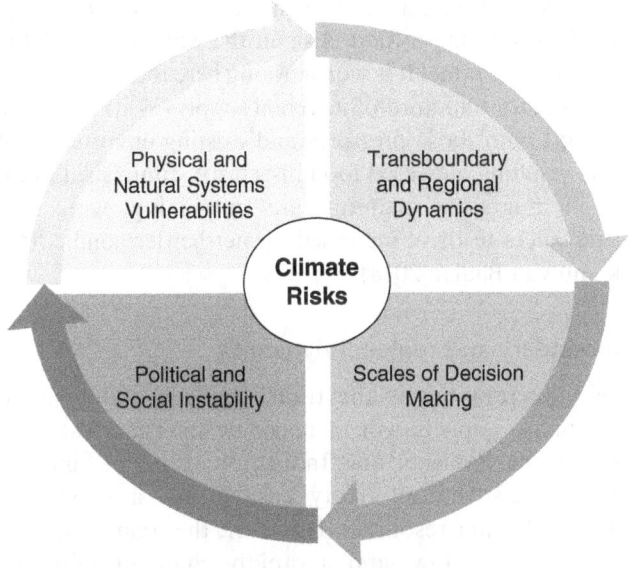

**FIGURE 6.1**   Framework for improving predictive capabilities based on four vulnerabilities.

*Source:* Author's creation.

This chapter begins with an overview of the four core dynamics, or vulnerabilities, identified in the framework to improve predictive capabilities for security risks posed by climate change (Figure 6.1) and provides examples of how these are playing out in country and region-specific examples, before applying the framework to a key region of US strategic interest – the northern Central American countries of Honduras, El Salvador, and Guatemala.

### Core dynamic: physical and natural systems

Climate change is increasingly interacting with existing physical and natural systems vulnerabilities around the globe. For example, Pakistan is a country that has been historically affected by severe floods and droughts, but the increase in frequency and severity of droughts and floods brought on by a changing climate are beyond the Government of Pakistan's capacity to respond and place the country's key industries – textiles and agriculture – at high risk (Reed and Bokhari 2022; YCC Team 2022). Similarly, in the Horn of Africa, recurrent drought and heavy, intermittent rains have been magnified by climate change, leading to longer and more frequent and severe droughts, devastating floods, and invasive desert locusts. Agriculture is a key pillar of the region's economy and populations' livelihood security. Reduced crop productivity as a result of heat, drought, and increased pest damage will further diminish food security in the region, which is already facing a food crisis in the wake of the global pandemic and Russia's war in Ukraine. It is worth noting here that East Africa relies on Ukraine and Russia for one-third of its cereal supply (Watkins 2022). As climate change interacts with those pressures and existing environmental change in the region, the resulting increased food insecurity, diminished livelihoods, and changes in migration patterns are combining with elite exploitation of local grievances to drive increased farmer-herder conflict (see Luizza 2019; Mobjork and van Baalan 2018).

### Core dynamic: transboundary and regional dynamics

A second category of vulnerabilities is the transboundary and regional dynamics with which climate impacts will increasingly interact. Take the Indus Waters Treaty between Pakistan and India, for example. Since its signing in 1960, the water-distribution treaty has become a symbol of coordination through shared water resources. However, the treaty does not account for variability in water flow, and a rapidly changing climate's impact on water availability will likely pose challenges to the treaty and exacerbate pre-existing tensions with India. In another example, Pacific Island countries (PICs) are caught in the middle of the great power competition between China and the United States. China is motivated politically and economically to be more engaged in the Pacific region. They currently use the Pacific to transport raw materials, support space missions with tracking ships housed in the Pacific, and resupply naval and scientific expedition ships at friendly ports on the islands. China's role as the largest lender in the region has primed its position to take advantage of a region that will be impacted routinely by large storms, sea level rise, and water scarcity. China is harnessing the region's need for capital and accruing debt from small island nations which cannot afford to pay back the loans. In many cases, China has extended economic relief only to 'flip it' for expansion of power. For example, China will build critical infrastructure with the stipulation that the host country pays it back within a certain time frame, or it will come under Chinese custody for an extended time frame – a 'loan to own' situation, where the inability to repay a loan results in the lender assuming ownership of the asset.

In addition to the impact of great power competition, the sovereignty of PICs is also under threat from rising tides. Much of the territory occupied by small island states may be lost completely to sea level rise, a trend which cannot, at this point, be stopped by any human intervention. When territory disappears, and some PIC residents are forced to relocate, two questions will arise: (1) how can identity and culture – both deeply tied to tribally owned land – be retained; and (2) whether it is possible for these countries to retain sovereignty in a global system which defines sovereignty, in part, by the land a country occupies? The issues of sovereignty and relocation for small island nations are issues that, while articulated, have not been addressed in the present (see Marinaccio 2022).

### Core dynamic: political and social stability

Political stability and strength of governance is a third category of vulnerabilities that can shape the conflict outcomes of climate-related disruptions. The key drivers of domestic insurgency in the Horn of Africa, for example, include the colonial legacy; government corruption and mismanagement; oppressive authoritarian regimes; restriction of movement; ethnic tensions; emergence of warlords with private armies; terrorism; weakened political stability; and fragile economies. Climate impacts interact with these existing vulnerabilities to create cascading security impacts. For example...years. with "In East Africa, for example, pastoralist (farmer-herder) conflicts have increased in recent years in part as a result of growing water insecurity. For example, during the 2011–2013 famine in Somalia, the Islamist insurgent group al-Shabaab used the famine to coerce and tax international aid agencies and withheld food from those it deemed uncooperative (Ro 2019).

### Core dynamic: the scales of decision making

Extreme weather events and climate change produce cascading, compound, and sequential impacts that include both short-term and long-term risks that are difficult to unpack and combat effectively. Governance of, and responses to, these events occur at multiple levels – local, national, regional, and global– with implications and feedback loops across all scales. The appropriateness of applying specific tools for prediction, resilience, and post-event action will vary across scales. Scales are necessarily discrete at their edges but have continuous and overlapping boundaries. If all levels of decision making and impacts are not considered, there will be opportunities missed for building resilience and mitigating vulnerability, as well as unintended consequences from actions taken or not taken at other scales.

In the Pacific, for example, several institutions are working to identify local, micro-solutions that will engage the community in building resilience. They have found that investing in local agriculture using traditional knowledge and native plants is an important source of resilience for PICs.

Co-production of knowledge and tools by local communities, governments, and experts is key to building resilience in the long term and ensuring that effective and applicable tools are being developed. Tools that are useful in theory are often developed, but lack practicality in a specific context, or are applied at the wrong scale of governance where the effect is diminished or negligible. Unwrapping the local context is also important for the inclusion of marginalised groups, for understanding conditioning factors and alternatives in specific cases, and for moving beyond generalisations drawn from large-N global studies of climate and security. If marginalised groups are not actively involved in the problem-solving process, the resulting solutions will often leave them behind.

## Climate-related security risks in the Northern Triangle

Latin America is experiencing a confluence of insecurity and migration challenges that are increasingly intertwined with climate change. High levels of ethnic, gender, and socioeconomic inequality are ubiquitous, and populism and authoritarianism are gaining traction. In particular, chronic conflict, violence, and weak governance are undermining human security in the Central American countries of the Northern Triangle. For decades, the region has experienced extreme rates of homicide and gang violence, failing democratic governance and ineffective institutions, pervasive corruption, and drug trafficking. More recently, it has also been propelled by the expansion of illicit networks, the limited capacity of cities and urban informal settlements to constructively absorb a growing number of migrants, and the compounding effects of the COVID-19 pandemic and recurrent natural disasters.

In the 2020 Preventive Priorities Survey, foreign policy experts ranked deteriorating economic and security conditions in the Northern Triangle, resulting in increased migration outflows from the region, as a top conflict to watch. While individuals have migrated from Guatemala, Honduras, and El Salvador for decades, it was in 2014 that the number of families and unaccompanied minors migrating to the US surged. Migrants from these countries are often fleeing local instability and violence, poverty, and drought. Climate change and extreme weather are deepening poverty and food and water insecurity in these countries, which will likely fuel increasing migration to the US border in coming months and years. And yet, environmental factors have been largely neglected in media and policy discussions, with experts tending to focus on the region's widespread criminal violence as the primary driver of migration and of the more recent surge. Within the last ten years, however, the region has experienced two multi-year droughts, including its most severe multi-year drought on record, and an uptick in the number and severity of hurricanes. This has driven both water and food insecurity. These climate and environmental factors are no doubt an important influence in the system of conflict–violence–migration in the region, and understanding how they intersect with other drivers is key to creating and implementing effective policy responses – both for climate-focused responses, as well as broader efforts to increase stability in the region.

To adequately understand and address the region's security challenges, it is critical to understand how the climate–migration–conflict nexus is being shaped by – and shaping – the existing vulnerabilities. Doing so through the framework of the four sets of vulnerabilities outlined above allows us to highlight both gaps and bridges and reveal how experiences of marginalisation and vulnerability – as well as opportunities for progress – are connected and mutually reinforcing.

### Physical and natural vulnerabilities

One only needs to look at the agricultural sector in Central America to understand how climate change will exist with the region's physical and natural vulnerabilities. Although agriculture has steadily declined as a share of GDP and source of employment in the Northern Triangle over the last 20 years, it remains important for the livelihood and food security of communities in the region (World Bank 2022a). It also contributes to more than 10% of GDP in each of the Northern Triangle (Savoy and Sady-Kennedy 2021). In Guatemala and El Salvador, agricultural activities account for around 13% and 12% of GDP, respectively. In Honduras, agriculture together with forestry and fishing contribute to 11% of GDP (World Bank 2022b). Coffee, a cash crop that has been cited as especially climate-sensitive, is the top agricultural export in all three countries and thus a notable contributor to GDP (Salazar et al. 2019). In Guatemala, coffee alone accounts for 2% of GDP.

Farming is also central to the day-to-day lives of many Central Americans, providing a key source of income and food, especially in rural areas (Seay-Fleming nd). Agriculture employs 31%, 29%, and 16% of the population in Guatemala, Honduras, and El Salvador, respectively (World Bank 2022c). And the cultivation of food commodities such as maize, beans, sorghum, and cassava - particularly important food sources for Central American populations - make up a large portion of agricultural land use (Salazar et al. 2019). Notably, the temperature-sensitive crops maize and beans in particular form part of a staple diet in the region (Pons 2021).

Across Central America, smallholder farmers play a significant role in agricultural production. This is especially true in Guatemala, Honduras, and El Salvador, which are home to more than 75% of the larger region's smallholder farmers (SG-SICA 2012). In Honduras, 70% of farming families are subsistence farmers. Many of these farmers are already under strain, with farming households making up over 66% of people below the poverty line in Guatemala, for example (de Janvry and Sadoulet 2010). A combination of factors, including limited access to land and economic opportunities through farming, has meant that over the last two decades, smallholder farmers in northern Central America have increasingly supplemented their farming income with off-farm income. Recurrent extreme weather events affecting El Salvador, Guatemala, and Honduras have also reduced adaptive capacity while further undermining societal and institutional resilience. Anecdotal evidence from 2018 suggests that major losses of agricultural crops such as maize and beans had a strong impact on migration flows from Central America to the United States, and the International Organisation for Migration's 2020 World Migration Report indicates that climate change in Central America appears to be having a profound impact on human mobility. Medium and long-term projections suggest immense additional pressure on agricultural crops that are key to the livelihoods of millions of low-income people. The region's vulnerability to climate change and its dependence on subsistence agriculture, combined with ongoing land pressures, severely affects the current and future context of both livelihoods and food security.

Latin America and the Caribbean is the second most disaster-prone region in the world. The COVID-19 pandemic and the 2020 Atlantic hurricane season – the most active Atlantic hurricane season and seventh costliest on record – have contributed to a complex social and economic scenario for Central America, specifically Guatemala, Honduras, and El Salvador. 2020 was one of the three warmest years in Central America and the Caribbean, and the second warmest year for South America. It saw 30 named storms, including hurricanes Eta and Iota (category 4 intensity), which made landfall in the same region in a quick succession, affecting 8 m people and incurring the loss or damage of 1 m hectares of crops. Simultaneously, drought was widespread across Latin America and the Caribbean. Importantly, investments in disaster risk reduction and humanitarian response to these acute crises could also present potential pathways to working with both formal and informal local institutions to make strides in strengthening governance and reducing corruption.

### Transboundary and regional dynamics

In a region plagued by high levels of poverty, inequality, insecurity, and fragile democratic institutions, it is the pervasiveness of chronic violence (both structural and criminal) that often has an enormous effect on individual decisions to move. While there are important differences among the countries and no single reason for the high rates of violence, there are also commonly shared and often intertwined conditions such as high homicide rates, pernicious gang violence, drug trafficking and economic despair which are coupled with rampant impunity and often impact every facet of life for many people in the region. At the same time, all three governments in the region struggle to provide functional legal, judicial, and correctional institutions. The populations of Honduras, Guatemala, and El Salvador are also notably young – the proportion of the population between the ages of 15 and 24 years old in all three countries hovers right around 20%, compared to 13% in the United States (UN 2023). The youth population has grown 51% since 2000, compared to just 16% in the United States in the same period.

Demographic trends influence economic sustainability in the region, which has direct implications for decisions to migrate. Notably, there is a dramatic difference between the limited number of formal economy jobs available compared to the burgeoning number of youths entering the workforce each year. As a result, many people must seek work in the informal economy or rely on irregular migration for employment reasons. Relatedly, urbanisation and population growth are further straining weak government infrastructure and service delivery, impacting physical security and wellbeing and taking an extra toll on more vulnerable populations, such as women and girls, who regularly lack access to public healthcare (Population Institute 2022). Notably, the increase in recent border crossings is almost entirely accounted for by families travelling together and unaccompanied minors, demonstrating a shift toward the migration of more at-risk and vulnerable groups.

In both media and policy discussions, migration tends to be securitised. Human mobility, however, is fundamentally rooted in political, social, and economic realities, and individual decisions to move are highly localised. In a region like Central America where outmigration is a predominant historical construct, influencing the entire social and economic fabric of society, mobility cannot be treated as a security risk without also acknowledging that the movement of people is now – and has always been – a critical component of regional adaptive and resilience capacity. Relatedly, securitisation of borders further amplifies historical inequities around land distribution and access coupled with less tangible constructs of embedded inequality, such as social identity and a sense of territorial belonging. Understanding local land tenure and property rights as a core concern in the region brings addressing the issues of livelihoods and food security as drivers of migration more constructively forward.

### Political and social stability

Governance challenges are an all-encompassing reality for Guatemala, El Salvador, and Honduras. All three countries have a history of autocratic rule and a chronic lack of investment in public institutions that have left the governments generally weak and open to corruption, as well as ill-suited to address the region's challenging socioeconomic and security conditions. These governance challenges have been further exacerbated by the increasing number of natural disasters. There is a pervasive distrust of formal government institutions, and an omnipresent sense of fear and hopelessness that contributes to citizens' real and perceived insecurity. Many Central Americans are unwilling to interact with the state, which affects the ability of the state to respond to citizens' needs, and also limits the ability of people to harness resources that could otherwise help them better address their own problems. At least in the near term, governments in the region are unlikely to become effective partners in addressing the drivers of insecurity and unsustainability fuelling migration trends. The international community must evolve its ability to better understand and support local interests more directly, while not subverting legitimate governmental authority in the process.

For example, targeted interventions in the region have found that when one community member has a positive experience with law enforcement, it can have a deep and enduring impact on their intentions to migrate, as well as their – and their community's – trust in government. Community policing, which increases positive engagement between police and citizens, can improve the safety perceptions of community members and, 'result in police being more responsive and accountable to local communities, enhancing trust between citizens and law enforcement entities, increasing the probability that citizens will report crime, and reducing police corruption and abuse' (USAID 2021).

### Scales of decision making

The international community often struggles to engage effectively with different levels of decision making, from local to regional. Unfortunately, the people and communities most directly affected by risks and stresses are frequently not consulted to determine their priority investments and to understand local sources of resilience. Donors are often constrained by requirements to work through national governments. Practitioners working to address climate–migration–security risks in Central America must therefore actively seek pathways to work more closely with the people whose needs and interests may not be sufficiently represented due to mass-scale mistrust in formal institutions, and in the context of endemic government corruption alongside weak institutional capacity.

From migration to climate vulnerability, it is important to move past looking at people as potential victims and instead as territorial actors who can contribute to, and enhance, system resilience. Indigenous knowledge and technologies must be leveraged and included in climate-risk analytics. In many instances in the region, communities are monitoring climate phenomena and the impacts of climate change but that information is not well represented or considered at higher scale. This type of traditional knowledge often includes attention to second- and third-order consequences, which more mainstream science can miss. Creating information that is as relevant and locally specific as possible enables all people, especially rural populations, to better use their traditional ways of communicating and coordinating for effective response.

It should be pointed out that where digital access and literacy are low, communication cannot rely solely on WhatsApp or social media more broadly, but must also help affected populations become part of the solution by sharing their wisdom in other ways as they move toward new technologies. Education in rural areas that brings in external information while elevating local knowledge and being attentive to local practices related to weather phenomena, natural disasters, and climate change is valuable for informing and engaging rural inhabitants on how to protect themselves, their livelihoods, and their crops. Given the governance challenges in the region, which often result in distrust and inaction in affected communities, solution finding for climate action will require coordinating more directly with *in situ* actors so that inputs to, and results from, studies and analytics produced at universities, research centres, or similar institutions can be informed by, and communicated to, local actors and communities in rural areas rather than relying upon government institutions as the interlocuters.

### Entry points for more effective action

Through enhanced understanding of how climate change will interact with the core dynamics outlined above, policymakers and practitioners working in support of the region will be better able to identify priorities and meaningful entry points for action. A critical component of all responses will include attention to fostering resilience – of people and systems – and seizing the opportunities for strengthening governance and improving human wellbeing in affected communities.

The terms 'threat multiplier' and 'risk multiplier' have increasingly become standardised concepts within the climate and security lexicon and, while accurate descriptors, they do obscure some of the complexity of climate impacts. Climate is not simply another layer on top of other factors that might result in instability or poor human development outcomes. Climate change *is* context and it influences as well as responds to all the factors around it.

With that in mind, there must be a multidimensional approach to risk management that incorporates understanding of both vulnerability and resilience and utilises reliable information on social, economic, environmental, institutional, and governance in early warning systems. There is a danger in modelling and predictive analytics of relying on climate change and its available data as a proxy for other vulnerabilities and variables that impact instability and security. At the same time, input from the dynamics of migration, food insecurity, water security, and other factors affected by climate – but beyond traditionally captured climate variables – needs attention when assessing climate risks.

### Data and early warning

Latin America and the Caribbean region are ill-equipped to address these challenges as multi-hazard risk-information systems and climate services in the region are underdeveloped. Support from governments and the science and technology community is critical to strengthening early warning alongside financial support for implementation. Simple provision of more equipment (i.e. radars) is not sufficient; the region needs smart support that considers the multidimensionality of how weather events impact the region. For example, the new Multi-Hazard Early Warning Systems (MHEWSs) checklist from the World Meteorological Organisation (WMO) helpfully addresses the operational aspect of improving predictive systems, including building better prediction models (WMO 2022). However, MHEWS does not account for vulnerable populations.

At the same time, it is clear that existing climate data and information are not fully utilised. Much of the science that is shared is just data, when ideally it should be analysed and communicated outward as information. It is also important to consider what type of climate information will be most useful to improve meaningful predictive capabilities as well as to inform prevention and response planning by various actors. For example, according to the US Department of Defense climate-risk analysis, one of the main risks of climate change in the region will be the need for more humanitarian assistance and disaster response support. This points to reconsidering what type of additional data and information collection might inform climate-risk analysis, and who should be involved in that process as well as in the responses to those analytics. Recognising the existing governmental challenges, information-gathering and sharing should be more often decentralised to the most effective local institutions and communities. This includes striving to incorporate as many different types and forms of information-gathering and communication as possible. For example, ground data can be improved by not just expanding a network of weather stations but by potentially placing/creating observation networks that are in the care of local communities. Dense observation networks can contribute to data collection, which can then inform more substantial data series for input into more accurate predictive models for the country or region.

Government budgets and poverty levels in Central American countries mean that even with access to robust information, resources for action may be constrained. To help address this, a wider-aperture on 'climate science' should be applied and non-traditional actors should be engaged in the process of data and information generation, communication and information sharing, and policy and decision making. Removing barriers to participation can range from addressing land tenure inequality to reducing limitations on the publication of information. To ensure that existing climate information and research are visible and available to those who can use it, a gap must be bridged between research and decision making by ensuring that more types of potentially useful climate data – including from communities and harnessed from local knowledge – are better utilised and translated into analytical and decision-making processes. This can be achieved in the region by, among other steps, actively promoting more scientific cooperation with a wide range of affected populations and experts to better capture existing and emerging observations, research, and knowledge.

Climate science is strongest on averages and variability because that's where data and models are best suited. But in terms of motivating strategic near-term action, these approaches are especially limited in their ability to show how climate-related factors may become more extreme in the short-term and trigger cascading effects, both of which are harder to predict based on averages. Climate information and climate services must become more accessible, user-friendly, and participatory to help inform pragmatic decisions and behaviour change at the ground level. In terms of improving the impact of climate-risk analysis, deep dives into locally grounded climate science as well as climate system linkages in particular geographies can help draw attention to more specific risks, points of vulnerability, and opportunities to shore up resilience that may emerge from complex connections across multiple scales of impact and systems from ecological to political to social.

### *Harmonising agendas for better regional coordination*

There remains a need to better harmonise climate-related agendas in the region, including the Sustainable Development Goals, the Sendai Framework for Disaster Risk Reduction, and the UN Framework Convention on

Climate Change. This would foster a more multidimensional view of the problems and illuminate entry points for action across different sectoral and policy agendas. At the country level as well as across borders, for example, supporting coordination and collaboration between climate, weather, and hydrological services could enable better service delivery as well as multi-hazard response because the approach would bring a wider range of institutional and local stakeholders together more consistently. This type of agenda coordination would also better centre the affected populations themselves, allowing their priorities – and risks as they understand them – to be better seen and addressed in more coherent and less siloed ways. The UN could play an important role in enabling this coordination in the region, especially if donors support the approach.

The value of collaboration is reflected in climate, migration, and peace efforts across the region. In the context of migration, the Comprehensive Regional Protection and Solutions Framework (MIRPS) is a central instrument to facilitate regional coordination around forced displacement. Born from the 2018 Global Compact on Refugees (UNHCR 2023a), MIRPS encompasses seven countries: Beliz, Costa Rica, El Salvador, Guatemala, Honduras, Mexico, and Panama. It involves improving mechanisms for reception and admission of displaced people, providing services directly to people on the ground, building resilience among host communities, and working towards long-term solutions in terms of livelihood generation and public policy at various levels of governance. MIRPS supports responsibility sharing among the seven countries to help ensure security for migrants, including asylum seekers, refugees, people in transit, and returnees.

### Fostering urban resilience

Rural-to-urban movements are a defining feature of migration in El Salvador, Guatemala, and Honduras. Rural migrants in the region often move to urban or peri-urban informal settlements where they have social networks. Yet, these settlements, which are the only viable destination for many migrants, tend to be characterised by high insecurity and low resilience. They are rapidly expanding across the region, and government institutions lack the capacity to keep up through sufficient social services and infrastructure. For example, according to GOAL, when hurricane Mitch hit, in 1998, there were 700,000 people living in Tegucigalpa, Honduras; today there are nearly 1.3 million (with approximately 600,000 people in informal settlements).

Given the limited essential services, such as early warning systems and housing, and insufficient employment options for low-skilled workers, risk is highly concentrated in these informal settlements. They are severely underserved by governments and are sometimes controlled by illicit networks. In other words, urban receiving communities are often underprepared to absorb migrants equitably, leading to heightened insecurity for everyone. But cities and city leaders are also close to the solutions. City leaders worldwide are demonstrating how cities can play a central role in facilitating peaceful mobility and realising the immense value that migration adds to receiving communities (Locke et al. 2022). These cities are supporting the integration of migrants by filling infrastructure gaps, investing in public health and employment opportunities, and shifting harmful narratives about mobility. Building resilience in urban and peri-urban settlements, including investments to make formal socioeconomic systems more robust and inclusive, must be prioritised as a crucial aspect of ensuring just and safe migration in the region.

### Addressing the digitisation of illicit networks

Central America has long been plagued by illicit networks that contribute to the region's insecurity and fuel unplanned and forced migration. Tellingly, in recent years, the global disruptions from the COVID-19 pandemic and the pervasiveness of digital connectivity have allowed existing illicit networks in the region to become more complex and potent, expanding their activities to cyberspace. The ubiquity of cell phones – even individuals who lack basic resources often have access to a cellphone – has made people more connected than ever. During the pandemic, cartels and gangs capitalised on cellphone and social media use to recruit young people, commercialise human smuggling services, spread fake news, and transfer monies. With the rise of cryptocurrency and digital wallets, moreover, illicit actors have grown increasingly nimble in how they develop and sell products. This digitalisation of illicit networks in the Americas has serious security implications; in some cases, it has even played a role in the destabilisation of democratic institutions (Realuyo 2022). Responding to the challenges posed by criminal groups will require addressing these new forms of communication and operation, including through regional and international cooperation.

### Prioritising vulnerable populations

Risks – whether they be climate-related, socioeconomic, or political – are distributed unequally across populations in the region. Women, youth, indigenous people, and other marginalised groups regularly bear the brunt of regional challenges like climate impacts and violence. Women and girls, for instance, tend to be disproportionately impacted by stressors because they lack sufficient access to public services, formal employment sector, political participation, land tenure, and other factors that enable resilience (Abdenur 2020). Moreover, the already-high levels of gender-based violence in Central America increased during the COVID-19 pandemic, further straining women's ability to cope with the socioeconomic and health implications of climate change. Youth also experience violence in unique ways through school, gangs, and street crime and are deeply impacted by a lack of employment opportunities, limited social mobility, and other broader conditions (UN 2022). In recent years, families and unaccompanied minors have made up an increasing proportion of Central American migrants, due to a combination of push and pull factors, including droughts, food insecurity, hopes of family reunification, and migration policies in the United States (Chishti and Hipsman 2016).

Despite their intentions, development interventions themselves can have lopsided impacts and deepen injustices if the unique experiences and needs of vulnerable populations are not centred. As awareness of this fact grows, there has been a rise in adaptation and resilience efforts that recognise and focus on such communities. At the same time, marginalised groups can be key agents for positive change as holders of knowledge about local priorities and contexts. Indigenous and local ecological knowledge and management strategies, for example, have proven beneficial for migration, mitigation, adaptation, development, biodiversity conservation, and resilience efforts across the globe (Veit 2019).

Youth engagement is also a promising channel for progress in the region. Spanos (2021) reports that 15–29-year-olds make up 28% of El Salvador's population and 30% of Guatemala and Honduras, while those under 15 years account for roughly 43% of people in Central America more widely. Young people experience the highest levels of unemployment in each country (Interpeace 2022), often driving their participation in the illicit or informal economy. As a result of this visibility, youth in Central America are often stigmatised in the context of security risks; yet they have been a central source of resilience and positive change, and their ongoing efforts should be supported and leveraged. In Honduras, youth played a key role in electing the country's first female president (Ernst, 2022) and have launched campaigns to promote transparency and accountability in elections (ASJ 2017). In Guatemala, youth activists led anti-corruption demonstrations in 2015 that helped oust the president and vice president (Flores 2019).

Importantly, prioritising the experiences and solutions of marginalised groups requires intersectional lenses – looking at how climate, migration, violence, and different forms of marginalisation interact, as well as the intersectionality of individuals' identities. Emphasising intersectionality allows us to identify the unique sets of influences that impact specific communities and support responses that account for this complexity. By centring indigenous voices, for instance, we can examine and adjust 'whose priorities' and 'whose solutions' international, regional, and national interventions are advancing. Intersectionality affords both a broader and more nuanced view, which can help us holistically and systemically think about the range of interests, problems, and opportunities we wish to affect.

### Developing responses that transform systems

In the region, governance issues, economic strain, and climate challenges intersect to drive vulnerability and influence peoples' decision to stay or to move. Since risks and opportunities are often dynamic and interdependent, it can be difficult to identify the inflection points that are best suited to outside support and to bring resources to bear that are not just sectoral (e.g. counternarcotics or disaster response).

Because these regional challenges are not isolated in technical silos, actors like the US government, including constituent agencies like the National Security Council, USAID, the State Department, and the Department of Defense, are increasingly using their policy and programming tools to develop more systems-focused solutions. For example, in this region, the United States is implementing a strategy to address the root causes of migration through a comprehensive set of initiatives focused on economic development, private investment, resilience and renewable energy, and governance. This work is part of the United States' broader effort to look at outcomes like forced migration, conflict, and natural disasters through a more inter-sectoral lens that understands causal and derivative connections.

The National Intelligence Estimate on Climate (NIA 2022) and Climate Migration Report (White House 2022) are two recent publications from the US government that have highlighted connections between climate, migration, and security. In this line of thinking, the Global Fragility Act (GFA), and its accompanying strategy, could also offer new tools within the US government to foster inter-agency coordination and encourage more comprehensive investments to address underlying causes of conflict and fragility, including in the Central American case issues like migration and climate change.

### Using scales of decision making

Given the interrelated and intractable nature of northern Central America's challenges and the challenges of working with national government institutions in particular, actors in the region can avoid decision paralysis by focusing more on scales of decision making. This way of identifying entry points allows for disaggregation along both timescales (timeframes of risks and responses) and levels of action (from individuals to communities to regional entities). In terms of timescales, the focus could be on longer-term solution finding, for example, by (1) strengthening resilience against slow-onset impacts of climate change and (2) investing in youth to prepare the next generation of leaders. On the other hand, in terms of levels of action, we can think about entry points at the individual scale, for example, migrant perceptions and motivations. Technologies like cell phones and social media have facilitated interpersonal connections, showcasing the lives of relatives who live elsewhere and allowing potential migrants to envision themselves living in those places, too. Against the backdrop of the COVID-19 pandemic and, increasingly, limited local economic opportunity, this connectivity has contributed to a migration pull into the United States. This shift means that the perceptions we have held about when and why a migrant makes the decision to move from the region may be inaccurate.

Since many observers were/are still unable to assess on-the-ground realities due to COVID restrictions on travel, field research on current dynamics is lagging. Integrating updated knowledge about individual perspectives, perceptions, and motivations into decision making can help address these individual-level factors and allow decision makers to better target interventions that enable and support safe migration, avoid forced displacement and poorly managed movement, and ensure the security of 'trapped populations', i.e. people who lack the resources or capability to move.

In the realm of resilience-building, thinking about scales of decision making also presents opportunities to localise support at the household and community levels. Despite being relatively underreported, small-scale events are most responsible for eroding household livelihoods. Working at a local scale can help development actors take into account these and other localised disruptions and support localised responses – distributed, localised energy resources, for instance, tend to be associated with faster community disaster recovery. But localised thinking need not be unidimensional – it can support the broader integration of services like water, health, and education systems. Moreover, localised work can help decision makers and practitioners elevate and leverage local sources of resilience. Outside actors may not always readily identify a community's adaptive behaviours and existing sources of resilience, especially when local communities do not frame their adaptive, resilience-building strategies as such. Yet, learning what these existing local approaches to resilience are and how they can be strengthened is a critical piece of addressing vulnerability in the region.

In short, localisation can narrow the scope of decision making and encourage effective policy interventions that are tailored to on-the-ground realities, tempering decision paralysis while still allowing for a comprehensive response at the scale of households and communities. And while these localised approaches focus on place-based work, they can and do help move the needle on complex regional challenges.

## Conclusion

This chapter has discussed the four core dynamics identified in the framework – physical and natural systems, transboundary and regional, political and social instability, and scales of decision making – to improve predictive capabilities for security risks posed by climate change. In providing an overview of how this framework applies to northern Central America, the chapter has also detailed how the climate–migration–conflict nexus is at once shaping and being shaped by existing vulnerabilities and, in turn, proposed some entry points for effective action.

## Acknowledgments

This chapter was informed by a series of workshops and expert discussions as well as a Wilson Center publication, "Addressing Climate Security Risks in Central America," authored by Cynthia Brady and Lauren Risi.

The project team that led the development of the climate risk framework includes Roger Pulwarty (NOAA), Sherri Goodman (Wilson Center), and Amanda King (Wilson Center).

The authors would like to express their appreciation to the many regional workshop participants who generously shared their insights and expertise and contributed both to the development of the case studies and the further refinement of the framework.

## References

Abdenur, A. E. 2020. Gender climate and security in Latin America and the Caribbean: from diagnosis to solutions. *Climate Diplomacy*. https://climate-diplomacy.org/magazine/cooperation/gender-climate-and-security-latin-america-and-caribbean-diagnostics-solutions (accessed 1 May 2023).

ASJ 2017. Young people lead advocacy for transparent elections in Honduras. *ASJ*. https://www.asj-us.org/stories/systems/young-people-lead-advocacy-for-transparent-elections-in-honduras (accessed 1 May 2023).

Chishti, M. and F. Hipsman 2016. Increased Central American migration to the United States may prove an enduring phenomenon. *Migration Policy Institute*. https://www.migrationpolicy.org/article/increased-central-american-migration-united-states-may-prove-enduring-phenomenon (accessed 1 May 2023).

de Janvry, A. and E. Sadoulet 2010. The global food crisis and Guatemala: what crisis and for whom? *World Development* 38(9): 1328–1339.

Ernst, J. 2022. Xiomara Castro poised to become first female president of Honduras. *The Guardian*. https://www.theguardian.com/world/2021/nov/29/xiomara-castro-declares-victory-in-honduras-presidential-election (accessed 1 May 2023).

Flores, W. 2019. Youth-led anti-corruption movement in post-conflict Guatemala: 'weaving the future'? *IDS Bulletin* 50(3). https://bulletin.ids.ac.uk/index.php/idsbo/article/view/3048/Online%20article

Interpeace 2022. Youth peace and security in the Northern Triangle of Central America. *Interpeace*. https://www.interpeace.org/wp-content/uploads/2018/04/2018-YPS-Central-America-v3-1.pdf (accessed 1 May 2023).

Locke, R., T. Albrecht, and J. Canney 2022. Harnessing the power of 'other': cities where human mobility is not a threat. *New Security Beat*. https://www.newsecuritybeat.org/2022/04/harnessing-power-other-cities-human-mobility-threat/ (accessed 1 May 2023).

Luizza, M. 2019. Urban elites' livestock exacerbate herder-farmer tensions in Africa's Sudano-Sahel. *New Security Beat*. https://www.newsecuritybeat.org/2019/06/urban-elites-livestock-exacerbate-herder-farmer-tensions-africas-sudano-sahel/ (accessed 1 May 2023).

Marinaccio, J. 2022. How Pacific climate diplomacy is changing. *Policy Forum*. https://www.policyforum.net/how-pacific-climate-diplomacy-is-changing/ (accessed 1 May 2023).

Mobjork, M. and S. van Baalan 2018. The nuts and bolts of a climate-conflict link in East Africa. *New Security Beat*. https://www.newsecuritybeat.org/2018/03/nuts-bolts-climate-conflict-link-east-africa/ (accessed 1 May 2023).

NIA 2022. National Intelligence Estimate: Climate change and international responses increasing challenges to US national security through 2040. *Office of the Director of National Intelligence*. https://www.dni.gov/files/ODNI/documents/assessments/NIE_Climate_Change_and_National_Security.pdf (accessed 1 May 2022).

Pons, D. 2021. Climate extremes, food insecurity, and migration in Central America: a complicated nexus. *UN Office for the Coordination of Humanitarian Affairs*. https://reliefweb.int/report/guatemala/climate-extremes-food-insecurity-and-migration-central-america-complicated-nexus (accessed 1 May 2023).

Population Institute 2022. Invisible Threads: Addressing the root causes of migration from Guatemala by investing in women and girls. https://www.populationinstitute.org/wp-content/uploads/2022/10/PI-3059-Guatemala-Migration-Report-WEB-Updated.pdf

Realuyo, C. 2022. How cryptocurrencies are empowering transnational criminal organizations and countries in Latin America. *Dialogo Americas*. https://dialogo-americas.com/articles/how-cryptocurrencies-are-empowering-transnational-criminal-organizations-and-countries-in-latin-america/#.Y8FxRS2l3sl (accessed 1 May 2023).

Reed, J. and F. Bokhari 2022. 'It's the fault of climate change': Pakistan seeks 'justice' after floods. *Financial Times*. https://www.ft.com/content/e69ece7d-11fb-4a8f-91ea-35b98d4b54db (accessed 1 May 2022).

Ro, C. 2019. Climate change is benefiting terrorists in Somalia. *Forbes*. https://www.forbes.com/sites/christinero/2019/10/27/climate-change-is-benefiting-terrorists-in-somalia/?sh=57a2a31c1016 (accessed 1 May 2023)

Salazar, M., T. S. Thomas, S. Dunston, and V. Nazareth 2019. Climate change impacts in El Salvador's economy: the agricultural sector. *International Food Policy Research Institute.* https://www.ifpri.org/publication/climate-change-impacts-el-salvadors-economy-agriculture-sector (accessed 1 May 2023).

Savoy, C. and T. A. Sady-Kennedy 2021. Economic opportunity in the Northern Triangle. *Center for Strategic & International Studies.* https://www.csis.org/analysis/economic-opportunity-northern-triangle (accessed 1 May 2023).

Spanos. K. K. 2021. Central America: what do the demographics tell us about the migrant surge? *George W. Bush Presidential Center.* https://www.bushcenter.org/publications/central-america-what-do-the-demographics-tell-us-about-the-migrant-surge (accessed 1 May 2023).

Seay-Fleming nd. Food insecurity in the Northern Triangle: leveraging agricultural policies and programs for the benefit of smallholders. *Wilson Center.* https://www.wilsoncenter.org/publication/food-insecurity-northern-triangle-leveraging-agricul-tural-policies-and-programs-benefit (accessed 1 May 2023).

SG-SICA 2012. Centroamérica en cifras, Datos de Seguridad Alimentaria Nutricional y Agricultura Familiar, a Diciembre 2011. *SICA.* https://www.sica.int/documentos/centroamerica-en-cifras-datos-de-seguridad-alimentaria-nutricional-y-agricultura-familiar-a-diciembre-2011_1_66467.html (accessed 1 May 2023).

UN 2022. Regional overview: Latin America and the Caribbean. *United Nations.* https://www.un.org/esa/socdev/documents/youth/fact-sheets/youth-regional-eclac.pdf (accessed 1 May 2023).

UN 2023. World population prospects. *United Nations.* https://population.un.org/wpp/Download/Standard/Population/ (accessed 1 May 2023).

UNHCR 2023a. The global compact on refugees. *UNHCR.* https://www.unhcr.org/the-global-compact-on-refugees.html (accessed 1 May 2023).

US AID 2021. Crime and violence prevention field guide. *US AID.* https://www.usaid.gov/sites/default/files/2022-05/PA00XGHG.pdf (accessed 1 May 2023).

Veit, P. 2019. Land matters: how securing community land rights can slow climate change and accelerate the sustainable development goals. *World Resources Institute.* https://www.wri.org/insights/land-matters-how-securing-community-land-rights-can-slow-climate-change-and-accelerate (accessed 1 May 2023).

Watkins, H. 2022. Horn of Africa facing worst hunger crisis in decades. American Red Cross. https://www.redcross.org/about-us/news-and-events/news/2022/horn-of-africa-facing-worst-hunger-crisis-in-decades.html (accessed 1 May 2023)

White House 2022. Report on the impact of climate change. *The White House.* https://www.whitehouse.gov/wp-content/uploads/2021/10/Report-on-the-Impact-of-Climate-Change-on-Migration.pdf (accessed 1 May 2022).

WMO 2022. Multi-hazard early warning systems: a checklist. *World Meteorological Organization.* https://public.wmo.int/en/our-mandate/focus-areas/natural-hazards-and-disaster-risk-reduction/mhews-checklist (accessed 1 May 2023)

World Bank 2022a. Agriculture, forestry, and fishing, value added. *World Bank.* https://data.worldbank.org/indicator/NV.AGR.TOTL.ZS (accessed 1 May 2023).

World Bank 2022b. Agriculture, forestry, and fishing, value added – Honduras. *World Bank.* https://data.worldbank.org/indicator/NV.AGR.TOTL.ZS?locations=HN (accessed 1 May 2023).

World Bank 2022c. Employment in agriculture – *Guatemala. World Bank.* https://data.worldbank.org/indicator/SL.AGR.EMPL.ZS?locations=GT (accessed 1 May 2023).

YCC Team 2022. Climate change made rain more extreme during Pakistan's deadly 2022 floods. Yale Climate Connections. https://yaleclimateconnections.org/2022/11/climate-change-made-rain-more-extreme-during-pakistans-deadly-2022-floods/ (accessed 1 May 2023).

# 7

# CLIMATE CHANGE, INSECURITY, AND ECONOMIC TRANSFORMATION

*Matthew Paterson*

## Introduction

The focus of this volume is on the broad security dynamics surrounding climate change: both the implications for conventional accounts of national and international security, but also the human security questions to do with food systems, migration, and so on. These are of course crucial to continue reflecting upon, not least for the security apparatus of many states that are charged with protecting their citizens from threats to their security, but for whom climate change poses rather unconventional sorts of threats. But in order to fully understand these security questions, they need embedding in a broader social and political analysis of both the nature of the climate change challenge and the drivers of policy and governance responses to this challenge. This chapter focuses on these two sorts of questions to help contextualise how security actors, discourses, and practices fit within this broader picture.

The overarching argument presented here is twofold. The first argument is that the world has come increasingly to understand the climate challenge as a sort of crude 'choice' between large-scale social collapse on a global scale and a radical and rapid transformation away from fossil fuels as the basis for the global economy (see Paterson 2021). We can no longer usefully understand climate change as a set of discrete impacts that can be disentangled and to which specific responses can be formulated, but must rather see it as an existential crisis for the current form that our societies take. In short, there will be some sort of massive multidimensional transformation brought about either by the totality of climate impacts or by the shift away from fossil fuels as the basis for civilisation.

DOI: 10.4324/9781003377641-9

The second argument is then to understand that the drivers for these transformations are in the dynamics of the global economy (Newell and Paterson 2010, Paterson and P-Laberge 2018). The global economy has developed fundamentally on the back of fossil fuels, especially coal and then oil. Economic activity and the policy and political support that enables that activity remains largely premised on access to, and mobilisation of, these fossil resources. Those controlling fossil fuel extraction, distribution, and consumption remain highly powerful within most political systems and have largely been successful in resisting even relatively modest climate policy in most countries. Access to, and control over, fossil fuel resources, in this instance mostly oil and natural gas, has also been highly important to geo-political dynamics since the early 20$^{th}$ century and remains so in several world regions. But these same political-economic drivers also hold the key to un-derstanding the transformational possibilities – both whether the world manages to avoid collapse by decarbonising rapidly enough, and how either collapse or decarbonisation unfold.

## Transforming understandings of climate as a global threat

In the late 1980s, when climate change first made it onto the international agenda, it was, even then, framed as a security threat. The first major international public conference on the subject was in June 1988, in Toronto, and entitled 'The changing atmosphere: implications for global security'. The statement of that conference was pretty stark: 'Humanity is conducting an uncontrolled, globally pervasive experiment whose ultimate consequences could be second only to a global nuclear war'. It then continues: 'The best predictions available indicate potentially severe economic and social disloca-tions for present and future generations, which will worsen international ten-sions and increase the risk of conflicts among and within nations' (Boyle and Ardill 1989, p. 247). Since then, there have been recurrent attempts to securitise climate change. However, to the extent that the effects of these efforts might have been intended to shift climate change up the political agenda, even justify emergency measures by governments along the lines of what the tradition of 'securitisation' arguments within security studies imply, these efforts have to be regarded as a failure (e.g. Oels 2013; McDonald 2012; 2021).

Despite the stark impression conveyed at the Toronto Conference, it is a mistake to read back onto these early developments our current under-standing of the nature of the challenge. In fact, at that point, the general consensus about the implications of climate change for world politics was much more modest compared to today's assessments.

Along with other developments, notably the politicisation of the huge drought in the United States in the summer of 1988 and the ongoing emer-gence of an institutionalised network of climate scientists coordinated by the

World Meteorological Organisation, the Toronto Conference led to the establishment of the Intergovernmental Panel on Climate Change (IPCC) in late 1988 (Paterson 1996). The IPCC produced its first assessment report in the summer of 1990. However, the way the climate change challenge was discussed in that report is radically different from the assessment we typically consider now. On the climate impacts side, the anticipated scale and pace of change were modest by comparison with contemporary assessments. And the possibility of a 'runaway greenhouse effect', where positive feedbacks in the climate system (release of methane from permafrost or under the seabed, for example) would lead to climate change getting beyond the possibility of human control and tip the earth's climate into something uninhabitable for human societies, was largely discounted by all but the most fringe climate scientists. And on the mitigation side, the IPCC's first report suggested that 60% cuts in $CO_2$ emissions (and differing levels of the other main greenhouse gases) would be necessary in order to 'stabilize atmospheric concentrations' of greenhouse gasses (GHGs) at current levels (Houghton et al. 1990, p. xviii).

The implications of this were important but by no means a threat to civilisation. The climate impacts would generate significant amounts of localised human insecurity – from sea level rise, increased extreme weather events, or disruption to food systems. And reducing global GHG emissions by 60% would be technically possible with concerted action in a relatively small number of sectors: decarbonising electricity generation and consumption, as well as road transport, in particular, would be more or less enough to achieve this sort of target. And it would be possible to pursue with minimal consideration of global inequalities in emissions since most developing countries – including at that point China – emitted underneath the emissions levels, in per capita terms, needed to keep global emissions within sustainable limits. But first gradually over the course of the 1990s and 2000s, and then much more rapidly since the late 2000s, this assessment has been radically reshaped (see Paterson 2021).

First, in the mid-2000s, policymakers and scientists started talking about temperature thresholds that need to be avoided to avoid catastrophic climate impacts, and the possibility of tipping points being reached generated by positive feedbacks in the climate system started to be taken much more seriously. A target of limiting warming to 2°C above preindustrial levels, initially proposed by the EU in 1996, became increasingly discussed during the mid-2000s as a way to focus attention on the necessary action and was included in the Copenhagen Accord in 2009 (Randalls 2010). Over time, other actors proposed 1.5°C instead and this became embedded in the Paris Agreement, albeit as an aspirational rather than firm target. These targets also became allied to a notion of an overall carbon 'budget': climate change is a 'stock' problem, not a 'flow' problem, i.e. a question of the total amount of GHGs emitted over time, not the amount in any given year. This has implications for the question of urgency (see below).

Second, the necessary cuts became more stringent. While in the IPCC's first report (Houghton et al. 1990), 60% cuts were deemed sufficient, during the 2000s 80% became much more commonly referred to, and by the time of the IPCC Fifth Assessment Report (2014) and its 'Global Warming of 1.5°C' report (2018), with the Paris Agreement sandwiched in between, this had shifted to a goal of reaching 'net zero' emissions by 'early in the second half of the 21st century' (IPCC 2014). The precise date is not given but most of the figures in the report indicate this has to be reached by 2070 at the latest (see e.g. IPCC 2014: figure SPM.5(a)).

Third, these cuts are ever more urgent. Part of this is simply because so few countries have actually started to reduce their emissions, and, globally, emissions have gone up by 60% since 1990, not down by 60% as the IPCC suggested then that they needed to. As Stoddard et al. (2021) put it, we simply have not 'bent the global emissions curve'. The downward slope of those emissions has to be correspondingly sharper to keep within an overall carbon budget. The later we leave it to start cutting emissions, the more that budget has been already used up. This is, therefore, part of the reason for the shift to thinking about 'net' emissions – in practice we have not only to get emissions to as near zero as possible, but also to take carbon out of the atmosphere directly. Indeed, some of the IPCC models that calculate what has to happen to keep within a 1.5°C limit show emissions have to become net negative at least for a period in the late 21st century (see van Beek et al. 2020).

But, fourth, the question of urgency is not only driven by these concerns about emissions levels and their trajectories but also that the onset of particularly worrying climate impacts has been more rapid than anticipated. Loss of ice sheets (Greenland, Antarctica, glaciers in various mountain ranges) and Arctic sea ice, frequency of highly damaging storms (Hurricanes Katrina, Sandy, and Harvey, all previously thought of as one-in-a-hundred-year events, for example), are the most visible phenomena to be accelerating faster than at least the position generated in scientific consensus-seeking processes like the IPCC. Those processes are always more conservative than the latest evidence, so projections in IPCC reports have been regularly overtaken by events.

At the heart of this shift in the assessment of the climate challenge is the forecast of some sort of 'civilisational collapse', or at the least a radical decline in the basic human security for people around the globe, with the collapse of food systems, critical infrastructure (transport, energy), and acceleration of climate-induced migration. This will, of course, be experienced in highly unequal ways given the existing inequalities and power relations in the global economy. But it is unlikely that even the most privileged will escape from this sort of widespread collapse of social provisioning systems (Chakrabarty 2017; MacGregor and Paterson 2021).

## Net zero as transformational

The flipside of this realisation of the acute potential for such socially catastrophic climate impacts is the increasing realisation that avoiding this sort of catastrophe entails the biggest purposive social transformation human societies have ever undertaken.

The shift in target from 60% to 100% (or even more) cuts entails a radical expansion in the range of economic and social activities that need to be transformed and also entails such transformations across the entire globe, not just in rich, high GHG-emitting countries (even if they still have to take the lead). No sector or country can (as many have over the course of the last 30 years) plea that they are 'only 2% of world emissions' and expect that to be taken as a serious argument. Thus, we not only have to address those bits of the economy which are both large percentages of emissions and/or relatively easy in technical terms – electricity generation and use, and land transport, in particular – but have to focus on the 'hard to decarbonise' – aviation, meat and dairy, steel, and cement, most notably (Bulkeley et al. 2022).

The 'net' in net zero also entails that a much more thorough, systematic approach to carbon sequestration is needed. Some of this focuses on the more innovative (but highly uncertain, especially at scale) approaches such as direct air capture of $CO_2$ or BioEnergy with Carbon Capture and Storage (BECCS). However, the bulk, certainly of what societies know how to do, entails radical shifts in land management, most notably in afforestation and peatland management, as well as in coastal areas to store 'blue carbon', notably in mangroves, tidal marshes, and seagrasses (Macreadie et al. 2019). But the implications of this are radical: to take one United Kingdom example, the government's Committee on Climate Change (2020) estimates that 22% of the UK's current farmland would have to be converted to forest to meet net-zero targets by 2050. Doing that while maintaining, or even expanding (given the United Kingdom is a considerable food importer) food production is an exceptional challenge.

This example also illustrates that the challenge of decarbonisation is multidimensional – not only technical in the narrow sense but also social, economic, cultural, and deeply political (e.g. Bulkeley et al. 2016). It entails radical changes in widespread social practices, to which there are powerful cultural value systems attached, and which are both highly unequal in how people are able to enjoy those practices and are embedded in long-established power relations. To take the land example, afforesting 22% of the UK's farmland while enhancing food security threatens most immediately the livelihoods of the landowners used to deriving income not only from farm produce but from land subsidies to farmers. Some of these are large landowners but many are smaller in scale. This pattern of land use has been sustained by forms of government subsidy based simply on acreage, which is long-established and has proved very difficult to shift despite two decades of attempts to move to more conservation-oriented subsidy regimes, reflecting the considerable power of agribusiness and landowner lobby groups within policymaking circles. Forests are not counted in the UK as farmland, so landowners risk losing considerable subsidies if they afforest their land: there are subsidies for afforestation but only to help plant the trees – land then gets rezoned as non-agricultural so you lose the annual subsidy. But it goes way beyond the question of farmers, in that the pursuit of this afforestation, food security, and net zero at the same time entails radical reductions in meat and dairy consumption. This then entails a fundamental challenge to established norms about what food *is*. While there is a current growth in vegetarianism and veganism, which is higher in the United Kingdom than in most countries, this also follows a cyclical pattern – there was a similar growth in the 1980s and 1990s for example, but not continued over time – this remains at pretty small percentages of the population, and the cultural attachment to meat and dairy remains, in general, extremely strong.

Each area that needs decarbonising has its own particular version of this multidimensional challenge. We can readily envisage these for questions of aviation, shipping, car driving, home cooking, and heating, for example. In each, there are powerful actors with interests in maintaining the status quo; significant distributional issues in the transition for a wide range of actors raising the issue of a 'just transition'; daily practices (not only of consumption but also in work itself) that need to change but which are highly valued; and often hugely complex socio-technical problems to solve. And all of these are intertwined to make decarbonisation even more complicated.

## Implications for global security and security institutions

The implications of climate impacts for global security are well understood by security actors and scholars and dealt with very well elsewhere in this volume. But the transformations that will be wrought as the world decarbonises also have considerable implications for global security and are worth detailing. Some of them are indeed underway. Three are particularly important to consider.

### Geopolitics of resources

First, and perhaps most obvious, are the geopolitics of the resources entailed in a so-called 'clean energy' economy (Paltsev 2016; Sivaram and Saha 2018; Jaffe 2021; but see Overland 2019 for a rebuttal). Fossil fuels have been both the target of geopolitical competition that has contributed to warfare, and the intrinsic means by which that competition and warfare have been pursued. Coal is relatively widespread across the globe but it was Britain's early deployment of coal, in particular in iron and steel production (including notably arms manufacture) in and around Birmingham, and innovation in manufacturing production through steam engines, mostly around Manchester, that was the basis for British imperialism and British global dominance during the 19th century and into the early 20th century (Satia 2019; Malm 2016a, 2016b). Coal and iron-based shipping gave the British Navy a decisive advantage over rivals. British advances in iron and steel production helped deliver this advantage but also underpinned improvements in gun manufacture for similar advantages in land warfare. Advantages were also derived from the financing of the triangular slave trade and colonialism more broadly (Satia 2019).

Oil and gas are much more unevenly distributed across the globe. As they became increasingly deployed – oil from the late 19th century onwards, gas from after World War II – their locations became the object of military rivalry, occupation, location of extraterritorial bases, destabilisation of governments, and a number of major wars. The Middle East is, of course, the central world region in this story. The interest and power-political manoeuvres of outside powers – Russia, Britain, France, and the United States, most obviously, and increasingly China as well – from the first decade of the 20th century onwards, cannot plausibly be explained without reference to securing access to, control over, and profit from oil resources (e.g. Klare 2001, 2007; Mitchell 2013; Frankopan 2015). The particular patterns of state building and interstate competition within the region itself can also not really be understood without understanding how oil made particular state elites rich, reinforced regional rivalries, undermined the importance of previously powerful but oil-poor states in the region, and so on. How we understand any of these specific conflicts may vary, but the centrality of oil to the overall pattern is undeniable.

Historically, natural gas has been less central to these dynamics. But as its role has grown within the overall global energy economy, so too has geopolitical conflict over its supply. It is worth noting that this shift to gas is at least in part stimulated precisely by the modest climate change measures to date in the last two decades, notably through displacing coal in electricity generation. Russia has become central to the geopolitics of gas, especially in its relations with both former Soviet republics now independent states and with European Union members. European countries, most importantly Germany, have become progressively more reliant on imported natural gas from Russia, which the Russian government under Putin has keenly used to gain leverage. Ukraine has become the principal target of the most aggressive forms of this, with control over natural gas forming part of the explanation for the Russian invasions of Ukraine in 2014 and then again in 2022 (Stulberg 2015; Riley 2022).

What are the geopolitical implications of the world shifting away from fossil fuels – assuming it does manage to do this? One of course is that for those oil-rich states, the future for them is highly uncertain. Oil and gas have provided almost the entirety of their livelihood for most of the last century, especially in the Arabian Peninsula. It has catapulted many of them into the category of high-income countries (if still 'developing countries' in UN diplomatic contexts), and their per capita carbon emissions are correspondingly the highest in the world – higher than the most high-emitting industrialised countries (the United States, Canada, and Australia). Oil has sustained highly autocratic political regimes in many of them (Mitchell 2013), by enabling them to use the resource rents from oil extraction to maintain their rule through mixtures of repression and lavish spending. Much of the coercive side of this has also been sustained by extensive military and arms trade relations with major arms-exporting countries in both the Cold War and post-Cold War eras. A decarbonising world undermines the rationales for extensive external intervention in the region. But at the same time, it undermines the capacity of many states in the region to sustain their current patterns of both domestic political rule and interstate relations in the region, or to sustain the sort of weapons purchases which are important to external arms manufacturers and the states they come from. Some states in the Gulf are clearly attempting to anticipate these shifts and transition to a renewable-based economy (UAE is the most advanced of these) but it is far from the dominant strategy of these states. Most of them, led by Saudi Arabia, have consistently led the charge against action on climate change within the UNFCCC (Kassler and Paterson 1997; Depledge 2008).

The second implication is based on thinking about the resources that are central to a decarbonising economy. Central to this is the investment in renewable energy technologies, in particular solar pV and wind, but also in the battery storage technologies that are crucial to the electrification of transport and to small-scale solar pV generation. Both of these are referred to as 'clean energy', and in their immediate electricity generation capacities this is perhaps reasonable. But they both rely extensively on a wide range of quite specific minerals. These include lithium, copper, nickel, cadmium, and rarer elements like indium, tellurium, neodymium, and so on, to a list involving around 80 key elements (there are about this number of such elements in a mobile phone). Some of these, like silicon, are widely distributed. But many are not. Lithium is perhaps the poster child of these minerals, critical to advanced batteries, in particular (Sterba et al. 2019). Australia is currently the world's largest lithium producer (around 60% of global production) and has about 15% of world's reserves. But Chile is even more central to the future of lithium, while it currently produces about 18% of the world's lithium it has a little over 50% of world's reserves (US Geological Survey 2020), and the 'lithium triangle' straddling the borders of Chile, Bolivia, and Argentina is widely regarded as crucial to a renewable energy-dominated future (e.g. Heredia 2020).[1]

Lithium has already become the object of intense investment, with the involvement of transnational capital, in ways that can be seen as analogous to the rush for external oil investment in the Middle East in the early 20th century. External involvement of American, Chinese, and other companies in the development of lithium extraction in the lithium triangle is considerable (Voskoboynik and Andreucci 2021). There is of course a substantial history of US intervention in Latin America to ensure regimes in the region are favourable to US interests, with the intervention to topple Allende in Chile in 1973 a paradigmatic case. The increasing strategic significance of lithium and other critical minerals, especially as US-Chinese competition for access, control, and profits, from those resources, seems to make such political intervention increasingly plausible, even while there is a countervailing process of attempts to 'onshore' lithium production in the United States and other Western countries (Riofrancos 2022). Alongside these geopolitical dimensions, there are also important socio-ecological aspects to these resources. The critical minerals for 'clean' energy are in practice anything but clean in their production. They have already entailed significant ecological destruction – of farmland and water resources, in particular – within the places from which they are extracted. This destruction is often in the places where highly marginalised communities, including indigenous peoples in many contexts, already live and thus have their livelihoods destroyed (Jerez et al. 2021), leading some to call these extractive sites 'sacrifice zones' (Lerner 2012; Zografos 2022). This socio-ecological destruction also generates significant resistance and conflict (Revette 2017).

### Fossil militaries

The second implication relates to the historical dependence of militaries on fossil fuels and the operational challenges of shifting away from them. Shifts to fossil energy sources, from sail to coal, coal to oil in ships, horse to tank and armoured car, train to car, the emergence of aviation, most notably, have been central to the development of warfare and the pursuit of military advantage since the early 19th century.

The early shift to coal, and then from oil, was central to the British navy's global dominance. The historian, A. J. P. Taylor (1969) famously made train timetables the explanation for the outbreak of World War I, since they created a rigidity in the logistics of mobilisation that made war inevitable once that mobilisation had begun. The emergence of the plane and the tank was important, if not decisive, in shifting World War I away from attritional trench warfare, but became central to World War II through the German strategy of blitzkrieg. Even now, advances in aviation, e.g. Unmanned Aerial Vehicles (UAVs), have been important in shaping the patterns of warfare that have enabled continued Western interventions and the acceleration of 'asymmetric warfare' – rendered legitimate domestically, at least in part, because of minimal losses to Western forces. All of these strategic developments have been underpinned by fossil fuel resources and their development.

The question then arises as to how do militaries – and the states that support and deploy them – redevelop strategy for a post-fossil world? This question is also intertwined with operational questions for militaries as climate impacts unfold and intensify: in what circumstances are they the appropriate agents for responding to climate-related disasters? Militaries have already been widely drawn into responding to extreme weather disasters (e.g. floods, wildfires, hurricanes, and droughts) because of their operational capabilities and hardware. However, many within the military worry this is a misuse of their resources and an inappropriate transformation of their mission, while many pro-climate advocates worry about such activity as heralding an inevitable militarisation of responses to climate change.

There is a fairly fundamental operational question at play in relation to the future of military operations, in that the energy density of fossil fuels is so much higher than all other sources. (Apart from nuclear, which has obvious limits in routine use for transport, and has never been used except for submarines and handful of other naval vessels.) The question of whether it is possible to pursue military operations with other sources is moot. Militaries are exceptionally high carbon in their operations, to the extent that variations in military spending per capita have strong correlations with per capita GHG emissions across countries (Jorgenson et al. 2010). There have long been efforts within militaries to improve energy efficiency, for cost and logistical reasons. Some of this has then triggered important technical developments enabling other low-carbon innovations (see below). But more recently some military establishments have begun to think more radically about a post-fossil fuel operational capability (Closson 2013). For example, the UK military has started to map out the possibility of the 'electrification of the battlefield' (British Army 2022; Emmett 2022). Such planning is in its infancy, but it seems difficult to imagine how post-oil military strategy remains the same as that dependent on oil. Conversely, of course, such remaining oil dependence by militaries may operate as a significant brake on weaning the global economy off fossil fuels.

### Militarised fossil economies

The third implication relates to the historical role of militaries and their institutional supports within states in socio-technical innovation (Johnstone and McLeish 2020). The flipside of the importance of shifts in energy sources to military operations is that military innovation has triggered repeated broader socio-technical shifts. Both World War I and World War II were important to the decisive shifts in transport technologies from trains, buses, bicycles, and horses to automobiles, in different ways. States drove demand for cars as tanks, armoured cars, and troop carriers became central to warfare, and car manufacturers expanded capacity to meet this demand; capacity that was then redeployed to civilian automobile manufacture after each war. Especially in planning for, during, and in the immediate decades after World War II, states went on a massive road-building programme, particularly in rapid long-distance routes (autobahns, autostrada, autoroutes, motorways, and interstate highways) that always had rapid troop movement as a primary political rationale (Paterson 2007, p. 117). Developments like these had enormous quantitative significance for the expansion of an automobile and oil-dominated economy and society. But they were also drivers of important technical improvements in engine performance, energy efficiency, fuel quality, and so on.

Military deployment of fossil fuels had, if anything, even more dramatic effects on innovation in shipping and aviation. In shipping, the military-driven shift to first coal and then oil had dramatic effects on the ability to expand global trade as it enabled bigger ships, faster movement of goods and people, and reduced reliance on wind direction and speed. Over time this then enabled developments like containerisation which, again, radically expand the potential for global trade. The extraordinarily rapid expansion of aviation – from the first, tentative, flight in 1903 to military deployment in 1911, to their centrality to World War II, and rapid expansion of commercial air travel from the 1920s onwards – can only be understood because of their military utility and the investment that thus poured into aircraft design, manufacturing capacity, and engine technology in particular. Many aircraft manufacturers (Boeing, Saab, etc.) remain essentially commercial-military hybrids, with strong connections to state military establishments. And, of course, the most recent development – of UAVs – remains a military innovation with enormous commercial spin-offs.

There are other areas where military investment and strategy have generated huge social and economic spin-offs. The emergence of the internet is perhaps the most famous of these. This is the result of the search for resilience in military installations and planning capacity. The US military pursued this in the 1960s through the capacity to have multiple sites of decision making, as well as flexible ways of moving and locating nuclear missiles, leading to the developments of multiple sites of computer capacity, linked, in a decentralised fashion, so that if one node was taken out, the others would still be able to continue operating unhindered. ARPANET was the Department of Defense programme that developed these networked computer labs, at both explicit parts of the US military establishment (the Pentagon itself, for example) and, initially, a small number and range of university and government research laboratories (MIT, Berkeley, etc.) A series of technical innovations – packet switching, protocols such as TCP/IP, server naming systems such as

DNS – were the result of this search for military security, but which became the central technologies that enabled the Internet that then rapidly expanded during the 1990s to become the basic infrastructure of the entire global economy, as well as a very substantial sector within it (Hafner and Lyon 1998).

Intertwined with the development of the internet, militaries (especially the US military) have also had significant impacts on innovation in low-carbon technologies. Laptop size, battery development, and small-scale solar pV are useful examples here. These became central questions for the US military as they developed what became known as 'network-centric warfare' during the 1990s (Dillon 2002). Wanting to be able to put small numbers of troops behind enemy lines, yet stay in effective communications (enabled by satellites and the internet itself), the need for units to have computers and radios, that could function over substantial periods of time, was a key challenge. The US military thus put substantial investment from the 1980s onwards into reducing computer size and weight, improving energy efficiency (for laptops – desktops continue to use much more electricity than their portable cousins, as it wasn't crucial how efficiency they are), improving battery performance and life, and improving small-scale (i.e. on the back of a rucksack) solar pV performance to be able the recharge of laptop batteries (Macneil 2017; Samaras et al. 2019). As Macneil (2017) shows, the role of the military has been reinforced by political dynamics in the United States, where federal climate legislation has been impossible to pursue but securing money for military-led energy innovation has been relatively easy. All of these developments have then fed into civilian technologies that are important for low-carbon transitions, with novel improvements in pV performance and radical improvements in battery life, for example, central to electric vehicle developments and renewable energy expansion.

## Conclusions

This chapter has sought to provide an understanding of the challenge of climate change that centres processes in the global economy as drivers of the origins of climate change, the patterns of response to it so far, and the potential dynamics of either how the world successfully abandons fossil fuels and/ or it responds to unfolding crises generated by cascading climate impacts. This enables us to return to how institutions and discourses of national security might understand their role. While for most practitioners and analysts in the field, climate change appears mostly via climate impacts and the security threats that might need to be managed, contained, or otherwise responded to, what I have tried to show is that understanding the role of militaries in the development and maintenance of a high-carbon global economy, and its potential role – both through military operations and broad socio-technical innovation processes – in the pursuit of a zero-carbon global economy, is also crucial.

From the point of view of the challenge of climate change as sketched above, this of course makes things enormously complicated. On the one hand, militaries remain extraordinarily high-carbon emitters and present a huge challenge to decarbonise effectively. On the other hand, the pursuit of military strategic advantage has been crucially important to a range of innovations in energy use that have for the most part accelerated the emergence of a high-carbon world. The question then becomes either are they likely to become similarly crucial to the pursuit of a low- or zero-carbon world and to the extent the world seeks to replicate (and intensify) the pace of innovation in the uptake of zero-carbon energy. If the military is not the trigger for the technical side of these innovations, how else may they be so rapidly spread across societies globally?

## Note

1 Note that this is reserves, not resources. 'Reserves' refers to those deposits where it is technologically and economically feasible to extract the mineral. 'Resources' refers to all known deposits. As prices go up, and exploration continues (it is still at an early stage for lithium since its utility to the global economy has been minor until recently), the geography of lithium reserves may change considerably.

## References

Boyle, S. and J. Ardill 1989. Statement issued by the participants at the World Conference on 'The changing atmosphere: implications for global security', June 1988. (In) S. Boyle and J. Ardill (eds) *The Greenhouse Effect: A Practical Guide to the World's Changing Climate*. London: Hodder & Stoughton, pp. 247–257.

British Army 2022. *Battlefield Electrification Approach*. Andover: Army HQ, Futures Directorate.

Bulkeley, H., M. Paterson, and J. Stripple 2016. *Towards a Cultural Politics of Climate Change: Devices, Desires and Dissent*. Cambridge: Cambridge University Press.

Bulkeley, H., J. Stripple, L. J. Nilsson, B. van Veelen, A. Kalfagianni, F. Bauer, and M. van Sluisveld 2022. *Decarbonising Economies*. Cambridge: Cambridge University Press.

Chakrabarty, D. 2017. The politics of climate change is more than the politics of capitalism. *Theory, Culture & Society* 34 (2–3): 25–37.

Committee on Climate Change 2020. *Land Use: Policies for a Net Zero UK*. London: Committee on Climate Change.

Closson, S. 2013. The military and energy: moving the United States beyond oil. *Energy Policy* 61: 306–316.

Depledge, J. 2008. Striving for no: Saudi Arabia in the climate change regime. *Global Environmental Politics* 8(3): 9–35.

Dillon, M. 2002. Network society, network-centric warfare and the State of Emergency. *Theory, Culture & Society* 19: 71–79.

Emmett, C. 2022. Analysis: British Army begins charge towards battery power. *IISS*. https://www.iiss.org/blogs/military-balance/2022/05/analysis-british-army-begins-charge-towards-battery-power (accessed 1 May 2023).

Frankopan, P. 2015. *The Silk Roads: A New History of the World*. London: Bloomsbury Publishing.

Hafner, K. and M. Lyon 1998. *Where Wizards Stay Up Late: The Origins of the Internet*. New York: Simon & Schuster.

Heredia, F., A. L. Martinez, and V. Surraco Urtubey 2020. The importance of lithium for achieving a low-carbon future: overview of the lithium extraction in the 'Lithium Triangle'. *Journal of Energy & Natural Resources Law* 38(3): 213–236.

Houghton, J.T., G. J. Jenkins, and J. J. Ephraums 1990. *Climate Change: The IPCC Scientific Assessment*. Cambridge: Cambridge University Press.

IPCC 2014. Summary for policymakers. (In) *Climate Change 2014: Synthesis Report*. Geneva: Intergovernmental Panel on Climate Change.

IPCC 2018. *Global Warming of 1.5°C*. Geneva: Intergovernmental Panel on Climate Change.

Jaffe, A. M. 2021. *Energy's Digital Future*. New York: Columbia University Press.

Jerez, B., I. Garcés, and R. Torres 2021. Lithium extractivism and water injustices in the Salar de Atacama, Chile: the colonial shadow of green electromobility. *Political Geography* 87: 102382.

Johnstone, P. and C. McLeish 2020. World wars and the age of oil: exploring directionality in deep energy transitions. *Energy Research & Social Science* 69: 101732.

Jorgenson, A. K., B. Clark, and J. Kentor 2010. Militarization and the environment: a panel study of carbon dioxide emissions and the ecological footprints of nations, 1970–2000. *Global Environmental Politics* 10(1): 7–29.

Kassler, P. and M. Paterson 1997. *Energy Exporters and Climate Change Politics*. London: Royal Institute of International Affairs.

Klare, M. T. 2001. *Resource Wars: The New Landscape of Global Conflict*. London: Henry Holt and Company.

Klare, M. T. 2007. *Blood and Oil: The Dangers and Consequences of America's Growing Dependency on Imported Petroleum*. London: Henry Holt and Company.

Lerner, S. 2012. *Sacrifice Zones: The Front Lines of Toxic Chemical Exposure in the United States*. Cambridge, MA: MIT Press.

MacGregor, S. and M. Paterson 2021. Island kings: imperial masculinity and climate fragilities. (In) P. M. Pulé and M. Hultman (eds) *Men, Masculinities, and Earth: Contending with the (m)Anthropocene*. Cham: Springer International Publishing, pp. 153–168.

MacNeil, R. 2017. *Neoliberalism and Climate Policy in the United States: From Market Fetishism to the Developmental State*. London: Routledge.

Macreadie, P. I., A. Anton, J. A. Raven, N. Beaumont, R. M. Connolly, D. A. Friess, J. J. Kelleway, H. Kennedy, T. Kuwae, P. S. Lavery, C. E. Lovelock, D. A. Smale, E. T. Apostolaki, T. B. Atwood, J. Baldock, T. S. Bianchi, G. L. Chmura, B. D. Eyre, W. Fourqurean, J. M. Hall-Spencer, M., Huxham, I. E. Hendriks, D. Krause-Jensen, D. Laffoley, T. Luisetti, N. Marbà, P. Masque, K. J. McGlathery, J. P. Megonigal, D. Murdiyarso, B. D. Russell, R. Santos, O. Serrano, B. R. Silliman, K. Watanabe, and C. M. Duarte 2019. The future of blue carbon science. *Nature Communications* 10(1): 3998.

Malm, A. 2016a. *Fossil Capital: The Rise of Steam-Power and the Roots of Global Warming*. London: Verso Books.

Malm, A. 2016b. Who lit this fire? Approaching the history of the fossil economy. *Critical Historical Studies* 3(2): 215–248.

McDonald, M. 2012. The failed securitization of climate change in Australia. *Australian Journal of Political Science* 47(4): 579–592.

McDonald, M. 2021. *Ecological Security: Climate Change and the Construction of Security*. Cambridge: Cambridge University Press.

Mitchell, T. 2013. *Carbon Democracy: Political Power in the Age of Oil*. London: Verso.

Newell, P. and M. Paterson 2010. *Climate Capitalism: Global Warming and the Transformation of the Global Economy*. Cambridge: Cambridge University Press.

Oels, A. 2013. Rendering climate change governable by risk: from probability to contingency. *Geoforum* 45: 17–29.

Overland, Indra 2019. The geopolitics of renewable energy: debunking four emerging myths. *Energy Research & Social Science* 49: 36–40.

Paltsev, S. 2016. The complicated geopolitics of renewable energy. *Bulletin of the Atomic Scientists* 72: 390–395.

Paterson, M. 1996. *Global Warming and Global Politics*. London: Routledge.

Paterson, M. 2007. *Automobile Politics: Ecology and Cultural Political Economy*. Cambridge: Cambridge University Press.

Paterson, M. 2021. Climate change and international political economy: between collapse and transformation. *Review of International Political Economy* 28(2): 394–405.

Paterson, M. and X. P-Laberge 2018. Political economies of climate change. *Wiley Interdisciplinary Reviews: Climate Change* 9(2): e506.

Randalls, S. 2010. History of the 2°C climate target. *WIREs Climate Change* 1(4): 598–605.

Revette, A. C. 2017. This time it's different: lithium extraction, cultural politics and development in Bolivia. *Third World Quarterly* 38(1): 149–168.

Riley, A., 2022. Gazprom set the Russian invasion of Ukraine in motion. *Atlantic Council*. https://www.atlanticcouncil.org/blogs/energysource/gazprom-set-the-russian-invasion-of-ukraine-in-motion/ (accessed 1 May 2023)

Riofrancos, T. 2022. The security–sustainability nexus: lithium onshoring in the Global North. *Global Environmental Politics* 23(1): 20–41. 10.1162/glep_a_00668

Satia, P. 2019. *Empire of Guns: The Violent Making of the Industrial Revolution*. Stanford: Stanford University Press.

Sivaram, V., & Saha, S. 2018. The geopolitical implications of a clean energy future from the perspective of the United States. (In) Scholten, D. (ed), *Lecture Notes in Energy, The Geopolitics of Renewables*. pp. 125–162.

Sterba, J., A. Krzemień, P. Riesgo Fernández, C. Escanciano García-Miranda, and G. Fidalgo Valverde 2019. Lithium mining: accelerating the transition to sustainable energy. *Resources Policy* 62: 416–426.

Stoddard, I., K. Anderson, S. Capstick, W. Carton, J. Depledge, K. Facer, C. Gough, F. Hache, C. Hoolohan, M. Hultman, N. Hällström, S. Kartha, S. Klinsky, M. Kuchler, E. Lövbrand, N. Nasiritousi, P. Newell, G. Peters, Y. Sokona, A. Stirling, M. Stilwell, C. Spash, and M. Williams 2021. Three decades of climate mitigation: why haven't we bent the global emissions curve? *Annual Review of Environment and Resources* 46: 653–689.

Stulberg, A. N. 2015. Out of gas? Russia, Ukraine, Europe, and the changing geopolitics of natural gas. *Problems of Post-Communism* 62(2): 112–130.

Taylor, A. J. P. 1969. *War by Timetable: How the First World War Began*. London: Purnell.

US Geological Survey 2020. *Mineral Commodities Summaries, January 2020.* Washington DC: US Geological Survey.

van Beek, L., M. Hajer, P. Pelzer, D. van Vuuren, and C. Cassen 2020. Anticipating futures through models: the rise of Integrated Assessment Modelling in the climate science-policy interface since 1970. *Global Environmental Change* 65: 102191.

Voskoboynik, D.M. and D. Andreucci 2021. Greening extractivism: environmental discourses and resource governance in the 'Lithium Triangle'. *Environment and Planning E: Nature and Space*: 25148486211006344.

Zografos, C. 2022. The contradictions of green new deals: green sacrifice and colonialism. *Soundings* 80: 37–50.

# SECTION II

# Defence and security implications

# 8

# TOWARDS A GREENER ALLIANCE

## NATO's energy efficiency and mitigation efforts

*Katarina Kertysova*

### Introduction

Russia's war against Ukraine has re-invigorated NATO and underlined its importance as the cornerstone of Euro-Atlantic security and defence. In response to Russia's actions, NATO has reinforced its eastern flank with a major new deployment of four battlegroups, agreed to develop a new force model, which will increase the pool of high readiness forces to 300,000, and a majority of NATO members have committed to investing more in defence. The alliance's collective defence assumed an even more important place in the new 'Strategic Concept', which was adopted at NATO's Madrid Summit in June 2022. While Russia's activities pose a real threat to NATO's member states, the alliance has not lost sight of the security risks emanating from transnational, actorless threats, such as climate change, which continue to shape the security environment.

The very same week that Russia launched its full-scale invasion of Ukraine, in February 2022, the Intergovernmental Panel on Climate Change (IPCC) released the IPCC Working Group II report, which gave the bleakest warnings yet of the impacts of climate change and the time humanity has left to limit global warming to 1.5°C (IPCC 2022). Despite the Russia's war on Ukraine, climate change is not going to go away. In line with its 360-degree approach to security – which commits alliance members to ensure collective defence against all threats from all directions – NATO needs to maintain focus on both strengthening Allied collective defence against the threat that Russia poses. Russia and addressing the security implications of a changing climate. As re-affirmed by the former NATO Deputy Assistant Secretary General for Emerging Security Challenges, Jamie Shea (2022, p. 11), 'even an event as catastrophic in humanitarian and political terms as Putin's invasion of Ukraine

DOI: 10.4324/9781003377641-11

cannot absolve NATO of its responsibility to keep its focus on the potentially even worse catastrophic effects of climate change'.

The Russia's war on Ukraine has also shown the necessity of weaning Allied militaries off of fossil fuels. With fuel depots and fuel supply infra-structure being attacked in Ukraine, the war is a stark reminder that fuel dependency is a significant military vulnerability. Moving away from fossil fuels is essential not only from a climate point of view but also for our combat survivability.

This chapter will first provide an overview of NATO's efforts to reduce reliance on fossil fuels and enhance the efficiency of allied armed forces to date. It will then zoom in on climate-related decisions that were taken at the Madrid Summit in June 2022. Finally, it will consider what more NATO can do to contribute to global mitigation efforts and outline ways in which the alliance can further support its members in their military emissions reduction efforts.

## NATO's climate security agenda

NATO first acknowledged the link between climate change and security in its 2010 Strategic Concept. The focus of the alliance's efforts has been, first and foremost, on adaptation to the negative effects of climate change on military installations, operations, equipment, and force readiness. In the area of mitigation, the need to reduce reliance on fossil fuels and improve the energy efficiency of military forces has been driven primarily by vulnerabilities associated with the provision of energy to front-line operations – the frequent targeting of NATO fuel supplies, the high cost of transporting fuel to the battlefield, and the risk to troops involved in fuel logistics – rather than explicit climate change considerations.

Over the past two years, references to mitigation and the need to embrace a net-zero transition have found their way into NATO's public statements. At NATO's Brussels Summit in June 2021, the NATO 2030 agenda and the Climate Change and Security Action Plan were adopted. In an important shift from earlier climate-related discussions within NATO, members of the alliance pledged to, 'significantly reduce greenhouse gas emissions from military activities and installations', tasking the Secretary General with the formulation of, 'a realistic, ambitious and concrete target for the reduction of greenhouse gas emissions by NATO political and military structures and facilities and assess the feasibility of reaching net-zero emissions by 2050' (NATO 2021a).

In line with this commitment, the Climate Change and Security Action Plan outlines mitigation as one of the four lines of NATO's climate change efforts – in addition to enhancing awareness, adaptation, and outreach efforts (NATO 2021b). As part of announced mitigation measures, NATO developed a mapping methodology to help alliance members measure their military emis-sions, which could contribute to determining voluntary cuts.

During an interview following his public remarks at the UN Climate Change Conference (COP26) in 2021 in Glasgow, NATO's Secretary General Jens Stoltenberg acknowledged that 'there [was] no way to reach net zero without also including emissions from the military'. Recognising that the alliance works by consensus, which continues to evolve, Stoltenberg (cited in John, 2021) added: 'this is the aim but of course I am dependent on agreement among 30 Allies'. The need to measure military emissions so as to reduce them more effectively was subsequently reiterated at COP27 in Sharm el-Sheikh, Egypt, where Stoltenberg also expressed his doubts that NATO can remain a 'fossil fuel Alliance in a world of renewables' (NATO 2022b).

## Current mitigation efforts: energy efficiency focus

Both the consumption and movement of fuel supplies have always been a critical yet challenging and dangerous part of any military operation. Large convoys of fuel transporters, which are required to provide energy on the battlefield, have often come under attack. It is estimated that between 2003 and 2007, 3,000 US soldiers were killed or wounded in attacks on fuel and water resupplies in Afghanistan and Iraq – about one casualty for every 24 fuel convoys (NATO 2015a). Large quantities of ammunition and equipment are moved around in military operations, which also necessitates additional fuel.

In 2011, recognising the vulnerability surrounding fuel supplies for deployed forces, NATO's Emerging Security Challenges Division started working on enhancing energy efficiency in NATO under the banner of 'smart energy' (NATO 2015b).[1] At the Chicago Summit in May 2012, alliance members unanimously agreed a position on smart energy, stating their determination to, 'work towards significantly improving the energy efficiency of [their] military forces' (NATO 2012). This commitment was reiterated in the Wales Summit Declaration in September 2014. Subsequently, NATO has been supporting projects that aim to reduce fossil fuel dependence in military camps, enhance renewable energy usage, and incorporate innovative technologies and approaches within military capability.

Following the Chicago Summit Declaration, a Smart Energy Team (SENT) was established, in 2013, with a two-year mandate to identify and spotlight best practices and opportunities for collaborative multinational smart energy projects. SENT was tasked with six concrete deliverables,[2] all of which were achieved. SENT's 'Comprehensive Report', published in 2015, concluded that a number of NATO and partner nations had successfully implemented smart energy technologies and established strategies, policies, and standards for smart energy use in the military. In spite of a desire and willingness for collaboration and knowledge sharing, the report acknowledged that there was, however, a lack of cooperation between defence, academia, and industry within, as well as among, NATO nations, with most national initiatives having been carried out in isolation. The report further identified that there was a lack of energy efficiency requirements in military procurement, and that policies, procedures, standards, and the overall knowledge and awareness on smart energy across the alliance were insufficient (NATO 2015b).

In addition to SENT, NATO's Science for Peace and Security (SPS) Programme also supported Serbian-led research into biofuel production from algae in 2017–2020, as well as the 'Camp Energy Efficiency' project, launched in 2018 in Canada, which sought to develop interoperable monitoring kits for energy data collection so as to identify and address wasteful energy consumption in deployed military camps (NATO 2017, 2018). Adding to the growing number of activities supported by the SPS Programme in the area of climate and security, a new multi-year project was initiated, in May 2022, that looks to exploit 'Big Earth' data analytics to improve alliance-wide understanding of security-relevant aspects of natural and human-made hazards (NATO 2022b, 2023a).

Since 2013, NATO has been holding 'Capable Logistician' exercises with the aim of testing and demonstrating various energy-saving technologies, as well as their interoperability, in the field. In 2013, the exercise was hosted by Slovakia, in 2015 by Hungary, in 2019 by Poland, and in 2023 by Lithuania (Lithuanian Armed Forces 2022). 'Smart energy' camps, showcased during the exercises, included solar panels and energy-saving LED lights, hybrid energy system grids, insulation for tents, intelligent power storage and management capabilities, and water purification systems (Michaelis 2017).

NATO's Green Defence Framework, adopted in February 2014, was an important step forward as it provided NATO staff and national experts with a broad basis for cooperation on green solutions for defence (NATO 2014, 2015b). The framework made several innovative suggestions to reduce the energy consumption of allied armed forces, as well as a proposal to apply 'green standards' across NATO's political and military structures and facilities (Figure 8.1). The momentum for mainstreaming green solutions was interrupted by Russia's illegal annexation of Crimea in March 2014, which took place only weeks after the framework was adopted. Indeed, NATO's renewed emphasis on collective defence is believed to have impeded a more systematic pursuit of the various innovative initiatives proposed in the framework (Rühle 2020).

Energy efficiency has also been reflected in NATO policies, namely, the 'Policy on Power Generation for Deployed Forces Infrastructure' and 'NATO Military Principles and Policies for Environmental Protection', which currently constitute two umbrella documents for further development of NATO standardisation in the area of smart energy (NATO 2015b). Furthermore, the NATO Energy Security Centre of Excellence, located in Lithuania, has devoted a large share of its activities to enhancing the energy efficiency of armed forces in the alliance. Since its establishment in 2012, the

## THREE PILLARS OF NATO'S GREEN DEFENCE FRAMEWORK

| Reinforcing efforts of NATO bodies | Facilitating Allies' efforts | External engagement: Improving NATO's 'green' profile |
|---|---|---|
| • Identifying a focal point within existing structures to improve overall coordination and streamlining of activities<br>• Development of 'green' accounting and benchmarks to measure progress<br>• Comprehensive database on energy consumption of NATO operations<br>• Factoring green defense into NATO training, education, and exercises | • Creating a platform for sharing of lessons learned, best practices, and nationally-developed 'green' technologies<br>• Incorporating questions about national Green Defence activities in the Defence Planning Capability Survey (DPCS)<br>• Exploring projects on 'green' capabilities and equipment<br>• Promoting environmental protection and developing STANAGs in this area | • Improving cooperation with partner nations and the private sector<br>• Better communication with the general public<br>• Leveraging various public diplomacy tools (including through relevant NATO CoEs)<br>• Coordinating with other international organizations to avoid unnecessary duplication, including through mutual briefings<br>• Using NATO activities as test beds for new technologies |

**FIGURE 8.1** NATO's three pillars of green defence.

*Source:* Author's creation.

Centre has generated knowledge, expert advice, and solutions for development of energy-efficient forces, as well as training courses that either concentrate on or at least feature 'smart energy' modules (Michaelis 2018).

The planned NATO accredited Centre of Excellence of Climate Change and Security (CASCOE), which Canada has offered to host, is also expected to support NATO's mitigation efforts. CASCOE started its accreditation process in 2022 and held its second establishment conference in February 2023. Once operational, as of fall 2023, the Centre could serve as a hub for the exchange of best practices and lessons learned, help address the above-mentioned lack of knowledge and awareness of the operational benefits of green solutions through courses for all levels of the NATO command structure, and make NATO's 'green' profile more visible through public communications campaigns and public events (Government of Canada 2022).

Even though NATO's primary objective has been to reduce reliance on fossil fuels when in the field and to make militaries in the alliance more energy efficient, the above-mentioned activities have also helped reduce $CO_2$ emissions from military operations and, as such, have contributed to global mitigation efforts.

### The Madrid Summit and the military energy transition by design initiative

The Summit in Madrid, in June 2022, marked a significant shift in NATO's environmental agenda, not least in relation to the alliance's emissions reduction efforts (Kertysova 2022). The Secretary General reaffirmed NATO's commitment to reducing military greenhouse gas emissions and announced concrete targets for NATO as an organisation: at least 45% reduction by 2030 and, in line with the Paris Agreement, reaching net zero by 2050 (Stoltenberg 2022). These targets only concern NATO's own installations and assets, such as the Airborne Warning and Control System (AWACS) surveillance planes and fleet of RQ-4D drones under NATO's Alliance Ground Surveillance (AGS) programme (SHAPE-NATO nd). However, whilst these reductions will only affect a small part of NATO-wide emissions, they are a powerful demonstration that NATO wants to lead by example and that the organisation takes climate change mitigation seriously.

Alliance members are responsible for their own policy decisions regarding greenhouse gas emissions, including their military emissions. However, with the exception of several member states that already produce such estimates, military emissions are neither counted nor reported – either due to a lack of data or because military emissions are treated as sensitive information (Goodman and Kertysova 2022). Even where military emissions are measured, the metrics are not consistent across the board. To help its members measure their emissions from military activities and installations, NATO developed a greenhouse gas emissions mapping and analytical methodology that has been shared with all capitals and published at the Vilnius Summit in 2023. This methodology draws on the best practices of alliance members and expertise residing in partner nations and other international organisations, including the EU. According to Richard Brewin, who leads on climate change and security in NATO's Emerging Security Challenges Division (ESCD), 2019 will serve as the baseline year against which NATO's military emissions reductions will be measured. As Brewin (cited in Kertysova 2022) has explained, 'the 45% emissions reduction target by 2030 was developed using the Science Based Target initiative and it applies to NATO assets and installations'.

To enable the sharing of best practices and lessons learned, NATO has also compiled and provided its members with a 'Compendium of Best Practices', which outlines examples of different awareness, adaptation, and mitigation measures that have been put into practice, and identifies which models can be replicated across the alliance (NATO 2022c). This compendium, first published in July 2022, continues to be updated to reflect ongoing progress.

Additionally, the Secretary General announced plans to develop a new 'military energy transition by design' initiative, (Stoltenberg 2022). Recognising the risk of divergence – in the event that alliance members pursue military emissions reduction goals at different speeds and ambition levels – this initiative will strive to ensure that a military transition away from fossil fuels does not impede interoperability of military forces nor does it negatively affect operational effectiveness (Rühle 2023).

In September 2022, NATO applied for observer status to the IPCC – the United Nation's body for assessing the science related to climate change – whose work NATO considers as a basis for its own adaptation efforts (IPCC Secretariat 2022). Under the IPCC governance structure, observer organisations may designate representatives to attend sessions of the IPCC and its working groups, suggest experts to review draft IPCC reports, and provide input into the work of the IPCC when requested. Once awarded, the observer status will enable NATO to share environmental data more effectively, coordinate its activities with key UN agencies, address existing gaps, as well as leverage its science programmes and bodies in support of the IPCC assessment process.

### Looking ahead: what more can be done within NATO's mandate?

It is important to bear in mind that NATO does not have the power to legislate and cannot impose binding emissions reduction targets on alliance militaries. In the words of the former Head of Climate and Energy Security Section, 'NATO is not a first responder to climate change. This role is played by other international bodies, in particular those who can set limits on $CO_2$ emissions' (Rühle 2020). NATO instead seeks to become, 'the leading international organisation when it comes to understanding and adapting to the impact of climate change on security', as the NATO 2030 agenda confirmed in 2021 (NATO 2021c). Even though the alliance does not seek to position itself as a 'first responder', it nevertheless has a range of tools in its toolbox that can support Allied emissions reduction efforts, both directly and indirectly.

Once national military emissions are known, with the help of the methodology NATO developed and delivered to member states at the Madrid Summit in 2022, NATO could set voluntary targets for their reduction. The Secretary General has already been invited to formulate a realistic target and assess how feasible it would be for alliance militaries to reach net-zero emissions by 2050 (NATO 2021c). For voluntary targets to work, they will need to be accompanied by a robust and agreed reporting mechanism.

NATO could also shift away from its single fuel policy (SFP) towards more sustainable alternatives, like biofuels or synthetic fuels. Under SFP, which is designed to simplify the logistic effort and enhance equipment interoperability, NATO forces aspire to use only one fuel on the battlefield – F-34 (Signorelli 2021). In an effort to reduce the negative environmental impacts of their military activities, some NATO members, such as Belgium, Canada, France, the Netherlands, and the United States, have already approved the use of synthetic fuels in their military platform (EDA 2017). The United States began the process over a decade ago with the 'Green Hornet' biofuel-powered flight, and the United Kingdom followed suit in 2020 when it changed its aviation fuel standards to allow substitution by sustainable fuels up to 50% in all its military aircraft (UK MOD 2020; US DOD nd). The Dutch Ministry of Defence aims to ensure that by 2030 all military aircraft will fly on a 20% addition of biofuels, and a 70% biofuel blend by 2050 (Biofuels International 2019). In addition to synthetic fuels, NATO could also lead the way in research and development of alternative propulsion systems for military applications (Goodman and Kertysova 2022).

Through the NATO Defence Planning Process (NDPP), which identifies the capabilities required by the alliance and ensures coherence in their development, NATO could develop green minimum capability requirements for each member to meet (Michaelis 2021). This could serve as an incentive for alliance nations to reduce their military emissions and shift to more sustainable technologies. While capability requirements set out in the NDPP do not constitute a legally binding commitment, they do drive national defence planning.

In addition, NATO has a successful track record as a standard setter. NATO can update existing policies and standardisation agreements (STANAGs), as well as introduce new ones. According to Susanne Michaelis (2018) former Science Officer at NATO's Emerging Security Challenges Division, new standards in the area of smart microgrids would constitute a 'quick win' for reducing the fuel consumption of field camps. Even though the alliance cannot mandate interoperability of national forces, units, and/or systems, STANAGs are binding on those alliance members that ratify and implement them. In general, NATO standards get a lot of support in the logistics community and are recognised and generally adhered to in manufacturing practices. This is due to the fact that STANAGs offer a bigger market (31 nations) and are a sign of quality for export (G. Stacey, pers. comm.).

At this moment, NATO has limited collective financial means in the civilian budget to do more on climate change. Given that NATO member states' commitment to spend at least 2% of GDP on defence is up for review in 2024, the NATO 2030 Young Leaders Group (NATO YLG 2022) proposed to reimagine what counts as a defence contribution moving forward. According to the group, the reformed 2% target after 2024 should also account for investments in areas such as climate security, economic resilience, or anti-hybrid warfare, helping the alliance achieve resilience across the board. The Defence Investment Pledge endorsed in 2014 also calls for alliance members to spend 20% of total defence expenditures on major new equipment and R&D (NATO 2023b). Some of this could be directed towards the development of sustainable technologies (both military and dual use). Shea (2022) has suggested setting up a NATO Green Fund that could help finance trials and demonstrations and assist less advanced alliance members in greening their militaries.

Technological innovation forms an important part of the mitigation process. Many of the technological solutions that can help our militaries lower their fuel use are readily available and do not require a market transformation. Deliveries by drones or 3D printing of weapons systems, components, and ammunition at the point of use already offer significant savings in terms of the logistics burden and fuel use on the battlefield (Lalkovič et al. 2022; Shea 2022). As mentioned, sustainable aviation fuels are also increasingly used. Furthermore, many alliance members are already electrifying their white fleets.[3] Improving energy efficiency of buildings and bases constitutes another 'low-hanging fruit'. Because the private sector drives much of the green innovation today, NATO members need to work alongside civilian agencies, private-sector companies, and research institutions when trying to reduce fossil fuel dependence in the military. Cooperation with civil society and the private sector is also important for NATO's own awareness about the technological breakthroughs and their societal impact.

Given their scale and buying power, defence departments have traditionally been effective at moving certain markets (Sivaram 2022). On the one hand, alliance militaries can scale up existing low-carbon technologies through their own procurement practices. On the other hand, they can create demand signals to enable research and development of next-generation technologies. The electronic chip industry, for example, started as a niche defence business in response to the military's demand for guidance computers in the post-World War II period. The innovations then extended to, and benefitted, other sectors. Indeed, by 1968, 75% of all the chips sold went to produce civilian goods, including hearing aids and corporate computers (Miller 2022; Sivaram 2022). The Secretary General's 'Climate Change and Security Impact Assessment Report' recognises that, 'a strong demand signal from the military establishments that they are moving towards new and cleaner technologies and energy sources can stimulate industry to create the necessary materials, for example, processing for biofuels and synthetic fuels' (NATO 2022d). Furthermore, NATO's 2023 Symposium on Climate Change and Capabilities, which gathered more than 150-member state and industry representatives to discuss ways in which innovative and green technology can be leveraged in the development of new military capabilities, is reflective of this military-industrial dependency in the climate security space (NATO 2023c).

While individual alliance members have control and influence over the procurement of military equipment, NATO as an organisation can lead with regard to commonly funded infrastructure (i.e. fixed installations which are necessary for the deployment and operations of the armed forces)[4] and the equipment and technologies that support NATO's command and control (C2). For example, sustainability and energy efficiency were integrated within the design of the new NATO Headquarters (SOM nd). The NATO Support and Procurement Agency (NSPA), the organisation's main logistics support and procurement agency, is also playing its part. Since its establishment in

2012, the agency has launched and implemented several cooperative activities that are environmentally sound and sustainable. Supporting NATO nations and partners in acquiring sustainable assets and capabilities that feature clean energy solutions lists among the agency's main areas of work (NSPA 2020; 2022).

At the Madrid Summit, NATO leaders agreed to establish a NATO Innovation Fund (NIF), the world's first multi-sovereign venture capital fund. The NIF intends to invest an initial US $1.1bn[5] to help early stage start-ups grow and to support NATO's technology needs, including in the areas of novel materials, energy, and propulsion. The fund is to become operational in 2023, with a 15-year timeframe (NATO 2022e). Also in 2023, NATO's Defence Innovation Accelerator for the North Atlantic (DIANA) is expected to launch its first pilot activities, reaching full operational capability by 2025. DIANA, which comes with more than 11 accelerator sites and 91 test centres in Europe and North America, will help military personnel work more closely with the alliance's technology companies, start-ups, and scientists to develop technological solutions to existing and future security problems shared across the Alliance (NATO 2022f). It has been announced that energy resilience will be one of the three priority areas of focus for DIANA's work, alongside secure information sharing and sensing and surveillance (NATO 2022a).

Another way in which NATO can help maintain momentum, bring in outside scientific expertise, and explore external ideas is through the SPS programme, which funds collaborative scientific activities among NATO members and partners. In 2022, 15% of proposals received by the SPS programme focused on environmental security, in addition to the 11% that revolved around energy security (NATO 2023a). In 2023, projects focusing on climate and energy security were given higher priority for funding, alongside those related to emerging and disruptive technologies (EDTs).[6]

### Discussion: measuring success, overcoming challenges, and setting expectations

Climate change mitigation in the armed forces is, without doubt, the most ambitious dimension of NATO's climate change and security action plan (Rühle 2021). As noted above, target setting for emissions reduction, for example, falls under the competence of individual member states, who find themselves at different starting points on the road to net zero. The lack of reporting on military emissions, with the exception of several members, such as Denmark, France, Germany, the Netherlands, Norway, the United Kingdom, and the United States, also impedes action. Even where military emissions are measured, the metrics are only starting to be harmonised across NATO. What is more, some NATO member states, especially those from Central and Eastern Europe, have expressed trepidation about the speed of energy transition, particularly in the defence sector (Kertysova 2022). In addition, there are fears in certain quarters that pressure to decarbonise defence might result in the loss of operational effectiveness. With the ongoing war in Ukraine, meeting higher emissions reduction standards has not been treated with the same sense of urgency as providing Ukraine with the means to defend itself and strengthening the collective defence of the alliance against the threat that Russia poses (Shea 2022).

At the organisational level, NATO has made considerable progress since the adoption of the Climate Change and Security Action Plan in 2021. Creation of a uniform methodology for measuring and analysing military greenhouse gas emissions is an important first step towards increased transparency and more ambitious mitigation efforts (Kertysova 2022; Wigell and Hakala 2022; Kõrts 2023). It not only enables a comparison of national military emissions, but it can also inform the formulation of voluntary emissions reduction goals at the level of individual member states (Rühle 2021).

In addition to providing its members with the tools to map and measure their military emissions, NATO strives to lead by example by reducing emissions of its own assets and installations. While these reductions will only affect a small part of NATO-wide emissions, they are a powerful demonstration that the organisation takes climate change mitigation seriously. NATO cannot, however, set binding targets nor enforce investments in clean military technologies. Individual alliance members have influence over the procurement of military equipment. NATO can nevertheless incentivise military decarbonisation through its standardisation agreements, by setting net zero and sustainable targets for defence planning, and/ or by shifting away from its SFP towards more sustainable alternatives. With its long and successful track record as a standard setter, NATO taking climate change mitigation seriously can have a normative influence on NATO members as well as partners worldwide.

To assess the progress that has been made, including but not limited to NATO's emission reduction and mitigation efforts, NATO set out to produce an annual 'Climate Change and Security Progress' report. The first one was delivered at the 2022 Madrid Summit (NATO 2021b). Because of its classification level, the progress report is treated as an internal resource and is unavailable for public scrutiny. Given that NATO's mitigation efforts are relatively recent, it is too early to draw conclusions on what has worked and what has not. It is, however, possible to infer that NATO has contributed to increasing awareness and sensitivity on these issues amongst alliance members and that it has helped build unity of effort and resulted in action on the part of member states, notably by those that previously paid little to no attention to their military emissions (Barberini 2022; Kertysova 2022).

NATO's Climate Change and Security Action Plan recognises that 'further work and sustained political ambition is needed to ensure that NATO is fully prepared to continue to deliver in a changing climate' (NATO 2021b). While existing scholarship welcomes NATO's mitigation efforts, there is a general recognition that more needs to be done to achieve a genuine impact on emissions (CEOBS 2021; Shea 2022; Wigell and Hakala 2022; Kõrts 2023). Because military emissions reporting has been voluntary, alliance militaries have been accused of being untransparent (Weir 2022). The non-reporting requirement has also made it difficult to track progress on mitigation and reinforce government accountability in the defence sector (Weir 2022; Wigell and Hakala 2022).

As noted earlier, for voluntary emissions reduction targets to work, they need to be accompanied by a robust and agreed reporting mechanism. In her latest report for NATO's Energy Security Centre of Excellence, Marju Kõrts (2023) calls for the UNFCCC reporting protocols to be strengthened and reformed to also include military emissions. According to her analysis, the reporting and reduction of military emissions must be, 'transparent, time-bound and measurable' (Kõrts, 2023, p. 6). As Shea (2022) has reported, the alliance has been under pressure from the NGO community to mirror the level of ambition and emissions cuts that were agreed at COP26. In his words, 'it will be important for [the measuring of military $CO_2$ emissions] to be rigorous, as it will be scrutinized carefully by the NGO community, who will no doubt be pushing for NATO to show transparency and accountability in publishing the results annually' (Shea 2022, p. 5). At the same time, Shea calls into question whether meeting the 2050 target will be achievable amid major ruptures in the international security environment that necessitate additional training, platforms, weapon systems, and – by extension – additional fuel.

On its road to net zero, NATO will need to overcome several challenges as well as set the right expectations. Firstly, NATO – as an alliance of 31 countries – works by consensus. Keeping the momentum, preventing diversion, and ensuring that climate security remains on top of NATO's agenda are key (Rühle 2023). Secondly, as alliance members shift away from fossil fuels, they need to be wary of not replacing one dependency for another – notably on China for critical minerals, which are essential to building and maintaining 'clean' technologies, such as solar panels and fuel cells, as well as modern weapons systems (Bazilian et al. 2023; NATO 2022d; Nugee and Clack, this volume). Thirdly, there is the challenge of military interoperability. While some members are already investing in green technologies, adjusting their fuel standards, and increasing the use of sustainable aviation fuel, others continue to operate legacy equipment. If alliance members move at different speeds as regards integration of cleaner military solutions, interoperability can be affected. However, through standard-setting and best-practice exchange, NATO is ideally positioned to help avoid divergences and ensure that members advance jointly (Rühle, 2023). Fourthly, as a result of Russia's war of aggression against Ukraine, fuel demand and consumption – as well as greenhouse gas emissions – are set to rise in the short term, given the fundamental shift in NATO's deterrence and defence posture. What is more, new energy-saving technologies – just like any other military equipment – have long lead times. As Rühle (2021) notes, 'they have to be developed, purchased, integrated into existing national forces, and they have to be made interoperable with the equipment of other Allies'. Thus, for a considerable period of time, armed forces will remain considerable consumers of fossil fuels. Alliance members need to be clear that energy transition in the military is a long-term project.

## Conclusion: increased ambition and action

Allied militaries generate considerable emissions in peacetime and even more when at war.[7] As this chapter has shown, NATO is moving beyond merely acknowledging the security implications of climate change and adapting its forces to extreme circumstances. The organisation has also stepped up its efforts to incentivise its members to reduce their military emissions and shift to sustainable technologies. At a time when war rages on NATO's doorstep, and collective defence is at the centre of members states' attention, the decisions taken in Madrid indicated that NATO was going to remain committed to its ambition to reduce military emissions and keep its focus on better understanding and addressing the impacts of climate change (Kertysova 2022).

While NATO – notably under Jens Stoltenberg's leadership – has formally embraced the need for alliance militaries to reduce their emissions, target setting falls under the competence of individual member states. NATO cannot set binding targets nor can it compel investments in or uptake of clean military technologies. It can only act within the framework of its mandate and purpose. Individual member states have control over the procurement of military equipment. Regardless, NATO has a range of tools in its box that can support alliance members in their emissions reduction efforts and, as Stoltenberg (2022) notes, it can itself strive to 'set the gold standard' by reducing emissions of its own assets and installations.

2022 was the eighth consecutive year of rising defence spending across NATO (NATO 2023a). The outbreak of the war has prompted alliance members to invest even more in defence, aiming to make 2% of GDP 'a floor, not a ceiling' (NATO 2022g). On the one hand, rising defence budgets are necessary to secure peace and stability in Europe, which is key to sustainable development. On the other hand, decisions to acquire new equipment present an opportunity to ensure that fuel and energy efficiency standards are factored in at the capability design stage, which will reduce logistical and fuel requirements in the future (NATO 2022d).

Military conflict itself is a significant driver of climate change and environmental damage. Effective deterrence, which is a core element of NATO's overall strategy, plays a vital role in preserving peace and preventing future conflicts and – by extension – future emissions. This is why operational considerations and military competence will remain of primary importance. In addition to boosting its deterrence, NATO needs to improve its early warning and strategic foresight capabilities to better understand the root causes of war and, where possible, address them (NATO YLG 2022). Prevention, after all, is less costly and more effective than dealing with the fallout.

## Acknowledgements

The author is grateful to Sir Graham Stacey, Sir Adam Thomson, and Timothy Clack for their review and comments on earlier drafts of this chapter. Parts of this chapter were originally published in *Decarbonized Defence: Clean Military Power in the Age of Climate Change* (Van Schaik et al. 2022) and 'Perseverance amidst crisis: NATO's ambitious climate change and security agenda after Madrid' (Kertysova 2022). The author wishes to thank the International Military Council on Climate and Security (IMCCS), European Leadership Network (ELN), and the editors for giving permission to reproduce these sections in the current volume.

## Notes

1 NATO's Smart Energy Team (SENT) defined smart energy as, 'the methods of providing energy to the user in a practical, effective, sustainable and environmentally responsible manner' (NATO 2015b, p. 4).
2 These six deliverables were: (1) project proposals; (2) a comprehensive report; (3) field trip assessments; (4) contributing a component to Exercise CAPABLE LOGISTICIAN 2013; (5) raising public awareness; and (6) establishing an information sharing internet platform (NATO 2015b).
3 The so-called military white fleet comprises vehicles that are used for administrative and non-operational purposes.
4 NATO's assets include airfields; signals and telecommunications installations; military headquarters; fuel pipelines and storage; radar warning and navigational aid installations; port installations; missile installations; forward storage sites; and support facilities for reinforcement forces. Such installations are financed collectively and may be used by each alliance member (NATO 1982).
5 These are public funds that participating nations can allocate, either from their existing defence budgets or established innovation funds.
6 The following topics were especially encouraged: assessment of current and future increased climate change-related biosecurity risks and mitigation options; climate change interactive scenario modelling (outlook to 2050/2050+); exploitation of innovative and low-carbon environmental technologies for operations; impact of increased variation in maritime salinity, acidity, and temperature on legacy and novel systems and technologies; impact of climate change on transboundary

security; strategic and critical resource management; nexus between terrorism and climate change; insurgency, terrorism and organised crime in a warming climate; exploitation of innovative sustainable energy systems in off-grid and on grid locations; and exploitation of hydrogen fuel cells.

7 Peacetime emissions relate to training and exercising, humanitarian assistance and disaster relief, military aid to civil authority, and tackling illicit activity (see Depledge 2023).

## References

Barberini, P. 2022. NATO green defence: from the 2014 green defence framework to the 2021 climate change and security action plan. (In) G. Iacovino and M. Wigell (eds) *Innovative Technologies and Renewed Policies for Achieving a Greener Defence.* New York: Springer (NATO Science for Peace and Security Series C: Environmental Security), pp. 7–16.

Bazilian, M. D., E. J. Holland, and J. Busby 2023. America's military depends on minerals that China controls. *Foreign Policy.* https://foreignpolicy.com/2023/03/16/us-military-china-minerals-supply-chain/ (accessed 1 May 2023).

Biofuels International 2019. Dutch ministry implements biofuel into military aviation. *Biofuels International.* https://biofuels-news.com/news/dutch-ministry-implements-biofuel-into-military-aviation/ (accessed 1 May 2023).

CEOBS 2021. Under the radar: the carbon footprint of Europe's military sectors. *Conflict and Environment Observatory.* https://ceobs.org/under-the-radar-the-carbon-footprint-of-the-eus-military-sectors/ (accessed 1 May 2023).

Depledge, D. 2023. Low-carbon warfare: climate change, net zero and military operations. *International Affairs* 99(2): 667–685.

EDA 2017. The use of alternative and synthetic fuels in the military. *European Defence Agency.* https://eda.europa.eu/docs/default-source/events/eden/phase-i/information-sheets/cf-sedss-information-sheet-use-of-alternative-and-synthetic-fuels.pdf (accessed 1 May 2023).

Goodman, S. and K. Kertysova 2022. NATO: An unexpected driver of climate action? *NATO Review.* https://www.nato.int/docu/review/articles/2022/02/01/nato-an-unexpected-driver-of-climate-action/index.html (accessed 1 May 2023).

Government of Canada 2022. Informing Design of a NATO Climate and Security Center of Excellence: Expert Engagement Workshop. Online, 26 April.

IPCC 2022. *Climate Change 2022: Impacts, Adaptation and Vulnerability. Contribution of Working Group II.* Cambridge, UK: United Nations. https://www.ipcc.ch/report/ar6/wg2/downloads/report/IPCC_AR6_WGII_FrontMatter.pdf (accessed 1 May 2023).

IPCC Secretariat 2022. Fifty-seventh session of the IPCC: admission of observer organizations. *Intergovernmental Panel on Climate Change (IPCC).* https://apps.ipcc.ch/eventmanager/documents/75/180820220419-Doc.%203%20-%20Observer%20Organizations.pdf (accessed 1 May 2023).

John, M. 2021. NATO chief: armies must keep pace with global climate efforts. *Reuters.* https://www.reuters.com/business/environment/nato-chief-armies-must-keep-pace-with-global-climate-efforts-2021-11-02/ (accessed 1 May 2023).

Kertysova, K. 2022. Perseverance amidst crisis: NATO's ambitious climate change and security agenda after Madrid. *European Leadership Network (ELN).* https://www.europeanleadershipnetwork.org/commentary/perseverance-amidst-crisis-natos-ambitious-climate-change-and-security-agenda-after-madrid/ (accessed 1 May 2023).

Kõrts, M. 2023. Climate change mitigation in the Armed Forces – greenhouse gas emission reduction – challenges and opportunities for Green Defense. *NATO Energy Security Centre of Excellence (ENSEC COE)*. https://enseccoe.org/data/public/uploads/2023/04/d2_climate-change-mitigation-in-the-armed-forces.pdf (accessed 1 May 2023).

Lalkovič, T., B. Hrozenská, and K. Kertysova (2022) Green defence toolbox. *Analytical Unit of the Slovak Ministry of Defence*. https://www.mosr.sk/data/files/4867_2022-b-06-green-defence-toolbox2.pdf (accessed 1 May 2023).

Lithuanian Armed Forces 2022. NATO and EU logistics experts held the initial planning conference ahead of Exercise Capable Logistician 2023. *Lithuanian Armed Forces*. https://kariuomene.lt/en/nato-and-eu-logistics-experts-held-the-initial-planning-conference-ahead-of-exercise-capable-logistician-2023/24679 (Accessed: 25 February 2023).

Michaelis, S. 2017. NATO smart energy: capable logistician 2015. *NATO*. https://www.nato.int/nato_static_fl2014/assets/pdf/pdf_2017_08/20170808_Smart-Energy-Ex-Capable-Logisti.pdf (accessed 1 May 2023).

Michaelis, S. 2018. How NATO is making progress in energy efficiency for military forces. *Eyvor Institute*. https://eyvor.org/how-nato-is-making-progress-in-energy-efficiency-for-military-forces/ (accessed 1 May 2023).

Michaelis, S. 2021. Smart energy: less fuel, more power'. (In) R. A. Kingham and O. Lazard (eds) *Sustainable Peace and Security in a Changing Climate: Recommendations for NATO 2030*. Brussels/The Hague: Environment & Development Resource Centre (EDRC), pp. 65–67. https://www.brusselsdialogue.net/news/sustainable-peace-security-in-a-changing-climate-recommendations-for-nato-2030 (accessed 1 May 2023).

Miller, C. 2022. History offers a guide to winning our growing 'chip war' with China. *The Washington Post*. https://www.washingtonpost.com/made-by-history/2022/10/04/history-offers-guide-winning-our-growing-chip-war-with-china/ (accessed 1 May 2023).

NATO 1982. Infrastructure and logistics. *NATO*. https://archives.nato.int/uploads/r/null/1/3/137761/0196_Aspects_of_NATO-Infrastructure_and_Logistics_ENG.pdf (accessed 1 May 2023).

NATO 2012. Chicago Summit declaration. *NATO*. https://www.nato.int/cps/en/natohq/official_texts_87593.htm?selectedLocale=en (accessed 1 May 2023).

NATO 2014. Green defence framework. *NATO*. https://natolibguides.info/ld.php?content_id=25285072 (accessed 1 May 2023).

NATO 2015a. NATO and its partners become smarter on energy. *NATO*. https://www.nato.int/cps/en/natohq/news_118657.htm (accessed 1 May 2023).

NATO 2015b. *Smart energy team (SENT) comprehensive report*. Brussels: NATO. https://www.nato.int/science/project-reports/Smart-Energy.pdf (accessed 1 May 2023).

NATO 2017. NATO-funded Serbian researchers develop biofuel from algae. *NATO*. https://www.nato.int/cps/en/natohq/news_146424.htm (accessed 1 May 2023).

NATO 2018. New NATO scientific project to reduce energy consumption of deployable camps. *NATO*. https://www.nato.int/cps/en/natohq/news_158964.htm (accessed 1 May 2023).

NATO 2021a. Brussels Summit communiqué. *NATO*. https://www.nato.int/cps/en/natohq/news_185000.htm (accessed 1 May 2023).

NATO 2021b. NATO climate change and security action plan. *NATO*. https://www.nato.int/cps/en/natohq/official_texts_185174.htm (accessed 1 May 2023).

NATO 2021c. NATO 2030: factsheet. *NATO*. https://www.nato.int/nato_static_fl2014/assets/pdf/2021/6/pdf/2106-factsheet-nato2030-en.pdf (accessed 1 May 2023).

NATO 2022a. NATO approves 2023 strategic direction for new innovation accelerator. *NATO*. https://www.nato.int/cps/en/natohq/news_210393.htm (accessed 1 May 2023).

NATO (2022x) High-Level Discussion on Climate Security with the NATO Secretary General Jens Stoltenberg at this year's United Nations Climate Change Conference (COP27), North Atlantic Treaty Organization. *(NATO)*. Available at: https://www.nato.int/cps/en/natohq/opinions_208773.htm (Accessed: 1 May 2023).

NATO 2022c. Compendium of best practices: factsheet. *NATO*. https://www.nato.int/nato_static_fl2014/assets/pdf/2022/7/pdf/0664-22_Climate_Change_Compendium_-_V3.pdf (accessed 1 May 2023).

NATO 2022d. The Secretary General's report: climate change and security impact assessment. *NATO*. https://www.nato.int/cps/en/natohq/news_197241.htm (accessed 1 May 2023).

NATO 2022e. NATO launches innovation fund. *NATO*. https://www.nato.int/cps/en/natohq/news_197494.htm (Accessed: 7 April 2023).

NATO 2022f. NATO sharpens technological edge with innovation initiatives. *NATO*. https://www.nato.int/cps/en/natohq/news_194587.htm (accessed 1 May 2023).

NATO 2022g. Pre-summit press conference. *NATO*. https://www.nato.int/cps/en/natohq/opinions_197080.htm?selectedLocale=en (accessed 1 May 2023).

NATO 2023a. The Secretary General's annual report 2022. *NATO*. https://www.nato.int/cps/en/natohq/opinions_212795.htm (accessed 1 May 2023).

NATO 2023b. Funding NATO. *NATO*. https://www.nato.int/cps/en/natohq/topics_67655.htm (accessed 1 May 2023).

NATO 2023c. NATO hosts symposium on climate change and military capabilities. *NATO*. https://www.nato.int/cps/en/natohq/news_211018.htm?selectedLocale=en (accessed 1 May 2023).

NATO YLG (Young Leaders Group) 2022. NATO 2030: embrace the change, guard the values. *NATO*. https://www.nato.int/nato2030/young-leaders/ (accessed 1 May 2023).

NSPA 2022. World Environmental Day 2022: NSPA clean energy endeavor. *NATO Support and Procurement Agency*. https://www.nspa.nato.int/news/2022/world-environment-day-2022-nspa-clean-energy-endeavor (accessed 1 May 2023).

NSPA 2020. NATO Support and Procurement Agency – Environmental Services. *Youtube*. https://www.youtube.com/watch?v=aim36g_POlQ&t=35s (accessed 1 May 2023).

Rühle, M. 2020. Scoping NATO's environmental security agenda. *NATO Defense College*. https://www.ndc.nato.int/news/news.php?icode=1426 (accessed 1 May 2023).

Rühle, M. 2021. NATO and the climate change challenge. *Internationale Politik Quarterly* [Preprint]. Available at: https://ip-quarterly.com/en/nato-and-climate-change-challenge (accessed 1 May 2023).

Rühle, M. 2023. Managing deep uncertainty in the energy and climate space: the importance of collaboration for geopolitics. *Westminster Energy Forum (WEF)*, London, 25 January. https://www.westminsterenergy.org (accessed 1 May 2023).

SHAPE-NATO (nd) NATO Assets. *Supreme Headquarters Allied Powers Europe (SHAPE)-NATO*. https://shape.nato.int/news-archive/2021/nato-assets (accessed 1 May 2023).

Shea, J. 2022. NATO and climate change: better late than never. *The German Marshall Fund of the United States (GMF)*. Available at: https://www.gmfus.org/news/nato-and-climate-change-better-late-never (accessed 1 May 2023).

Signorelli, G. 2021. Military aspects of energy security. *NATO Energy Security Centre of Excellence (ENSEC COE)*. https://enseccoe.org/data/public/uploads/2021/10/d1_military-aspects-of-energy-security.pdf (accessed 1 May 2023).

Sivaram, V. 2022. Jumpstarting demand for climate solutions: the first movers coalition and US national security, 13 January. *Youtube.* https://www.youtube.com/watch?v=dLjrHDuTIT0 (accessed 1 May 2023).

SOM nd. NATO headquarters. *Skidmore, Owings & Merrill.* https://www.som.com/projects/nato-headquarters/ (accessed 1 May 2023).

Stoltenberg, J. 2022. Opening Speech by NATO Secretary General Jens Stoltenberg. *High-Level Dialogue on Climate and Security, NATO Public Forum.* https://www.nato.int/cps/en/natohq/197168.htm?selectedLocale=en (accessed 1 May 2023).

UK MOD 2020. Sustainable fuels to power RAF jets. *GOV.UK.* https://www.gov.uk/government/news/sustainable-fuels-to-power-raf-jets (accessed 1 May 2023).

US DOD nd. Green Hornet. *US Department of Defense.* https://www.defense.gov/Multimedia/Photos/igphoto/2002002645/ (accessed 1 May 2023).

Van Schaik, L., D. van der Meer, A. Ramnath, and K. Kertysova 2022. *Decarbonized Defence: Clean Military Power in the Age of Climate Change.* International Military Council on Climate and Security (IMCCS). https://imccs.org/decarbonized-defense-the-need-for-clean-military-power-in-the-age-of-climate-change/ (accessed 1 May 2023).

Weir, D. 2022. NATO won't say how it will count its carbon emissions. *Conflict and Environment Observatory (CEOBS).* https://ceobs.org/nato-wont-say-how-it-will-count-its-carbon-emissions/ (accessed 1 May 2023).

Wigell, M. and F. Hakala 2022. Towards a greener defence: an introduction. (In) G. Iacovino and M. Wigell (eds) *Innovative Technologies and Renewed Policies for Achieving a Greener Defence.* Springer (NATO Science for Peace and Security Series C: Environmental Security), pp. 1–6.

# 9
# THE EVOLVING CLIMATE CHANGE THREAT

## UK defence preparations

*Richard Nugee and Timothy Clack*

## Introduction

Climate change threatens our way of life. This should not need to be noted – the climate science is clear. The global threat is here today and amplifies in the future, particularly if we do nothing. The effects of extreme weather, loss of biodiversity, and destruction of soils, combined with predicted population growth and scales of consumption, will change the face of the planet in a way that will make our current lifestyles and social arrangements difficult, if not impossible, to sustain. Living will cost far more. China, for example, predicts that it will cost hundreds of billions of US dollars if climate change continues unchecked (Maizland 2021).

But why is this a defence issue? Surely, defence is about fighting, deterring, and mitigating immediate threats and adversaries? Such a conventional approach to threats has them correspond to human activity, with an identifiable 'guiding hand' to oppose. Such a framework had utility in the past but is increasingly outdated. Indeed, climate has emerged as a national security issue; one that will become increasingly dominant as an underlying, and, occasionally, direct threat issue.[1] Governments look to provide security for their citizens. For example, the purpose of UK defence is 'to protect the people of the United Kingdom, prevent conflict, and be ready to fight our enemies' (UK Parliament 2022). This does not change because of the emergence of a new or novel threat. Instead, strategies, plans, and equipment must, as always, be adapted to the new threat (Clack and Nugee 2022).

The distinction with climate change is that this also includes adapting to a harsher, climate-impacted environment – a new defence operating environment. In the United Kingdom, climate change also demands a reassessment

DOI: 10.4324/9781003377641-12

of the fundamentals which have been taken for granted for at least 100 years, and take new unknowns into account which have, potentially, existential consequences. If nothing is done, at the very least the freedom of manoeuvre, strategically to tactically, will be steadily eroded and diminished. So, to remain at the forefront of operational capability, it is imperative that defence understands the future and adapts to it.

Climate change poses a number of threats to UK defence. This chapter will attempt to explain some of these threats and how defence can prepare and overcome them.

## National security

The Integrated Review of 2021 described climate change as the UK government's number one international priority and declared climate change a threat to humanity (UK Government 2021). Indeed, the then UK Prime Minister stated at the UN Security Council on 23 February 2021 that 'it is absolutely clear that climate change is a threat to our collective security and the security of our nations' (PM Office 2021). His words were echoed by the former US President, Barack Obama who, on 8 November 2021 at COP26 in Glasgow, stated that 'climate change poses a national security threat for the United States and for everyone else' (Obama 2021). This assessment was reinforced, in the United Kingdom, by a Joint Intelligence Committee report that stated climate change was a direct threat to the UK's national security and economic resilience. Moreover, when the World Economic Forum declared the top ten Global Risks for 2021–22, these included geo-economic confrontation, social cohesion loss, and livelihood crises, partly as a result of climate change, along with extreme weather and natural resource crises (WEF 2022).

In order to be designated a national security risk, climate change has to manifest itself in ways that will damage the citizens of the United Kingdom, either as a threat to life or to our livelihood, i.e. the way we live. This is happening and increasing. Here it must be recognised that climate change is not a singular threat and there is no single measure that will preserve our security. Indeed, the impacts of climate change will – and in some respects already do – manifest in ways that affect the whole society from the economic and social to political and legal. Climate change has been described for some time as a 'threat multiplier', which implies that rather than being a cause of insecurity in its own right, it contributes to, and exacerbates, existing threats – to a point that simmering tensions might boil over into conflict. But it is more than that, as it is also a shaping threat, determining not only the strength of the threat (via a multiplier effect) but also driving the way the threat manifests itself in different environments and conditions. These differences will shape the influence and ferocity of the impact of climate change. The implication is that in order to deliver resilience different responses will be needed in different contexts. Defence will have to be supremely adaptable to cope with threats that are driven by a scale of unpredictability, such that even the most sophisticated modelling capability struggles to assess it accurately. Moreover, while the most severe effects of climate change may not be felt in the United Kingdom and northern Europe (IPCC 2021), they will be felt acutely by respective overseas territories, allies, and trade partners. As such, they will impact the United Kingdom significantly.

## Threat multiplication

### *Infrastructure and costs*

Climate change, and the international response to it, will manifest itself in many different ways. Perhaps the most obvious of these are the direct threats from a global weather system which is impacted by the warming of the seas particularly (Met Office 2022; see also Otto 2020). The result, as we have seen unmistakably in the last couple of years, is extreme weather events; typified perhaps by the extraordinary fires in Australia, California and the Mediterranean (Boyle 2020). Another example is the heat dome in northwest United States and Canada, described as a once in a 1000-year event, made 150 times more likely by climate change (Thompson et al. 2022). Then there were the destructive floods in Germany, with downpours 20% heavier than normal which caused over 200 deaths (Else 2021). In China's Henan Province, an estimated year's rainfall fell in two days in December 2021, with over 200 mm of rain falling in one hour, causing over 300 deaths, requiring the evacuation of 815,000 people, and otherwise affecting 14.5 m people around the province (Zhang et al. 2021). In May 2023, the Shabelle River in central Somalia broke its banks and submerged the town of Beledweyne, washed away crops and livestock, and displaced a quarter of a million people, even as regionally the Horn of Africa faces its fourth consecutive year of drought, due to seasonal rains in Somalia and upstream Ethiopia (The Guardian 2023).

The IPCC (2021) has made it clear that climate change is already affecting every region on Earth and that the resultant unpredictable meteorological conditions will continue. These, in turn, will place ever-increasing demands on finite public funds, with inevitable consequences for the ability of states to respond to other priorities or fund other capabilities. In certain parts of the world, these storms and floods are powerful enough to lead to a breakdown in law and order, which can require the extensive stabilisation efforts of security forces. The financial cost of such destruction is significant. Estimated costs of at least RMB2.7bn (US$393.1mn) were reportedly faced by China linked to the Henan Province flood (WMO 2022). Moreover, the United

States has assessed that the cost of extreme weather events between 2014 and 2018 was about US$400bn. But there is also a penalty in terms of the clearing up, much of which is done by militaries around the world. According to almost all relevant studies, the need for Humanitarian Assistance and Disaster Relief (HADR) is likely to grow, both at home and abroad (see Evans and Lewis, this volume). Where once the United Kingdom would send a naval platform to the Caribbean to support victims of the hurricane season perhaps once every five years, now it is an annual deployment. Each such naval commitment is also longer in duration, which correlates with evidence that the hurricane season itself may be lasting longer (Gibbens 2020). Defence will have to take this into account in its planning and procurement. (In this context it can be noted that there is also genuine opportunity, e.g. for engagement and diplomacy, collaborative training and operating, and shared equipment development.)

Military bases, airfields, ports, and training areas are also susceptible to powerful climactic events. Consequently, defence must think hard about the resilience of its own fixed assets, the effects of a rising sea level on their naval bases, and impact of hotter, drier weather on their training areas across the world. In the United Kingdom, for example, draining some of the training areas to make it easier to train may no longer be sensible; it makes them more likely to catch fire in the heat of the summer and the water runoff adds to already full rivers, potentially exacerbating local flooding. In 2020 and again in 2022 training time was lost – and military effort redeployed – due to fires (caused by training activity) on the British Army's principal training area of Salisbury Plain (BBC 2022). Moreover, having had fast jets severely damaged by intense hail storms and bases by flooding, the Pentagon has recently devoted significant effort to looking at the flood, hurricane, and fire risks posed at all their bases (Roblin 2020).

### Scarcity and conflict

The other direct threat is from the environmental degradation and permanent resource scarcity that results. Decline of forests, reduced biodiversity, expansion of deserts, rising sea level and increased oceanic acidification, and expansion of disease zones and widespread drought – all of which we already see evidence of – will have a profound effect. One of the key areas will be a reduction of food production from the traditional 'grain baskets' of the world, such as Canada and South America. The global trade in staple foods is changing and becoming increasingly vulnerable, which is creating higher food prices (see Benton et al. this volume), and may further result in changes to trade relations across the world. As a nation which imports the majority of its food, these global trends hold the potential to impact significantly the UK economy and way of life.

There is evidence from the 17th century of climate shifts causing changes in food production, which, in turn, precipitated conflict. Historians have linked the 'General Crisis' of that century to the coinciding 'Little Ice Age', which reduced temperatures by 1°C on average after a long period of restricted agricultural growth, with far-reaching consequences for the stability of parts of the world (Hessayon and Taylor 2022; Lanchester 2019). Between 1618 and 1707, as food supplies reduced and vulnerability to disease increased, there was an uptick of wars around the world. Conflict proliferated in Europe, e.g. the Thirty Years War and English Civil War, in China, e.g. the collapse of the Ming Dynasty, and in India, e.g. the Mughal-Maratha Wars. Although a direct causal link between climate and war during this period remains unclear, the evidence is considerable that climate configured the context, exacerbated threats, and – in concert with other factors – amplified the propensity for violence.

Inevitably, as resources become scarcer, they are hoarded by those who have control or access to them. An obvious example here is dams, which not only store water but hold potential for hydro-electric power generation. Despite dams offering prospects for green energy production, the effect of restricting water flows could – and often does – have dramatic downstream consequences. From the tensions linked to the Grand Ethiopian Renaissance Dam and Kalabagh Dam in Pakistan to those involving small holdings and pastoral rangelands in Kenya, the potential for increasing conflict by those suffering restrictions of water downstream is significant (Heggy et al. 2021; Mustafa et al. 2017; Agade et al. 2022). In various parts of the world, e.g. Marsabit and Mandera in Kenya, control over access to wells has already led to a reported uptick of incidents of local violence. Such incidents will only become more likely as water becomes scarcer as a result of climate change.

### Insecurity drivers

Climate change reinforces insecurity. In Autumn 2021, *CBS News* and *The New York Times* published assessments that climate change had strengthened the position of the Taliban in Afghanistan during the Western campaign there (Korte 2021; Sengupta 2021). The reports asserted that conflict had been amplified and agriculture undermined as a result of repeated droughts and the rapid melting of winter snow. With agriculture failing, farmers were often left with two choices – both undermining the ISAF's counterinsurgency campaign – to grow poppies for the opium trade or seek an income by working directly for the Taliban (US$5–10 per day). This pattern of climate-induced local jeopardy being exploited by malign actors is evident in other areas as well. Andrew Harper, the Special Advisor to the High Commissioner for Climate Action at the UN Refugee Agency has suggested that Boko Haram, for example, gained footholds in the Lake Chad Basin after droughts in 2017, and Islamic State (ISIL) took advantage of the extreme droughts in Iraq and Syria to consolidate their support base.

After living in the Sahel for six months, Shwartzein (2021) identified four critical areas that escalate insecurity. Firstly, he observed that climate change became a 'stoker of resentment', as some communities were worse affected than others, which exaggerated the difference between the 'haves' and the 'have-nots'. It has been recognised elsewhere, of course, that assorted structural inequalities are driven by climate change in multiple contexts (Burzynski et al. 2022; Deivanayagam et al. 2022; Islam and Winkel 2017). Secondly, Shwartzein (2021) also identified that the climate change impacts, such as droughts, failing harvests, and economic impoverishment, created conditions where radical movements, such as ISIL in Iraq or Boko Haram in Sahelian West Africa, could influence the poor and desperate. These violent and extremist organisations (VEOs) and others have proven themselves experts at exploiting weaknesses in the local government capacity and turning their followers and sympathisers against the authorities who are seen as incompetent, illegitimate, corrupt, and/or otherwise exacerbating negative conditions (see also Nett and Ruttinger 2016; UNSC 2021). Christophe Hodder, the UN climate security and environmental advisor to Somalia, and others have described eloquently how drought led to Somali farmers moving to cities, where they were recruited by al-Shabaab (C. Hodder, pers. comm.). When aid was able to get through to local communities and irrigation and water management solutions improved conditions, recruitment proved more challenging. In response and demonstrating that they understood the relationship, al-Shabab began to attack irrigation systems (Brown et al. 2021; Marshall 2021).

Thirdly, Shwartzein (2021) identified that the dislocation caused by the scale of those leaving rural environments for urban centres shattered local cohesion. It also robbed village communities of the critical mass required for viable micro-businesses and communal agricultural practices. This, fourthly, was made worse by the fact that climate change – the weather – was beyond people's control. This powerlessness – a 'despondency effect' – was compounded by the community, family, and tribal structures becoming less effective and at times breaking down. This, in turn, resulted in further displacement.

It is reported that, in 2022, 23.7 m people were internally displaced because of climate and disasters and 14.4 m through conflict and violence (IDMC 2023). As much conflict is shaped and exacerbated by climate, the scale of the impact of climate change on migration is clear. It is important to be sympathetic to those exposed to forms of climate insecurity, particularly vulnerable groups, such as IDPs and refugees. It would be remiss, however, not to point out that large-scale displacement has, historically, often resulted in the exploitation of those on the move and had a perceived or real destabilising political effect in receiving theatres (see Chirodea 2018; Karamanidou 2015; Norwood 2020; Lewis and Dwyer 2014). The dangers of disposed and desperate peoples becoming radicalised prior to departure or whilst *en route* are also apparent. Wherever climate migrants go at scale, they hold the potential to exacerbate pre-existing local tensions. In one sense this is obvious, they are going to places where people already live.

Migration is an adaptation strategy. Throughout history, it has been the means to escape conflict, persecution, and exploitation and has been precipitated by a multitude of factors. There is often a perception of impermanency to the situation. In the case of war, for example, migrants and receiving nations often presuppose that return will be possible after the conflict abates. Climate change is different because, if the place of origin becomes uninhabitable, the possibility of return is remote.

Narratives indicating that migration is one of the main threats to European security are increasingly apparent in political and public discourse. There is a logic to these: as the equatorial belt heats up to levels where it is physiologically impossible to live for long periods, in line with current modelling, the numbers will grow of those moving to areas where temperatures will be less extreme and subsistence viable. Globally, the numbers displaced could be vast. Indeed, the UN Secretary General Antonio Guterres and the Institute for Economics and Peace have reported, respectively, that global migration due to the climate crisis could hit 1bn and 1.2bn by 2050 (Farhoud 2022; Henley 2020). Moreover, with forecasted figures of displacement varying from 25m to 250m in Africa alone, it is conceivable – and understandable – that a proportion of transborder migrants would consider the UK a sensible destination. The scale of such potential migration holds both opportunity and challenge for receiving nations. However, it is important to note that destabilisation is not inevitable. Since the 1960s due to planning and management, the city of Nouakchott, Mauritania, for example, has grown from 9,000 to 1.3 m without significant tensions.

### Balancing the role of defence

The defence and the security sectors cannot – nor should they – intervene everywhere to solve these potential areas of conflict. Nonetheless, they have the potential to act to support the building of local resilience and provide local and specific expertise. They can also support the activities of other government departments and NGOs on the ground, including through the collection and assessment of information, deployment of equipment and human power, and in the provision of security envelopes as and when required. Moreover, as the then UK Defence Secretary described at COP26, one of defence's roles must be to prevent conflict – it is part of its very purpose – and that includes conflict exacerbated or triggered by climate change.

Here it is worth acknowledging the very real fears that some countries and communities have about any defence organisation being involved in climate change response. In many parts of the world, militaries are considered malign influences who, the people fear, will simply use climate change as yet another excuse to dominate the environment for their own benefit or on behalf of powerful elites. In 2018, a survey in Burkina Faso and Mali reported that 75% of villages considered national defence and security forces to be a threat to peace and security. For these villages, protection from local threats and the predations of defence and security actors came from elsewhere. Indeed, 62% were in favour of local armed self-defence militias, whilst 50% saw VEOs as protection from the state security forces (see Venturi and Toure 2020). There is also much literature on internal and external perceptions of intervention in recent decades linked to various theatres, such as Iraq, Afghanistan, and Somalia (e.g. Hesse 2015; Karlborg 2014; Pitchford 2011; Stollenwerk 2018). The point is that overt defence activities in climate-destabilised theatres may be unwelcome, even if they are offered with the best of intentions. Thus, any intervention to prevent conflict must be delivered both carefully and sensitively in order to ensure that the outcome is genuine resilience and support. Moreover, any defence force engaged in such missions must be clear about its mission and operate at the highest of professional standards.

### The technology threat (and opportunity)

Climate change is another cause of conflict albeit one that has the potential for extreme impact. It exacerbates existing threats, frictions, and fault lines. But there is another area of strategic geopolitical impact that needs to be highlighted and carries with it the potential for significant re-alignment and destabilisation: the transition away from fossil fuels to green energy solutions.

There have been energy transitions in the past, but one – the move to oil over a century ago – had particularly massive geopolitical effects for the world and shaped military history. The effect of converting the Royal Navy from Welsh coal to oil on the eve of World War I, for example, conferred advantage (Brown 2003). The need for oil before and during World War II impacted all sides and shaped events. The Japanese need for oil added impetus to their imperial designs. Moreover, Hitler's push for the Caucasus and the determination of the British to maintain the security of the Suez Canal were also driven by oil considerations. Interest in the Middle East over the last 50 years also shows the clear importance of oil in Western geo-strategic thinking (Colgan 2013). In both war and peace, access to oil is a strategic consideration for all states that depend on it.

The current transition away from oil will be no less impactful but has two unique properties that make it much more challenging than that which has gone before. The first is that the requirement is for an energy replacement.

In the past, new energy sources were added to a pre-existing mix. Thus, the intent is to replace global reliance on fossil fuels very significantly, perhaps ultimately to zero. In contrast, coal was still used in the transition to oil. The second is the timeframe. The world has set itself a target of net zero emissions by 2050 – fewer than 30 years from now. By comparison, it took oil 50 years to establish market penetration and decades longer for market dominance. The scale and pace of the required transition are therefore exceptional. Both these unique factors are very likely to increase the instability that will be caused by the transition away from fossil fuels.

As the world pivots to renewable energy, there will be significant winners and losers. The International Energy Agency estimates that states whose GDP and/or exports are dominated by fossil fuels will have to adapt to US$7tn loss of income by 2040 (Walker 2021). Some states are already adapting and are reasonably well prepared, diversifying, and creating new jobs in the renewable industry and elsewhere. Some have yet to adapt but are working towards finding solutions. These adapting states still face transition risks despite preparations. But the states likely to suffer the most are those that are already fragile and most reliant on fossil fuels or fail to see the need to adapt. The lesson from history is that states experiencing rapid economic decline – and the conditions that go with it, including (hyper)inflation, drop in standards of living, and overwhelmed services – often also experience internal unrest, domestic instability, political and sectional violence, and even conflict with neighbouring states (Collier and Hoeffler 1998; Miguel et al. 2004; Rodrik 1999).

The move to renewables is also uneven. The materials to power the renewable energy industry appear, at the moment, to be concentrated in certain parts of the world – rare earths in China, cobalt in the Democratic Republic of Congo, and lithium in Chile. The control of such resources holds the potential to create new power bases and opportunities. In the 1970s, the Organisation of Petroleum Exporting Countries (OPEC), for example, was able to alter the economies of countries far outside their own, as the power invested in the source of energy had a strategic geopolitical impact. The upshot is that the reliance on fossil fuel-producing countries – and the alliances and relationships that have been forged to ensure continuity of supply – may need to be replaced by a similar reliance on other countries, particularly China, in order to ensure the supply of raw materials for renewable energy generation. As the world has discovered increasingly, the extraction of minerals for renewables is far from straightforward. The cost of a tonne of lithium, for example, went from US$9,600 to over US$50,000 in the 12 months from January 2021. This is as much to do with the environmental damage of extraction and local resistance as it is demand and the concentration of deposits. In Chile, for example, protestors have curtailed output and caused some mine closures (Sherwood 2019).

The effects of the pivot to renewables and quest for reductions in fossil fuel dependence are driving research and development beyond existing energy solutions. Importantly, many in industry are already changing their research and development to embrace the opportunities and respond to challenges. This has significant consequences for defence. The ADS Group, the trade body for the UK defence industry, wrote to the UK Chancellor and Defence Secretary in December 2021 calling on national governments to follow a number of principles to support sustainability within defence. In an attempt to ensure that they would not be penalised by the markets for appearing to ignore sustainability and wider ESG considerations when dealing with defence, one stated explicitly, 'Any possible exclusion of defence activities from regulations on sustainability should mirror only the exclusions set out in international treaties signed, ratified, and adopted by the UK and European national governments'. Another principle noted that 'There is no sustainability without security, no security without defence capabilities, and no defence capabilities without defence industries'. Through such communications, industry has indicated strongly that they recognise they are going to have to become more sustainable to survive, with the implication that the speed at which society and the banks are demanding changes to technology, particularly renewable energy technology, has the potential to leave defence behind.

This is a complex situation for UK defence. There is no single solution to the issue of removing fossil fuel propulsion systems for air, land, or maritime equipment. Thus, while alternatives are developed, defence will have to choose what type of energy is optimal for each type of equipment. There are, of course a few elements where defence is unique and these should be researched using the defence R&D budget. Outside these, however, defence should act as a 'fast follower' or 'early adopter' of the settled view of industries and sectors for whom this challenge is existential. The maritime and air industries, for example, must find an optimal energy source that is financially viable and which delivers the same capability – even if it involves different supply chains. Over the coming years, there will be significant debate, experimentation, trials, and failures in the pursuit of the optimal energy for the requirement, including hybrids, ammonia, hydrogen, synthetic fuels, or electric drives and storage. If it can avoid doing so, defence should not decide prematurely or unilaterally rather it should wait for the clear determination of other industries. Only that way will it be possible to avoid the potential for stranded assets in the future and ensure retention of interoperability with allies and overseas nations. There will also remain a role for fossil fuels even in 2050. As such, defence should not look to retrofit existing propulsion systems, e.g. aircraft carrier engines, as that would be prohibitively expensive. Instead, it should concentrate on the design of new equipment and alternative fuels as they become viable and take advantage of proven new technologies at the first opportunity.

There are significant tactical and operational advantages with a move to renewable energy beyond the obvious compensations of reduced emissions. Small electric aircraft, ideal for short passenger journeys, are already a reality, and this sort of technology is recognised by the Royal Air Force (RAF) as being ideal for training pilots and air cadets. Electricity for such aircraft can derive from renewable sources or perhaps generators powered by sustainable aviation fuel. As the technology develops further, there will be improvements in speed – a recent successful breaking of the electric air speed record was achieved by the United Kingdom (see Rolls-Royce 2022) – and the endurance of batteries, allowing greater range and/or more passengers. This technology, translated into uncrewed air systems would offer greater endurance on task. Moreover, with the RAF looking to achieve 80% of their training synthetically in the future, the opportunities for greater tactical experimentation out of sight of adversaries, while at the same time saving money (and emissions) on fuel, maintenance, and wear and tear, is genuinely capability enhancing. Furthermore, sustainable aviation fuel, although not yet viable economically, is on the horizon and is likely to deliver the same capability whilst reducing emissions (OEERE 2022).

The British Army and industry partners have already co-delivered a hybrid Jackal and are developing hybrid systems for both the Foxhound and logistic vehicles, such as MAN SV (British Army 2021). The hybrid Jackal has the ability to run off-road on battery power for in excess of two hours before re-charging via the use of its onboard engine. While the tactical advantages of a silent, emission-free vehicle with reduced thermal profile are obvious in terms of stealth, the demonstration programme is showing other significant benefits. The weight of the battery and careful placing in the design, for example, creates a lower centre of gravity and a more balanced vehicle. Moreover, with electric drives on each wheel, the vehicle offers more control and manoeuvrability. The vehicle also demonstrates better off-road performance, including more effective slope ascent and descent, and – particularly useful in urban environments – the ability to rotate 360 degrees on a single point. Fuel consumption is also reduced, by as much as 20% in a battlefield scenario, and more in an urban setting. The implications for the crew are also positive. With no noise in battery mode, there is increased situational awareness; and with no vibration and noise from an electric motor, there is less crew fatigue. Importantly, with electric drives, there is far less driver training required – it is described as intuitive, straightforward, and 'like driving a golf buggy' – giving both confidence to the driver and making it less likely that the vehicle will get into difficulty. Cumulatively, the effect is greater operability in terms of range, duration without resupply, and 'fightability'.

In some of the buildings that are being developed for the UK Defence training estate, such as at Nesscliff Training Camp in Shropshire (MOD 2021), developers have been required to look at every material they have used and, in order to reduce embedded carbon, have changed designs and materials. The net effect has been to reduce cost and carbon, both in the construction and operation. The financial savings have been recycled into the buildings – to make them carbon net negative – and into the wider training estate. In the case of Nesscliff the savings were used to electrify the infantry ranges; creating an enhanced environment and improving operational capability.

The lesson from these threats and opportunities is that defence and industry must work in partnership on these issues to improve operational effectiveness, build resilience, enhance attractiveness to recruits, and reduce emissions. By taking the lead on some of these technologies and using them as arenas for experimentation, not only will defence and military actors become significantly better at setting requirements in the future, but they will also de-risk future technologies and innovate capabilities. In the case of the UK and in line with the new Defence Industrial Strategy (DIS), momentum should be exploited for innovation around (some of) the new technologies available.

### A threat to freedom of manoeuvre

Freedom of manoeuvre is based on the ability to understand the environment, at home and abroad, and act accordingly. It is defence's duty, of course, to provide the best and most militarily capable armed forces possible and to be prepared to deploy wherever they are sent in the world. Due to climate change, the world is changing. Indeed, what might have been described previously as 'once in a 100-year event' has become almost the routine. The 'new normal' is a climate-changed world. As various chapters in this volume describe, heating is affecting the ice caps, temperature of the sea, and ambient heat. All of these have an impact on military capability, and the freedom of manoeuvre to operate around the globe.

Militaries are obligated to be able to operate in a world where the environmental envelopes built into equipment design requirements are already being pushed to their limits. Climate change will worsen this situation. Indeed, rising surface sea temperatures will force either operational compromises or new design criteria to be adopted. Over the next 10–15 years, it is assessed that in the Gulf – where the United Kingdom has only recently opened a new military base East of Suez – the summer surface sea temperatures could reach 36°C frequently, and 38–40°C occasionally. Such temperatures will forcibly slow ships as the sea – currently used to cool the engines – would act as a thermal blanket and push engines beyond their running range.

Mostly caused by the melting of ice sheets and glaciers, the rate of mean sea level rise for the period 2006–15 was exceptional over the last century (NOAA 2022). If the Western Antarctic ice sheet melts, which models indicate will happen if temperatures increase by 2°C (Hulbe 2020), then mean sea level could rise by as much as 3.3 m. Similarly, the Greenland ice sheets may start to melt far more rapidly if the temperature rise reaches 1.5–2°C. Over the long term, there is thus the potential for cumulative sea level rise of over 7 m. Changes such as this will take time but will have very momentous effects on the understanding of littoral strikes and the abilities of navies to operate. Here we should highlight recent studies that have shown ice sheets retreating much faster than the upper limits of models (Batchelor et al. 2023; Graham et al. 2022).

Climate change is affecting the Arctic nearly four times faster than elsewhere, with the risk that large parts of the Arctic ice cap melt in 15–20 years (Rantanen et al. 2022). One of the results would be the 'opening up' of Arctic trade routes. This has not been missed by states who are looking to exert control and influence. In March 2019, Russia proposed that all foreign warships navigating the Northern Sea Route (i.e. in international waters) were to be restricted. This blatant Russian challenge to the freedom of the seas was signed into law in December 2022 (Overfield 2022). Similarly, in January 2018, China declared itself a 'near-Arctic' state, so that it was better positioned to configure the 'polar silk route'. If freedom of manoeuvre is to be maintained alongside competing nations, navies need to be able to operate within the same environment. Forewarning the competition to come, the Russians, Chinese, and Canadians are all preparing their fleet to be able to deal with the disruptive ice state between open water and ice pack for which icebreakers are required. Action includes the hardening of ship's hulls. There have been calls for other states with interests in the Arctic, such as the United Kingdom, to do the same. Russia is also refurbishing many of their Arctic bases and has increased their regional military activity, with surface ships, icebreakers (of which they now have at least 40), submarines, and ground exercises (Clack and Nugee 2022). In response, NATO has reacted with increased preparedness. This has seen the US 2nd Fleet being given responsibility for the Arctic and the staging of exercises, such as Exercise Cold Response, in 2022, which involved 35,000 soldiers from 28 nations and two aircraft carriers from the United States and United Kingdom (NATO 2022). Militaries operating more in the Arctic also need to be adapted to the increased exposure in that region to the phenomenon of space weather, which affects navigation, timing, radar, and communications systems.

The effects on land and in the air will also be significant. As a warm year in the 2000s will broadly become a typical year in the 2040s, training days will be lost as temperatures exceed thresholds making it 'too hot to train'. Indeed, a recent study by the UK Met Office identified that in temperate climates the number of training days lost due to temperature is likely to rise by between 75% and 150% by 2040 (Sheridan 2021). In Cyprus, where the UK have a tri-service military headquarters, the projection is that all training will be lost in

August to heat. In the air, higher temperatures result in reduced air density, which results in less effective aerodynamic lift. The effect on take-off runs, in terms of reduced payloads and fuel capacity, and reduced climb rates means that increased runway length and increased engine thrust may be necessary (Coffel and Horton 2017). In the absence of such adjustments, operational reach may be diminished.

Climate change is also likely to affect the transmission of infectious diseases which, in turn, will potentially hamper deployments, in particular HADR operations. For example, the growing spread and geographic reach of diseases transmitted by arthropod vectors, such as malaria, Zika, and dengue fever, could create challenges for force survivability, increasing the need for individual medical assistance, vaccinations, and personal protective equipment. Medical preparation for deployments to areas with a risk of exposure to mosquito-borne diseases could also become more costly. Similarly, higher rainfall could increase the dissemination of infectious agents in water sources – raising the risk of personnel exposure to waterborne diseases – whilst higher temperatures could increase the growth and survival of infectious agents. Climate change can impact air quality and is likely to lead to an increase in concentrations of surface ozone, an urban air pollutant responsible for human respiratory problems, which can also damage crops (US EPA 2022).

Many aspects of climate change could have a potential impact on intelligence, surveillance, and reconnaissance (ISR) activities and related sensors. Radio communications could be affected by extreme weather events and increased rainfall. Moreover, the presence of clutter in a radar sweep can reduce the resolution of data acquired and lead to misidentification. Optical sensors are also susceptible to a number of climate effects. For example, passive infrared (IR) sensors are unreliable in harsh weather conditions; IR, ultrasonic, and laser are susceptible to moisture, dirt, and temperature; and millimetre electromagnetic radio waves – also known as extremely high frequency (EHF) band – are degraded by atmospheric conditions such as snow, cloud, dust, smoke, and fog. The increased atmospheric turbulence caused by climate change also imposes a fundamental limit to electro-optical/infrared (EO/IR) capability, particularly at the long-range required for standoff ISR. With implications for underwater acoustic sensors, melting ice is changing acoustic propagation in the world's oceans, and changes in ocean chemistry and glacial melt are altering the surface layer and thermohaline circulation of the oceans (Wunsch 2002). The loss of sensor capability may restrict freedom of manoeuvre.

It is also worth noting that the ever-greater reliance on electricity and connectivity to optimise renewable energy usage – a critical part of moderating demand and reducing surges in power – will lead, inevitably, to greater vulnerability to cyberattack. The Critical National Infrastructure (CNI) of various states, the United Kingdom included, could be particularly vulnerable to targeting as nations become increasingly interconnected.

If freedom of manoeuvre is to be maintained, militaries must adapt to a changing, more challenging world. In short, the environmental norms the UK military has adapted to operate under are changing, and further adaptation is required to maintain operability.

## A threat to the licence to operate

Defence does not operate in isolation. The political, social, and legal contexts are always centrally important to and shape outcomes. The same is true for military activities linked to climate change. Governments around the world, for example, are pledging increasingly to reach net zero greenhouse gas (GHG) emissions in the short to medium term. Indeed, via the Climate Change Act, the United Kingdom has enshrined in law its commitment to do so by 2050 (GOV.UK 2019). The UK government will thus be breaking the law if the target is not realised. This has a number of implications for the United Kingdom and its military. As the MOD is responsible for 50% of central government emissions, it will have to act to reduce its emissions to enable government to reach its target. The government is increasingly legislating for emission-reduced end states in many areas from vehicles and heating to land use and biodiversity preservation. Defence is not – and will not be – exempt from much of this legislation and so will have to act. Already, through the requirements of the Committee on Climate Change and the new Environment Act, the MOD has external requirements that must be met (GOV.UK 2021).

Reducing emissions will also confer advantages for the UK military in terms of being able to operate globally. It is conceivable that many host nations may become more demanding on the emission profiles of any support and intervention. States may limit the ability to deploy if it is recognised that by so doing, militaries would be adding further stress to the environment, i.e. the licence to operate may be restricted. After the Hunga Tonga-Hunga Ha'apai volcanic eruption (Proud et al. 2022) and despite the need, for example, the local Tongan government laid very strict restrictions (due to COVID-19) on the aid that was brought from overseas. It is also possible that certain states will refuse the opportunity for the UK military to train on or near their territories if they are not considered environmentally responsible and sustainable.

At the same time, the single services need to ensure that they remain attractive to quality recruits. With climate change recognised as a public concern by more than 85% of the UK population – and the majority of those aged 18–24 disagreeing with the statement that 'there is too much concern for the environment' – a failure to act (and to be seen to be acting) will ensure militaries are increasingly out of touch with the population base it protects and from which it recruits (GOV.UK 2022; see also Milfont et al. 2021).[2]

While this may not be of great concern today for certain parts of the defence force, such as the combat arms, this will become more relevant as public concern for the effects of climate change escalates. Additionally, the cadet movement in the United Kingdom, which acts as a strong interface between the public and the military – and is the source of a number of recruits – is likely to be increasingly shaped by the climate change mitigation response.

Despite the fact the ruling was later overturned, there is potentially still a significant legal impact from the UK Court of Appeal's decision not to allow the third runway at London Heathrow on the grounds that the government had not considered its climate commitments (Carrington 2020a, 2020b). The implication is that the MOD should consider its net zero commitment as part of its decision-making processes. Procurement decisions, for example, might reasonably be expected to balance carbon emissions against costs and technology requirements against procurement rules. If it does not, it might face legal action based on measurable and attributable environmental impacts.

## Opportunities and prospects

From the perspectives of defence and security, climate change is not exclusively a threat. There will be some opportunities for certain states in certain contexts. An enhanced understanding of climate change on the different domains of warfighting and the development of technologies to facilitate adaptation to changed environments and mitigation of risks can drive the emergence of a more capable force. Whether directly (e.g. adoption of new technologies) or indirectly (e.g. opportunities to become more self-reliant and resilient), financial and personnel savings can then be recycled to further enhance operational capability.

As described above, there are opportunities for equipment and training to be optimised to enhance operational capabilities. There is, however, a greater prize to be won by thinking differently about sustainability and resilience. The potential for recycling water and reducing waste and adoption of 3D-printing techniques for replacement replacement parts reduces logistical resupply requirements and adds to combat effectiveness. Sustainability ensures that combat troops, including infantry, armour, and aviation, that have hitherto been required to protect logistic chains in contested environments could be redeployed to the front line.

There are significant advantages to militaries becoming increasingly self-sufficient in operations, e.g. through the generation of power via sustainable and portable equipment and the harness of renewable energy. Energy efficiency can, of course, serve as a force multiplier by increasing the range and endurance of forces in the field and reducing combat forces diverted to protect energy supply lines. (These are crucial and vulnerable to both asymmetric and conventional attacks and disruptions.) The cost of fuel supply in terms of both lives and funding would be reduced substantially through enhanced levels of self-sufficiency.[3] Hybrid generators, connected to a microgrid, and solar panels, for example, have shown a 70% reduction in fuel consumption with the energy optimised for demand rather than supply.

There is also considerable potential advantage linked to the operational use of small and micro nuclear reactors (Andres and Breetz 2011). Such technologies are air portable and can be deployed to remote localities to generate reliable power. They can also be installed rapidly and, due to being fuelled and defuelled in factory settings and passive safety measures, do not require a large team of specialised operators. Whilst the risk of attack might make the deployment of these reactor technologies to warzones inappropriate, remote but stable environments, such as Cyprus, Diego Garcia, and the Falkland Islands would seem viable. In other localities, conditions would be key in any determination. During the Afghanistan campaign, with ISTAR and other protections in place, a small modular reactor could have been used to power Camp Bastion, for example. There are, however, factors limiting the technology and its deployment.[4]

The creation of deployed operational bases that are more self-reliant will reduce the impact on local communities (Gordon 2010; Ucko 2013; see also Dardia et al. 1996) and further reduce supply chain vulnerabilities. This would also reduce the environmental impact of a deployment. This is important as it is possible, even likely, that UK defence, for example, will be required to deploy to areas most vulnerable to climate change-related stress. The economic and resilience benefits of becoming more efficient on the base estate, through adoption of renewable and self-generating power (e.g. solar, wind, tidal, nuclear micro-reactors, and deep geothermal) or through efficiencies associated with building monitoring systems or adoption of LED lights, for example, will give a return on investment that will rapidly ensure that more money can be spent elsewhere on defence equipment and personnel. Such developments will also insure against the potential overload of the national grid in the future as it moves exclusively to renewables (National Grid 2023).

Ensuring that sustainability and emission mitigation is made central to the design of new military equipment at every stage of the development, will facilitate innovation. Indeed, throughout history, militaries have been responsible for significant technological developments. These, in turn, have catalysed wider technological and societal transformations. For example, the military invented or at least oversaw the rapid development of drones, the Internet, GPS, digital photography, jet engines, caterpillar track, blood banks and transfusion techniques, autoinjector ('EpiPens'), feeding cylinder, and spotlight (Clack and Selisny 2021; Virilio 2005; 2009). In today's climate change context, the reduction of weight and size through, for example, transitioning to remote controlled or even semi-autonomous equipment, confers greater opportunities for renewable energy solutions to be found. These will both reduce cost and logistic drag and, in so doing, enhance capability. More widely, the adoption of a concept of a circular economy for all military equipment and disposables, where possible, should offer opportunities for revenue generation. In many cases, recycling and selling on equipment will both lengthen their use and reduce waste; recycling materials will also generate finance and possibilities for novel use. The methane stored in grass cut on military establishments, for example, can be converted into a fuel to drive the mowers that cut the grass in the first place.

Taking a lead – and being seen to take a lead – on climate action confers other advantages. In a vicious double effect, communities in conflict zones contribute to the effects of climate change both by significantly increasing their emissions and by being distracted away from trying to mitigate or find solutions to the climate crisis. Through reducing the incidence of conflict, deploying to build resilience and support the efforts of other militaries to reduce their own emissions, and delivering capacity and resilience building for states most vulnerable and affected by climate change provides prospects to reduce threats, increase partnerships and alliances, and build skills for military and adjacent personnel. Such activities can also promote good governance and economic opportunity in host nations. From the home nation perspective, the approach can deliver security of upstream interests, enhance political influence, and support trade and commercial opportunities.

### Conclusion

Defence must prepare for the threats and opportunities of climate change. As this chapter has demonstrated, this does not have to be a zero-sum game between 'green' or 'capable'. There are many opportunities when adopting a sustainable approach which can result in a more capable force. These can be direct, such as in the adoption of new technologies, or indirect, such as developing pathways to greater self-sufficiency and resilience which confer economic savings which can themselves be recycled into operational capability. UK defence also has the opportunity to act as a global leader in understanding the climate-security nexus, and in acting to build resilience and engage in conflict prevention. With its hard and soft power capabilities, defence is well-positioned to adapt. Adaptation and a level of organisational change are at the core of the defence enterprise – this is how, throughout its existence, it has delivered against the challenges of changing character and environments of war.

By embracing a responsive climate change and sustainability posture, UK defence has significant potential to gain competitive advantage in a number of key areas. Early response – political, military, and commercial – holds the potential to amplify rewards from the strategic to the tactical. The states that are prepared for the future, including importantly the near term, have a significant advantage over those that are not. A key challenge for UK defence will be modernising the force for operations in a complex, climate-changed environment whilst, simultaneously, complying with the UK government's evolving environmental regulations.

If militaries fail to adopt an approach to sustainability and ignore the effects of climate change, there is the potential for significant operational, political, social, economic, and legal disadvantages. There are also, of course, disadvantages arising from loss of competitive edge, poor investment choices, and costs of later retrofit. Crucially, if UK defence does not adapt and reduce its emissions while it can, it is conceivable that in order to meet its legal requirements (i.e. to be net zero by 2050), the only option in the future will be to reduce operational capability. This is unacceptable and would leave future generations exposed.

## Notes

1 Many commercial enterprises are also directly threatened by climate change or the effect on the environment of their products. Indeed, for some, the move to an environmentally sustainable world would be existential, e.g. fossil fuel companies.
2 A number of studies have suggested that the ability of commercial employers to recruit and retain their talent is becoming influenced increasingly by their so-called 'sustainability credentials' (e.g. HSBC 2021).
3 As an example, during the ISAF campaign in Afghanistan fuel in Helmand cost over 40 times more than at home, with the US reportedly spending US$500 a gallon.
4 That noted, deploying such reactors to warzones exposes them to attack. As evidenced in Afghanistan, Iraq, Mali, and Somalia recently, forward operating bases are attacked with indirect fire (IDF) and person and vehicle-borne IEDs. The risk of overrun is also present (al-Shabaab has overrun KDF camps in Somalia in recent years, for example). Moreover, if attacked or if there is a reactor accident, troops at the base and local civilians risk exposure and contamination. The conventional measures to avert kinetic damage from IDF, such as burial or enclosure with hesco barriers, could undermine the safety and performance of reactors, which rely on air flow to reduce overheating. Climate change is, of course, relevant here as the military is increasingly asked to conduct operations in hot environments.

## References

Agade, K. M., D. Anderson, and E. A. Owino 2022. Water governance, institutions and conflicts in the Maasai rangelands. *The Journal of Environment & Development* 31(4). 10.1177/10704965221123390

Andres, R. B. and H. L. Breetz 2011. Small nuclear reactors for military installations: capabilities, costs, and technological implications. *National Defense University Strategic Forum*. https://ndupress.ndu.edu/Portals/68/Documents/stratforum/SF-262.pdf (accessed 1 May 2023).

Batchelor, C. L., F. D. Christie, D. Ottesen, A. Montelli, J. Evans, E. K. Dowdeswell, L. R. Bjarnadottir, and J. A. Dowdeswell 2023. Rapid, buoyancy-driven ice-sheet retreat of hundreds of meters per day. *Nature* 617: 105–110.

BBC 2022. BBC. MoD fire on Salisbury Plain: helicopter will be used to tackle flames. *BBC News.* https://www.bbc.co.uk/news/uk-england-wiltshire-62153403 (accessed 1 May 2023).

Boyle, L. 2020. How the climate crisis is affecting wildfires around the world. *The Independent.* https://www.independent.co.uk/climate-change/news/wildfires-australia-california-amazon-climate-change-b1775230.html (accessed 1 May 2023).

British Army 2021. Army hybrid vehicles power forward. *British Army.* https://www.army.mod.uk/news-and-events/news/2021/07/army-hybrid-vehicles-power-forward/ (accessed 1 May 2023).

Brown, O., A. Farhan, and C. Hodder 2021. On the front line of climate change, Somalia needs help. *Chatham House.* https://www.chathamhouse.org/2021/12/front-line-climate-change-somalia-needs-help (accessed 1 May 2023).

Brown, W. M. 2003. The Royal Navy's fuel supplies 1898–1939: the transition from coal to oil. Unpublished PhD thesis, King's College London.

Burzynski, M., C. Deuster, F. Docquier, and J. de Melo 2022. Climate change, inequality, and human migration. *Journal of the European Economic Association* 20(3): 1145–1197.

Carrington, D. 2020a. Top UK court overturns block on Heathrow's third runway. *The Guardian.* https://www.theguardian.com/environment/2020/dec/16/top-uk-court-overturns-block-on-heathrows-third-runway (accessed 1 May 2023).

Carrington, D. 2020b. Heathrow third runway ruled illegal over climate change. *The Guardian.* https://www.theguardian.com/environment/2020/feb/27/heathrow-third-runway-ruled-illegal-over-climate-change (accessed 1 May 2023).

Chirodea, F. 2018. Migration: a factor of development or destabilisation? *Eurolimes* 23/24: 5–14.

Clack, T. and L. Selisny 2021. From Beijing bloggers to Whitehall writers: observations on the 'invisible war'. In T. Clack and R. Johnson (eds) *The World Information War: Western Resilience, Campaigning, and Cognitive Effects.* London: Routledge, pp. 259–279.

Clack, T. and R. Nugee 2022. Climate change is creating security threats around the world – and militaries are responding. *The Conversation.* https://theconversation.com/climate-change-is-creating-security-threats-around-the-world-and-militaries-are-responding-173668 (accessed 1 May 2023).

Coffel and Horton 2017. How hot weather – and climate change – affects airline flights. *The Conversation.* https://theconversation.com/how-hot-weather-and-climate-change-affect-airline-flights-80795 (accessed 1 May 2023).

Colgan, J. D. 2013. *Petro-Aggression: When Oil Causes War.* Cambridge: Cambridge University Press.

Collier, P. and A. Hoeffler 1998. On economic causes of civil war. *Oxford Economic Papers* 50(4): 563–573.

Dardia, M., K. F. McCarthy, J. D. Malkin, and G. Vernez 1996. The effects of military base closures on local communities. *RAND.* https://www.rand.org/pubs/monograph_reports/MR667.html (accessed 1 May 2023).

Deivanayagam, T., S. Selvarajah, J. Hickel, R. Guinto, P. de Morais Sato, J. Bonifacio, S. English, M. Huq, R. Issa, H. Mulindwa, H. Nagginda, C. Sharma, and D. Devakumar 2022. Climate change, health, and discrimination. *The Lancet.* https://www.thelancet.com/journals/lancet/article/PIIS0140-6736(22)02182-1/fulltext?rss=yes (accessed 1 May 2023).

Else, H. 2021. Germany's deadly floods. *Nature News*. https://www.nature.com/articles/d41586-021-02330-y (accessed 1 May 2023).

Farhoud, N. 2022. One billion people may be climate crisis refugees by 2050 as UN warns 'clock is ticking'. *The Mirror*. https://www.mirror.co.uk/news/world-news/one-billion-people-climate-crisis-28443433 (accessed 1 May 2023).

Gibbens, 2020. Hurricanes are lasting longer, staying stronger, over land. *National Geographic*. https://www.nationalgeographic.com/environment/article/hurricanes-lasting-longer-climate-change-study-finds (accessed 1 May 2023).

Gordon, S. 2010. The United Kingdom's stabilisation model and Afghanistan: the impact on humanitarian actors. *Disasters* 34(3): 368–387.

GOV.UK 2019. UK becomes first major economy to pass net zero emissions law. *GOV*.UK. https://www.gov.uk/government/news/uk-becomes-first-major-economy-to-pass-net-zero-emissions-law (accessed 1 May 2023).

GOV.UK 2021. World-leading Environment Act becomes law. *GOV.UK*. https://www.gov.uk/government/news/world-leading-environment-act-becomes-law (accessed 1 May 2023).

GOV.UK 2022. The children's people and nature survey for England. *GOV.UK*. https://www.gov.uk/government/statistics/the-childrens-people-and-nature-survey-for-england-summer-holidays-2021-official-statistics/the-childrens-people-and-nature-survey-for-england-summer-holidays-2021-official-statistics (accessed 1 May 2023).

Graham, A. G., A. Wahlin, K. A. Hogan, F. O. Nitsche, K. J. Heywood, R. L. Totten. J. A. Smith, C-D. Hillenbrand, L. M. Simkins. J. B. Anderson, J. S. Wellner, and R. D. Larter 2022. Rapid retreat of Thwaites Glacier in the pre-satellite era. *Nature Geosciences* 15: 706–713.

Heggy, E., Z. Sharkawy, and A. Z. Abotalib 2021. Egypt's water deficit and suggested mitigation policies for the Grand Ethiopian Renaissance Dam filling scenarios. *Environmental Research Letters* 16(7). 10.1088/1748-9326/ac0ac9

Henley, J. 2020. Climate crisis could displace 1.2bn people by 2050, report warns. *The Guardian*. https://www.theguardian.com/environment/2020/sep/09/climate-crisis-could-displace-12bn-people-by-2050-report-warns (accessed 1 May 2023).

Hessayon, A. and D. Taylor 2022. The original climate crisis – how the little ice age devasted early modern Europe. *The Conversation*. https://theconversation.com/the-original-climate-crisis-how-the-little-ice-age-devastated-early-modern-europe-178187 (accessed 1 May 2023).

Hesse, B. J. 2015. Why deploy to Somalia? Understanding six African countries' reasons for sending soldiers to one of the world's most failed states. *Journal of the Middle East and Africa* 6(3–4): 329–352.

HSBC 2021. Are sustainability credentials your biggest hiring advantage? *HSBC*. https://www.business.hsbc.com/en-gb/insights/sustainability/are-your-sustain-ability-credentials-your-biggest-hiring-advantage (accessed 1 May 2023).

Hulbe 2020. How close is the West Antarctic ice sheet to a 'tipping point'? *Carbon Brief*. https://www.carbonbrief.org/guest-post-how-close-is-the-west-antarctic-ice-sheet-to-a-tipping-point/ (accessed 1 May 2023).

IDMC 2023. Global report on internal displacement 2022. *Internal Displacement Monitoring Centre*. https://www.internal-displacement.org/global-report/grid2022/ (accessed 1 May 2023).

IPCC 2021. Sixth assessment report. https://www.ipcc.ch/assessment-report/ar6/ (accessed 1 May 2023).

Islam, N. and J. Winkel 2017. Climate change and social inequality. *UN Department of Economic and Social Affairs.* https://www.un.org/esa/desa/papers/2017/wp152_2017.pdf (accessed 1 May 2023).

Karamanidou, L. 2015. The securitization of European migration policies: perception of threat and management of risk. (In) G. Lazaidis and K. Wadia (eds) *The Securitization of Migration in the EU: Debates Since 9/11.* Basingstoke: Palgrave, pp. 37–39.

Karlborg, L. 2014. Enforced hospitality: local perceptions of the legitimacy of international forces in Afghanistan. *Civil Wars* 16(4): 425–448.

Korte, C. 2021. How climate change helped strengthen the Taliban. *CBS News.* https://www.cbsnews.com/news/climate-change-taliban-strengthen/ (accessed 1 May 2023).

Lanchester, J. 2019. How the Little Ice Age changed history. *The New Yorker.* https://www.newyorker.com/magazine/2019/04/01/how-the-little-ice-age-changed-history (accessed 1 May 2023).

Lewis, H. and P. Dwyer 2014. *Precarious Lives: Forced Labour, Exploitation and Asylum.* Bristol: Policy Press.

Maizland, L. 2021. China's fight against climate change and environmental degradation. *Council on Foreign Relations.* https://www.cfr.org/backgrounder/china-climate-change-policies-environmental-degradation (accessed 1 May 2023).

Marshall, W. 2021. Is climate change fuelling al-Shabaab's resurgence in Somalia? *Global Risk Insights.* https://globalriskinsights.com/2021/09/is-climate-change-fuelling-al-shabaabs-resurgence-in-somalia/ (accessed 1 May 2023).

Met Office 2022. Effects of climate change. *UK Met Office.* https://www.metoffice.gov.uk/weather/climate-change/effects-of-climate-change (accessed 1 May 2023).

Miguel, E., S. Satyanath, and E. Sergenti 2004. Economic shocks and civil conflict: an instrumental variables approach. *Journal of Political Economy* 112(4): 725–753.

Milfont, T. L., E. Zubielevitch, P. Milojev, and C. G. Sibley 2021. Ten-year panel data confirms generation gap but climate beliefs increase at a similar rate. *Nature Communications* 12. 10.1038/s41467-021-24245-y

MOD 2021. Lt Gen Richard Nugee opens Nesscliff NETCAP buildings. *GOV.UK.* https://www.gov.uk/government/news/lt-gen-richard-nugee-opens-nesscliff-netcap-buildings (accessed 1 May 2023).

Mustafa, D., G. Gioli, M. Karner, and I. Khan 2017. Contested waters: subnational scale water conflict in Pakistan. *United States Institute of Peace.* https://www.usip.org/publications/2017/04/contested-waters-subnational-scale-water-conflict-pakistan (accessed 1 May 2023).

National Grid 2023. Future developments. *National Grid.* https://www.nationalgrid.com/national-grid-ventures/what-we-do/future-developments (accessed 1 May 2023).

NATO 2022. Dispatch from the field: exercise Cold Response 2022 wraps up in Norway. *NATO.* https://www.nato.int/cps/en/natohq/news_194434.htm (accessed 1 May 2023).

Nett, K. and L. Ruttinger 2016. Insurgency, terrorism and organised crime in a warming climate. *Climate Diplomacy.* https://uploads.guim.co.uk/2017/04/20/CD_Report_Insurgency_170419_(1).pdf (accessed 1 May 2023).

NOAA 2022. Climate change: Global sea level. *Climate.gov.* https://www.climate.gov/news-features/understanding-climate/climate-change-global-sea-level (accessed 1 May 2023).

Norwood, J. S. 2020. Labor exploitation of migrant farmworkers: risks for human trafficking. *Journal of Human Trafficking* 6(2): 209–220.

Obama, B. 2021. COP26 climate speech transcript. *Rev.* https://www.rev.com/blog/transcripts/barack-obama-cop26-climate-speech-transcript (accessed 1 May 2023).

OEERE 2022. Sustainable aviation fuels. *Office of Energy Efficiency and Renewable Energy.* https://www.energy.gov/eere/bioenergy/sustainable-aviation-fuels (accessed 1 May 2023).

Otto, F. 2020. *Angry Weather: Heat Waves, Floods, Storms, and the New Science of Climate Change.* London: Greystone.

Overfield, C. 2022. Russia's Arctic claims are on thin ice. *Foreign Policy.* https://foreignpolicy.com/2022/12/20/russia-arctic-claims-territorial-internal-waters/ (accessed 1 May 2023).

Pitchford, J. 2011. The 'Global War on Terror', identity, and changing perceptions: Iraqi responses to America's war in Iraq. *Journal of American Studies* 45(4): 695–716.

PM Office 2021. PM Boris Johnson's address to the UN Security Council on climate and security, 23 February 2021. *GOV.UK.* https://www.gov.uk/government/speeches/pm-boris-johnsons-address-to-the-un-security-council-on-climate-and-security-23-february-2021 (accessed 1 May 2023).

Proud, S., A. Prata, and S. Schmauss 2022. The January 2022 eruption of Hunga Tonga-Hunga Ha'apai volcano reached the mesosphere. *Science* 378(6619): 554–557.

Rantanen, M., A. Yu, A. Lipponen, K. Nordling, O. Hyvarinen. K. Ruosteenoja, T. Vihma, and A. Laaksonen 2022. The Arctic has warmed nearly four times faster than the globe since 1979. *Communications Earth & Environment* 3(168). 10.1038/s43247-022-00498-3

Roblin, S. 2020. The US military is terrified of climate change. It's done more damage than Iranian missiles. *NBC News.* https://www.nbcnews.com/think/opinion/u-s-military-terrified-climate-change-it-s-done-more-ncna1240484 (accessed 1 May 2023).

Rodrik, D. 1999. Where did all the growth go? External shocks, social conflict and growth collapses. *Journal of Economic Growth* 4(4): 385–412.

Rolls-Royce 2022. The 'Spirit of Innovation' officially breaks speed record and becomes the world's fastest all-electric vehicle. *Rolls-Royce.* https://www.rolls-royce.com/media/press-releases/2022/20-01-2022-the-spirit-of-innovation-officially-breaks-speed-record.aspx (accessed 1 May 2023).

Sengupta, S. 2021. A new breed of crisis: war and warming collide in Afghanistan. *The New York Times.* https://www.nytimes.com/2021/08/30/climate/afghanistan-climate-taliban.html (accessed 1 May 2023).

Sheridan, D. 2021. Climate change stopping soldiers from training, warns military's green tsar. *The Telegraph.* https://www.telegraph.co.uk/news/2021/09/16/climate-change-stopping-soldiers-training-warns-militarys-green/ (accessed 1 May 2023).

Sherwood, D. 2019. Chile protestors block access to lithium operations. *Reuters.* https://www.reuters.com/article/us-chile-protests-lithium-idUSKBN1X42B9 (accessed 1 May 2023).

Shwartzein, P. 2021. Why water conflict is rising, especially on the local level. *New Security Beat.* https://www.newsecuritybeat.org/2021/03/water-conflict-rising-local-level/ (accessed 1 May 2023).

Stollenwerk, E. 2018. Securing legitimacy? Perceptions of security and ISAF's legitimacy in Northeastern Afghanistan. *Journal of Intervention and Statebuilding* 12(4): 506–526.

The Guardian 2023. 'The city was underwater': quarter of a million Somalis flee flooded homes. *The Guardian*. https://www.theguardian.com/world/2023/may/17/city-underwater-quarter-of-million-somalians-flee-homes-floods (accessed 17 May 2023).

Thompson, V., A. Kennedy-Asser, E. Vosper, Y. T. Eunice Lo, C. Huntingford, O. Andrews, M. Collins, G. Hegerl, and D. Michell 2022. The 2021 western North America heat wave among the most extreme events ever recorded globally. *Science Advances* 8(18). 10.1126/sciadv.abm6860

Ucko, D. 2013. Beyond clear-hold-build: rethinking local-level counterinsurgency after Afghanistan. *Contemporary Security Policy* 34(3): 526–551.

UK Government 2021. Global Britain in a competitive age: the integrated review of security, defence, development and foreign policy. *Her Majesty's Stationary Office*. https://assets.publishing.service.gov.uk/government/uploads/system/uploads/attachment_data/file/975077/Global_Britain_in_a_Competitive_Age-_the_Integrated_Review_of_Security__Defence__Development_and_Foreign_Policy.pdf (accessed 1 May 2023).

UK Parliament 2022. The integrated review, defence in a competitive age and the defence and security industrial strategy: government response to the committee's second report. *UK Parliament*. https://publications.parliament.uk/pa/cm5803/cmselect/cmdfence/865/report.html (accessed 1 May 2023).

UNSC 2021. Security in the context of terrorism and climate. *United Nations Security Council. UN Web*. https://media.un.org/en/asset/k1r/k1rw3gy2mo (accessed 1 May 2023).

US EPA 2022. How climate change may impact ozone pollution and public health through the 21st Century. *United States Environmental Protection Agency*. https://www.epa.gov/sciencematters/how-climate-change-may-impact-ozone-pollution-and-public-health-through-21st-century (accessed 1 May 2023).

Venturi, B. and N. A. Toure 2020. Challenges and choices in the Sahel. *Foundation for European Progressive Studies and Instituto Affari Internazionali*. https://www.iai.it/sites/default/files/venturi_toure_en.pdf (accessed 1 May 2023).

Virilio, P. 2005. *The Information Bomb*. London: Verso.

Virilio, P. 2009. *War and Cinema: The Logistics of Perception* (trans. P. Camiller). London: Verso.

Walker, A. 2021. Shift to green energy 'could cost oil states $13 trillion' by 2040. *BBC News*. https://www.bbc.co.uk/news/business-56017415 (accessed 1 May 2023).

WEF 2022. Global risk report 2022. *World Economic Forum*. https://www.weforum.org/reports/global-risks-report-2022/ (accessed 1 May 2023).

WMO 2022. Economic losses from extreme weather rocket in Asia. *World Meteorological Organization*. https://public.wmo.int/en/media/press-release/economic-losses-from-extreme-weather-rocket-asia (accessed 1 May 2023).

Wunsch, C. 2002. What is thermohaline circulation? *Science* 298(5596): 1179–1181.

Zhang, J., I. Hilton, and J. Turner 2021. What's China doing to fight climate change? *BBC World Service*. https://www.bbc.co.uk/programmes/w3ct1hsp (accessed 1 May 2023).

# 10

# MARITIME RESPONSE TO CLIMATE CHANGE

*Sherri Goodman, Pauline Baudu, and Rachel Fleishman*

## Maritime security in the age of climate change

Maritime security has traditionally been focused on protecting vessels tran-siting the maritime environment. However, the advent of climate change requires a broader conceptualisation. For the purpose of this analysis, mar-itime security encompasses the safety and health of people and assets at sea, security of land-based populations subject to climate-induced sea-level rise and extreme weather threats, corresponding threats to marine life, and deg-radation of the oceans' capacity to sequester atmospheric carbon. Climate change impacts range from the physical and social to military and security.

### Physical impacts

Rising greenhouse gas (GHG) levels in the atmosphere are warming air, water, and land-based environments worldwide, intensifying climatic phe-nomena. Warmer air carries more humidity, making storms larger and more powerful. As storms intensify, mean precipitation escalates, increasing flood risk. Melting glaciers feed into the system, further strengthening river flows and air moisture content and amplifying both storms and floods. When glacial melt reaches the sea, it increases the volume of ocean waters and further compounds sea-level rise and its cascading effects, including erosion and salination of low-lying agricultural areas. Warming seawater absorbs carbon dioxide, leading to ocean acidification endangering marine health and marine ecosystems. Sea-based species either migrate towards cooler water – towards the poles – change sizes and reproduction patterns or begin to die off. One recent study suggests that marine species are disappearing from their

DOI: 10.4324/9781003377641-13

habitat twice as fast as their land-based counterparts (Pinsky et al. 2019). Such studies are a foreboding signal of the ocean's rapidly deteriorating health. But the oceans also play a crucial role in protecting humanity from global warming. Oceans help reduce climate change by storing large amounts of carbon dioxide; in fact, an estimated 83% of the global carbon cycle takes place through marine waters (BCI 2019).

### Societal impacts

For coastal communities, sea-level rise is a growing danger to lives and livelihoods, as well as to the governments supporting them. Storm strength, storm surge, and over-use of local freshwater sources further exacerbate the challenge. At least 267 m people are at risk of inundation today, with almost 60% of those in Asia (Hooijer and Vernimmen 2021). Warming and acidifying seas are also decimating fish stocks, a key source of protein for over 3bn people (Our Ocean 2016). Dwindling fish stocks, combined with economic inequalities and weak governance, can prompt illegal, unreported, and unregulated (IUU) fishing, as fishing communities struggle to maintain their trade. Coastal agricultural areas face a similar challenge. Salinisation, which affects 20% of irrigated cropland today, could increase to over 50% by 2050 (Singh 2021). In these contexts, farmers are often left to choose between limited local employment options and migrating to another region altogether.

### Military and security impacts

Naval installations face the same physical impacts as other important coastal infrastructures. Sea-level rise endangers coastal military infrastructure, subject to flooding and loss of access. Decision makers will face billion-dollar decisions in the coming years as infrastructure from dry docks and ports to base housing and offices contend with impending floods. At the same time, extreme-weather events increase the demand for naval forces in humanitarian assistance and disaster response (HADR) missions. The United States and other navies regularly contribute naval forces and hospital ships to rescue missions in response to natural disasters worldwide.

Climate change is a threat multiplier, exacerbating resource scarcity, taxing government capacity, and fuelling existing societal tensions that can lead to human suffering, forced displacement, and conflict. According to the Stockholm Climate Security Hub (Barquet et al. 2020), 'for small island states and the 70% of the most climate-vulnerable countries that also happen to be among the most politically fragile countries, this is bound to be a matter of state and collective security'. The physical destruction wrecked by climate events, compounded by the associated competition for scarce resources, and loss of the roads, bridges, power plants, water delivery, and other infrastructure critical for the functioning of modern society, strains states, particularly those already stressed by under-development. It also provides conditions and incentives for the critically poor to engage in the maritime crimes of IUU fishing, piracy, and human trafficking. Climate impacts are also expected to amplify drivers of population movements, including cross-border migrations through increasingly dangerous waters.

Competition for sea-based minerals and hydrocarbons is another maritime flashpoint. While there has not yet been a shooting war over sea-based oil rigs or cobalt deposits, economic dependence is growing upon oceans for fossil fuel deposits, renewable energy sources, and the minerals essential for the green energy transition. The location of deposits in contested territorial waters further complicates matters, as does the absence of comprehensive and widely accepted governance regimes. Finally, sea-level rise impacts marine territorial boundaries, many of which are already disputed.

This analysis explores maritime climate security: the threats to maritime security that emanate from or are amplified by the climate crisis. Illustrations are drawn from three regional case studies: the Caribbean, South China Sea, and Arctic. Recognising naval forces' responsibility to protect the marine environment and the challenges climate change represents to security and defense forces, the chapter also discusses mitigation, adaptation, and conservation of the marine environment through a military lens. The recommendations proffered can inform not only military forces, but also political decision makers with the ability to create ocean governance regimes tailored to the challenges of the 21st century.

## Navigating troubled waters from the High North to the Pacific

Climate change and coastal ecosystems degradation impact maritime security differently around the world. As shown in Figures 10.1 and 10.2,[1] impacts are shaped by: (1) the physical impact of climate change on the regions; (2) the dependence of local communities on affected ecosystems for food security or incomes; (3) the resources available to local communities to prepare for, and adapt to, climate impacts; and (4) governance capacity to manage or invest in these resources. This chapter illustrates the scope of climate impacts on maritime security through three regional case studies: communities' classic struggles due to sea-level rise and altered marine production on Caribbean islands; multinational competition for control over resources in the South China Sea; and human security threats and increased rivalry due to melting ice in the Arctic.

### *Case study 1: the Caribbean Sea*

*Political geography*. The Caribbean Sea is a basin of the western Atlantic Ocean of approximately 2.78 m km$^2$, which includes continental countries,

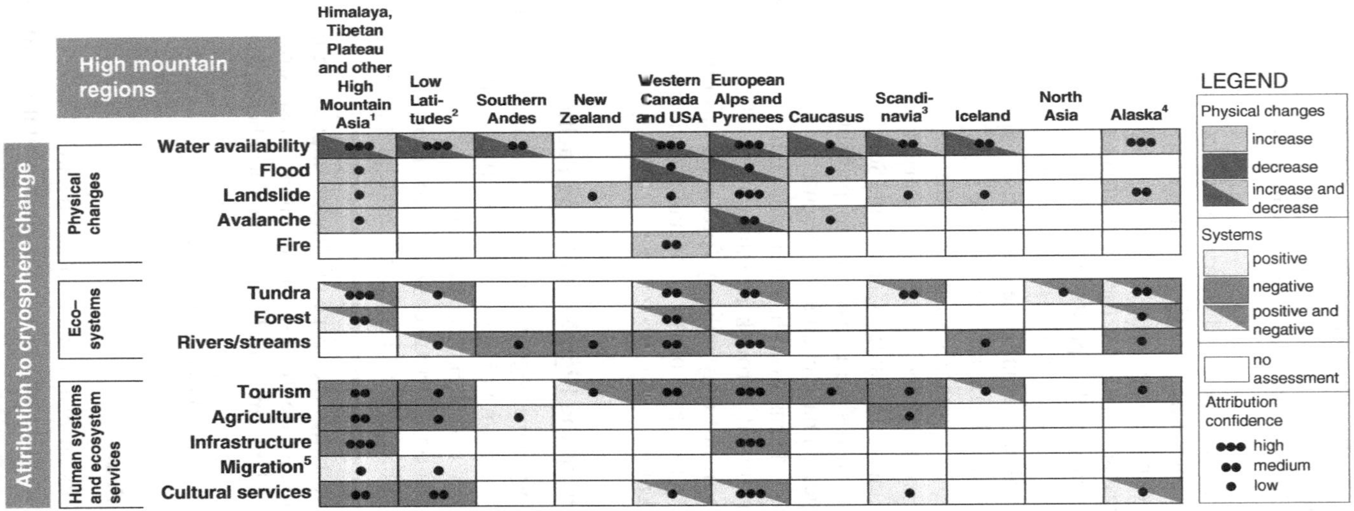

**FIGURE 10.1** High mountain regions.

*(Figure 2.8 from Hock et al., 2019: High Mountain Areas. In: IPCC Special Report on the Ocean and Cryosphere in a Changing Climate [H.-O. Pörtner, D. C. Roberts, V. Masson-Delmotte, P. Zhai, M. Tignor, E. Poloczanska, K. Mintenbeck, A. Alegría, M. Nicolai, A. Okem, J. Petzold, B. Rama, N. M. Weyer (eds.)]. Cambridge University Press, Cambridge, UK and New York, NY, USA, pp. 131–202. from 10.1017/9781009157964.004.*

[1] includes Hindu Kush, Karakoram, Hengduan Shan, and Tien Shan; [2] tropical Andes, Mexico, eastern Africa, and Indonesia;
[3] includes Finland, Norway, and Sweden; [4] includes adjacent areas in Yukon Territory and British Columbia, Canada;
[5] Migration refers to an increase or decrease in net migration, not to beneficial/adverse value.

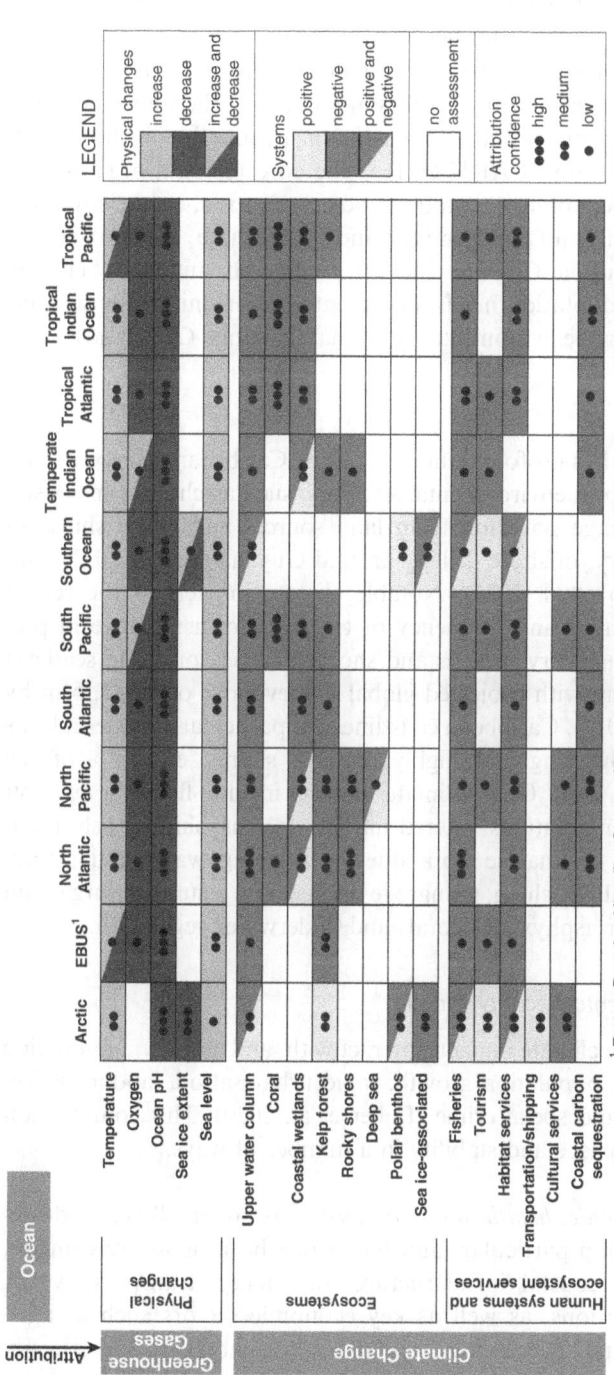

**FIGURE 10.2** Oceanic and cryosphere changes.

*(Figure 5.24 from Bindoff et al., 2019: Changing Ocean, Marine Ecosystems, and Dependent Communities. In: IPCC Special Report on the Ocean and Cryosphere in a Changing Climate [H.-O. Pörtner, D.C. Roberts, V. Masson-Delmotte, P. Zhai, M. Tignor, E. Poloczanska, K. Mintenbeck, A. Alegría, M. Nicolai, A. Okem, J. Petzold, B. Rama, N.M. Weyer (eds.)]. Cambridge University Press, Cambridge, UK and New York, NY, USA, pp. 447–587. 10.1017/9781009157964.007.*

insular nations, and dependent territories. It is bounded to the south by the coasts of Venezuela, Colombia, and Panama; to the west by Costa Rica, Nicaragua, Honduras, Guatemala, Belize, and Mexico's Yucatán Peninsula; to the north by Greater Antilles Islands of Cuba, Hispaniola, Jamaica, and Puerto Rico; and to the east by the north-south chain of the Lesser Antilles. According to United Nations data, about 43.5 m people live in the region. The Caribbean Community (CARICOM) is the core regional organisation and aims to promote economic integration and cooperation, ensure equitable sharing of the benefits of integration and coordinate foreign policy. CARICOM has set up a Climate Change Centre to anticipate climate impacts, respond to adaptation needs, and create opportunities for sustainable development (Carribean Community Climate Change Center nd).

### Climate impacts

Global climate change has profound impacts on the Caribbean Sea ecosystem. It also interacts with local environmental stressors, such as changes in coastal land and sea use, sewage pollution from land sources and cruise ships, an intense tourism industry, offshore drilling around Guyana (Juhasz 2021), and overfishing (Seminario et al. 2021). Notable climate impacts in the region include increased intensity and frequency of tropical storms; change in precipitation patterns: longer dry seasons and shorter wet seasons; and sea-level rise and coastal erosion, with projected global sea level rise of 0.5–2.15 m by 2100 (Simpson et al. 2011). Caribbean coastlines are particularly vulnerable, as they include many low-lying and highly erodible shores, e.g. 80% of the Bahamas is below sea level. Other climate impacts include flooding and salt intrusion; rising sea temperatures; altered marine streams; shifting fish distributions; and impacts on marine flora due to warming water and stream changes including coral bleaching, mangrove decline, and a surge in sargassum seaweed, which further asphyxiates corals and underwater vegetation.

### Climate and environmental security threats

In the Caribbean, these climate impacts interact with, and amplify, pre-existing vulnerabilities, such as population growth, rapid urbanisation, labour market inequalities, poverty, and social crime (Fuller et al. 2020). This combination can threaten regional peace and stability in a number of ways.

- *Human risks: economic, health, and food insecurity.* In small island developing states (SIDS) in particular, disaster-related hazards severely impact the predominantly coastal infrastructure for energy, transport, water, and telecommunications, as well as key economic sectors such as agriculture. Storms and decreasing freshwater availability have already induced crop failure in the region. Sea surface temperature and ocean acidification also threaten biodiversity and fisheries. The massive expansion of sargassum seaweed disturbs fishing boat engines and emits gasses harmful to coastal populations (UN FAO 2022). These factors negatively affect (or impair) the health of local populations, as well as the region's economies (Fuller et al. 2020). The tourism sector, which contributes as high as 40% to the economies of SIDS, is already threatened by rising sea levels, and touristic assets risk further atrophy due to coral bleaching, alteration of marine biodiversity, and overall aesthetic degradation.

- *Risks of political instability and social unrest.* Such impacts on livelihoods and key economic sectors imperil food security and economic development, and increase unemployment, poverty, and debt-GDP ratio. This may lead to increased competition over resources, social unrest, and political instability, for example, through undermining governments' ability to provide essential services, including security and disaster relief. One example was Puerto Rico in 2019, when, building on pre-existing tensions, locals demanded Governor Ricardo Rosselló's resignation due to the slow recovery after Hurricane Maria (Fuller et al. 2020).
- *Rise in social violence and crime in coastal communities.* Natural disasters, sea-level rise, and loss of livelihoods will intensify displacement and exacerbate existing urban security challenges. Climate-induced loss of livelihoods such as fishing will increase opportunities for illegal coping mechanisms and criminal activity within territorial waters, like IUU, piracy, and trafficking, but also gang activity, kidnapping, and organised crime (Bolaji 2020). Here it is worth noting that organisations policing IUU have hitherto been hindered by a lack of domain awareness and naval capacity in monitoring Caribbean waters (Seminario 2021). In addition, the combination of natural disasters and forced displacements have been found to correlate with increasing sexual and gender-based violence in the region (Fuller et al. 2020).

### Case study 2: the South China Sea

*Political geography*

The South China Sea is a vast marine expanse (3.6 m km$^2$) bordered by Brunei, Indonesia, Malaysia, the Philippines, Singapore, Thailand, Taiwan, Vietnam, and China. An estimated US$5.3tn in goods – over 60% of global maritime trade – sail through its waters each year. The region's young and increasingly educated population is leading the shift to manufacturing and services. Growing domestic demand, international trade, and foreign direct investment together fuel one of the most economically vibrant regions of the world. Southeast Asian economies grew at an impressive 5.6% in 2022, and are projected to grow at 4.8% in 2023 (OECD 2023). While Southeast Asian countries differ in government systems and economic development, the political and economic construct of ASEAN serves to align the interests of its ten Member States. This alignment, while far from perfect, often positions ASEAN in a delicate relationship with its enormous neighbour to the north. Chinese claims of over 80% of the South China Sea are demarcated by the so-called 'nine-dash-line'. Their contravention of the UN Convention on the Law of the Sea (UNCLOS) has led to disputes with Brunei, Indonesia, Malaysia, the Philippines, Thailand, and Vietnam. At issue is not only the enforcement of international law, but also the right of free passage, and valuable fishing, subsea mineral, and oil and gas rights.

## Climate impacts

The 2021 Intergovernmental Panel on Climate Change Sixth Assessment Report provides a clear indication of the extent and diversity of climate impacts affecting the South China Sea region. Among these are increasingly frequent and intense extreme heat events, with levels of heat and humidity that threaten human health and survival; greater mean precipitation, inducing flooding; tropical cyclones which are predicted to decrease in number but increase in intensity; compound meteorological heat and drought events, water shortages to ecological systems and agricultural areas; sea-level rise, inducing coastal flooding exacerbated by land subsidence; marine heatwaves of increasing intensity, with potential feedback loops into storm strength and marine environmental degradation; and oceanic carbon absorption, leading to ocean acidification, coral bleaching, and the concomitant degradation of marine plant and animal life.

## Climate security threats

Overlapping political fault lines and climate-vulnerable zones reveal serious maritime climate security hot spots. On land, coastal megacities including Bangkok, Jakarta, and Hong Kong confront both domestic political turmoil and lack of preparedness for impending climate impacts. Inaction on adaptation and resilience increases the lethality of storms and floods, compounding the risks to human health and compromising agricultural regions like the Mekong Delta. The resulting challenges to health and safety, food security, and infrastructure integrity will tax political stability and social cohesion in ways that are both predictable and preventable. At sea, Chinese assertion of dominance over marine areas and resources provides the backdrop for multiple conflict vectors with potential for rapid escalation. For example, China is projecting dominance and taking control of oil and gas reserves within the nine-dash line (Kuo 2018). Countries which dispute Chinese claims but rely on Chinese trade and investment are left in a difficult position – one that reportedly cost Vietnam US$1bn in 2020 (Hayton 2020). Similarly, in the recent past armed Chinese fishing fleets have occupied a Philippine reef (Chavez 2021), attacked fishing boats in Vietnam (Bengali and Bao Uyen 2020) and illegally trawled rich fishing grounds near the Natuna Islands, Indonesia. The latter activity destroying the seabed and thus compromising the very resources at the centre of the dispute (Beech and Suhartono 2020). Ongoing Chinese militarisation of small islands and features in the South China Sea portends preparation for such cold conflicts to escalate into hot war – unless steps are taken to de-escalate with confidence-building measures and governance structures.

### Case study 3: the Arctic

#### Political geography

The Arctic is a 14 m km$^2$ territory delimited by the Arctic Circle and spread across three continents. It includes the Arctic Ocean surrounded by the northern territories of eight countries: Canada, Russia, the United States (via the state of Alaska), Norway, Denmark (via Greenland), Iceland, Sweden, and Finland. Russia is geographically dominant in the Arctic. The Kola Peninsula is home to its Northern Fleet, including strategic submarines (Østhagen 2020). For the past decade, Russia has been expanding its military presence and up-grading its northern naval infrastructure to enable greater control over sea lanes (Boulègue 2019; Clack and Nugee 2022). The main governing body is the Arctic Council (AC), which convenes the eight Arctic nations to address issues related to indigenous communities, environmental concerns, scientific research, and sustainable development. The AC has initiated agreements covering oil spill response and search and rescue (Maddox 2021), which are supported by the Arctic Coast Guard Forum (Arctic Council 2021), while shipping is managed by the IMO Polar Code (2017). Military and security issues are explicitly excluded from the AC agenda and were discussed in other forums until Russia's invasion of Crimea in 2014 (Conley and Melino 2016). In addition, most of the AC work was paused following Russia's full invasion of Ukraine in February 2022. At the time of writing this chapter, the AC was studying how to resume its work under the new Norwegian chairmanship.

#### Climate impacts

The Arctic is warming at least three times faster than the rest of the globe, and its unique ecosystems are among the world's most vulnerable to climate change. Direct climate impacts in the Arctic include new temperature extremes, which contributed to wildfires in Siberia in 2020; decline in Arctic sea-ice extent and thickness (Arctic sea-ice extent has decreased by an average of 3.5–4.1% per decade since the 1980s); thawing of permafrost; intensification of the hydrological cycle; biodiversity changes; higher ocean primary productivity; and oceanic acidification – much faster than the global ocean – with implications for algae, zooplankton, and fish (Guy et al. 2021; NOAA Arctic Program 2021). Furthermore, the melting ice generates destructive feedback loops: thawing permafrost releases high amounts of methane, which accelerates global warming. Similarly, melting sea ice exposes larger parts of a dark ocean which absorbs more heat, further aggravating global warming. Climate change also has important geo-economic impacts in the Arctic. The melting sea ice is expected to lead to an increase in navigable surfaces in the medium term (see Marten, this volume). The Northwest Passage (NWP) and the Northern Sea Route (NSR), the main trans-Arctic lanes, are substantially shorter in distance than the ones currently used, and thus expected to be cheaper by shipping companies and investors - although navigation constraints due to the harsh weather and environment seriously limit their safety and commercial viability. In addition, increased shipping brings ship strikes, noise and light pollution, as well as chemical pollution, and marine debris impacting the region's marine biodiversity (Maddox 2021). Arctic ice melt is expected to allow relatively easier access to natural resources, including hydrocarbons, minerals (e.g. those crucial to the green energy transition), fishery resources, and water. It is estimated that the Arctic contains 13% of the world's undiscovered oil resources and 30% of undiscovered natural gas resources, including in deep waters (Gautier et al. 2009).

### Climate security threats

Climate change and biodiversity loss in the Arctic maritime domain are increasingly shaping the security environment.

- *Human security threats.* Climate-altered maritime dynamics have crucial impacts on human security, of which Indigenous communities are on the front lines. threats. Isolated coastal municipalities, especially in the US and Canadian Arctic, face existential threats from coastal erosion, sea-level rise, saltwater intrusion, storm surges (against which sea ice is no longer acting as an effective natural buffer), as well as endangered critical infrastructure and ecosystems due to permafrost thaw (Maddox 2021). Climate change is also a threat to human health, with biological and environmental risks emerging from expanding toxic algae and from the collapse of industrial, nuclear, and oil and gas infrastructure built on thawing permafrost. In addition, surging commercial and touristic activity in Arctic waters increases risks of accidents and casualties in maritime areas that already confront extreme weather conditions and the tyranny of distance: long response times, limited communication capabilities, and inadequate support infrastructure (Maddox 2021).

- *Military and hard-security threats.* As the sea ice melts, competing interests add strain to global power relationships and challenge maritime governance frameworks. Russia has been hyper-securitizing its economic assets for Arctic resource exploration, and its provocations have included a strong maritime component, increasing the likelihood of escalation due to accidents at sea or misunderstandings. Furthermore, the loss of Russia's natural defensive barriers due to melting sea ice is expected to further affect its strategic decisionmaking and force posture in the region. Climate change is also likely to amplify opportunities for malign actors to engage in the Arctic by instrumentalizing societal vulnerabilities. For example, climate change may empower Chinese hybrid activities by justifying a deeper scientific footprint to monitor climate impacts in strategically-significant locations. Lastly, climate effects are challenging military readiness at sea, as they impact missions and coastal installations. They will put a growing strain on security forces to address a surge in concurrent emergencies such as search-and-rescue and environmental-disaster response, due to the growing volume of maritime traffic in more challenging sea conditions.

## Maritime responses to climate change

### *Mitigation in maritime security*

The IPCC Sixth Assessment Report published last year (IPCC 2022a) offers the most sobering assessment to date of the actions necessary to avoid uncontrollable escalation of the climate crisis. The analysis confirms that the world has exceeded its carbon budget – or maximum level of allowable carbon dioxide emissions – to limit warming to 1.5°C as per the Paris Agreement. The record breaking temperatures on both land and sea during the summer of 2023, with forecasts portending more heat trapped in ocean basins to come, presages turbulent weather patterns ahead. Furthermore, a recent study has demonstrated that in the absence of radical emission curbs, marine systems will experience a shift that could rival previous mass extinction events, while a reversal of emissions trends would diminish extinction risks by more than 70% (Penn and Deutsch 2022). The magnitude, breadth, and interconnectedness of the climate crisis is, in essence, the business case for mitigation. The only way to achieve the 1.5°C limit is for emissions to peak before 2025, then radically reduce (at least 43% by 2030, based on 2019 levels). Furthermore, the scale and scope of the action required have become quite clear. With current pledges insufficient and the range of options narrowing quickly, some strategies are non-negotiable: curbing fossil fuel use; accelerating the renewable energy transition; electrifying buildings, vehicles, and manufacturing; and capturing carbon dioxide already in the oceans and atmosphere. Such efforts should involve all maritime stakeholders, including navies, ship-owners, and fisheries. This chapter focuses from here on the role of navies and coastguards in mitigation efforts.

Militaries have largely and deliberately been excluded from the mitigation conversation. This precedent was set during negotiations of the 1997 Kyoto Protocol, when the United States argued successfully that only emissions from installations and non-tactical vehicles were to be included in carbon emissions inventories, with emissions from military operations omitted. The 2015 Paris agreement took a step towards reversing this trend when it made military reporting voluntary. It is now obvious to most nations that exempting militaries from accounting for, and reducing, emissions entirely is unwise from both a climate and a national security perspective. Today's climate crisis presents an 'all hands on deck' moment, and world militaries are not excepted. While no comprehensive accounting of the emissions of armed forces exists, attempts are underway (Military Emissions Gap 2021). One conservative estimate puts the US military's carbon footprint as larger than that of Portugal or Denmark, with the US Navy and Marine Corps responsible for about 37% of emissions.

Around the world, work to both measure and reduce military emissions is getting underway. The UK's Royal Air Force committed to achieving net zero by 2040, while the US Army committed to net zero by 2050, with a 50% reduction by 2030. Meanwhile, NATO has developed and recently released, in July 2023, a Greenhouse Gas Emissions Mapping and Analytical Methodology.

### A plan for transformation

Leadership in climate readiness is an investment in mission assurance. Becoming less reliant on increasingly outdated infrastructure, such as fuel truck convoys, increases operational capacity and resilience both in rescue missions and on the battlefield. Funnelling military procurement monies into distributed energy, advanced energy efficiency, new fuels, and digitally enabled building and transport systems also accelerates their testing and scale economics, making them cheaper for government and private sector actors (Figure 10.3). From a military perspective, decarbonisation at sea is an enterprise-wide undertaking that incorporates at its core a range of viable mitigation strategies:

- *Rapid scaling back of fossil fuel use.* According to the International Maritime Organisation (IMO 2020; IPCC IMO 2021), rising vessel traffic has translated into increased shipping emissions, even as maritime vessels become more carbon efficient. This poses a conundrum for world navies,

**Many options available now in all sectors are estimated to offer substantial potential to reduce net emissions by 2030. Relative potentials and costs will vary across countries and in the longer term compared to 2030.**

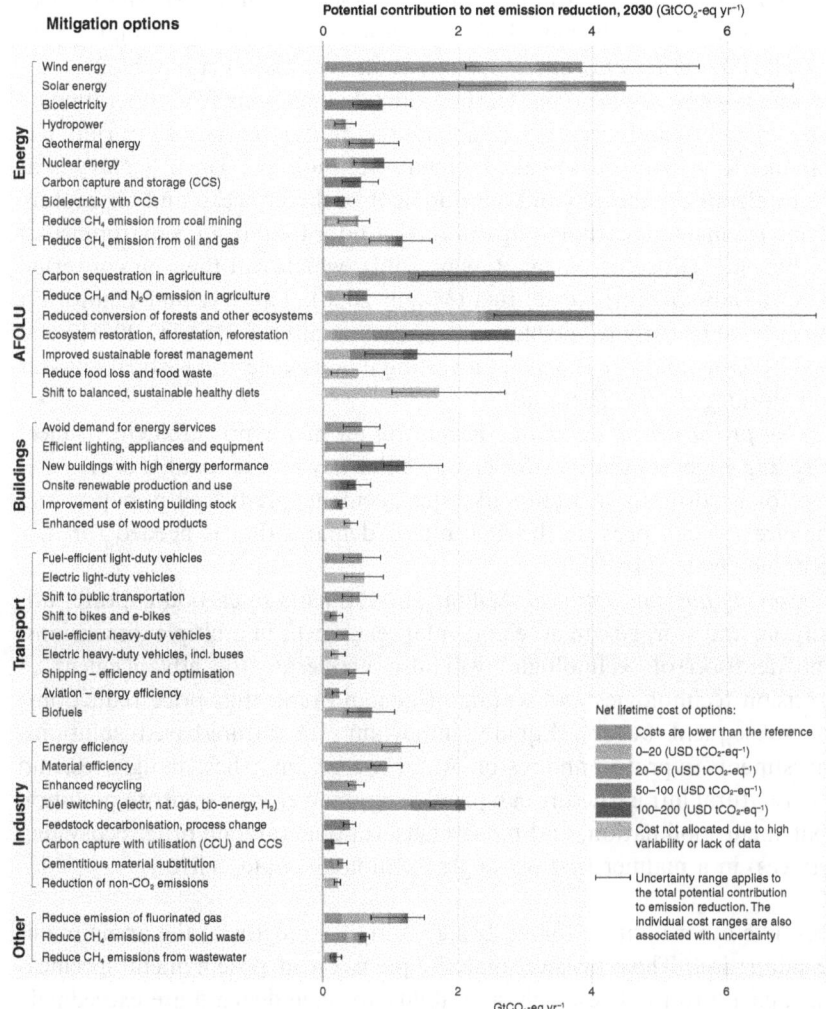

**FIGURE 10.3** Options available to reduce net emissions by 2030.

(Figure SPM.7 from IPCC (2022b)).

whose mission is to protect people and vessels at sea. Research is underway to step up presence while reducing carbon footprint. One set of promising technologies is Prepare Ships (Copernicus nd), a project developed by Swedish research and corporate entities to improve both the safety and efficiency of shipping.

- *Energy efficiency and renewable energy at naval installations.* Efforts to reduce energy use at US Navy bases have been ongoing for years (CNIC nd). Under the Biden Administration's leadership, however, the level of ambition has risen sharply. The Pentagon is now turning to industry to source 100% renewable energy to power its facilities (Vergun 2022).
- *Tackling operational carbon.* Carbon dioxide emissions from operations – vehicles, ships, and aircraft – dominate the military carbon footprint. The response is a focus on so-called 'climate technology'. The US Navy is at the forefront of research and development of direct ocean and air capture of environmental carbon (Operational Endurance from Environmental Carbon nd; Office of Naval Research nd), which can then be converted directly onboard into vessel fuel (Magill 2020). The UK and Danish air forces have recently trialled low- or no-carbon aircraft (NATO 2021b) and the US Army and Air Force are investing in a new electric vertical take-off and landing aircraft (Beta nd).
- *Upping procurement standards.* Requiring all major procurement to meet 'bleeding edge' standards – for energy efficiency and renewable energy but also for hard-to-abate sectors like steel, cement, plastic, pulp, paper, and chemicals – can provide the guaranteed demand that is needed for new technologies to take root.
- *Carbon capture and removal.* Military investments in carbon capture, utilisation, and storage can accelerate market growth in multiple dimensions. The iteration of technologies will also accelerate the advancement of precision technologies and scaling of demand (and thus price reductions) for subsequent buyers. Equally important are nature-based solutions. Investments to protect and restore water sources, marshes, mangroves, and other natural infrastructure can provide ongoing erosion control, pollution abatement, pollination, and biodiversity support (among other ecosystem services) in a manner that no single technology could deliver.

Navies, like other military forces, bring multiple strengths to the arena of climate mitigation. Their incisive analysis, precision in procurement specifications, capacity to fund research, and ability to scale demand are exceedingly valuable at this critical juncture of the climate crisis. As militaries start investing in clean energy and climate technologies through their innovation portfolios, such as those managed by the Defense Innovation Unit and Defense Advanced Research and Projects Agency in the United States, there is prospect to start seeing climate technologies develop at the speed of climate change itself.

## Security and conservation in the maritime environment

Oceans hold invaluable natural conservation assets, from their biodiversity to their carbon absorptive properties. Oceans are also the source of vast economic riches, from fish to hydrocarbons, and from clean energy materials to next-generation pharmaceuticals. The emerging Blue Economy is defined by the World Bank as the 'sustainable use of ocean resources for economic growth, improved livelihoods, and jobs while preserving the health of ocean ecosystems' (Leiva 2020).

Navies have come to recognise their environmental stewardship responsibilities. The Convention on the Prevention of Marine Pollution by Dumping of Wastes and Other Matter 1972 (or the 'London Convention') is one of the first global conventions to protect the marine environment from human activities. Its objective is to take all practicable steps to prevent pollution of the sea. Currently, the Convention has 87 State Parties (IMO nd). In response to its requirements, the US Navy developed a plastics waste processor – that today operates aboard all naval vessels – to compact plastic trash for safe disposal on land.

Navies have also improved their protection of marine mammals, such as whales. To avoid harming these, the US Navy conducts pre-exercise screening of waters and often reroutes activity to avoid ship strikes. Sonar and other sounds in the ocean are a perennial concern. The United States and other navies conduct extensive research on their impact on marine mammals and make adjustments, particularly in shallow waters, to reduce harm. The US Navy has even redesigned the propeller on some vessels to avoid contact with the slow-moving manatee in shallow coastal areas.

With accelerating climate change in the world's oceans, the main conservation concerns for navies are fish stocks, biological resources, mining and energy extraction, and support to marine protected areas (MPAs). All of these concerns occur within the context of growing competition for ocean resources, as well as increasing demand for the conservation of these same resources.

Compounded by warming seas, IUU fishing poses a direct threat to food security and socio-economic stability in many parts of the world. Developing countries that depend on fisheries for food security and export income are most at risk from IUU fishing. For example, total catches in West Africa are estimated to be 40% higher than reported catches (NOAA Fisheries nd). Many crew members on IUU fishing vessels are from poor and underdeveloped parts of the world, and work in unsafe conditions on boats pursuing multiple illegal activities. One notable example is *Sea Breez 1*, which was intercepted by the Indonesian Navy in 2018 after eluding capture by Interpol and multiple other national jurisdictions (Maritime Executive 2018).

As the Blue Economy grows, interest increases in harvesting resources from the ocean. Ocean resources are, for example, of interest to the pharmaceutical industry, mining and fossil fuel extraction sectors, and the green energy transition is catalyzing interests too. States whose economies depend on fossil fuels, such as Russia, are likely to step up their oil and gas extraction efforts from Arctic coastal areas before they become stranded assets. Seabed mineral mining creates a sustainability paradox: cobalt, manganese, nickel, and copper from such deposits could power the clean energy economy at the cost of ocean health. However, as extraction technology improves, this might be done more sustainably. In this context, it is likely that navies will play various roles and should prepare.

Ocean governance for conservation and sustainability is fragmented. The main framework is the UN Law of the Sea Convention (UNCLOS), which the United States has not ratified. Other important conventions include the Convention on Biological Diversity and the International Maritime Organisation. Navies also have some forms of ocean governance, such as the Incidents at Sea Agreement (INCSEA). This 1972 agreement between the US and Soviet navies was designed to reduce the chance of an accident at sea and prevent escalation in the event one occurred (US Department of State 1972). It was supplemented by the Agreement on the Prevention of Dangerous Military Activities 1989 (see Campbell 1991), establishing a similar high-level forum to discuss ways of avoiding dangerous military confrontations over land and territorial waters. While both agreements are dormant today, they could form the basis for confidence-building measures among the US, Russian, and other navies to avoid incidents, and simultaneously support conservation of the environment. Here we can also note that conservation in the marine environment is benefitting from growing interest in establishing MPAs, supported by the 2020 intergovernmental High Ambition Coalition initiative to protect 30% of land, freshwater, and ocean areas by 2030 (O'Shea et al. 2021). This was taken up by the Biden Administration and reiterated during the One Ocean Summit in Brest, France, in February 2022. One key question will be the extent to which navies will be called upon to enforce MPAs and the so-called '30 × 30 goal'. To do so, navies need further training on working with local communities to respect local practices and customs and to leverage local communities' domain awareness of IUU and other encroachment on MPAs (J. Tasse and S. Kabbej, pers. comm.).

Overall, the past 25 years has witnessed growing conservation stewardship by some navies. Quieter vessels, more efficient fuels and power sources, and improved domain awareness technologies all further mitigation and conservation objectives. The question remains, however, whether the pace of conservation efforts will catch up on the rapid pace of ocean degradation and pollution.

### Adaptation in maritime security

Another line of response to climate and environmental challenges lies in adaptation and resilience. Adaptation consists of managing the climate impacts that cannot be avoided through mitigation. Resilience is the social, economic, and ecosystem capacity to cope with a hazardous event or disturbance (IPCC 2014). Both require adjusting natural and human systems in anticipation of, and in response to, the changing climate. The earlier and the more efficiently adaptation efforts are anticipated and implemented, the better they are able to safeguard human security and reduce risks. This chapter identifies some policy priorities and showcases a few key adaptation efforts. It acknowledges, however, that the list is far from exhaustive and will only grow in scale and complexity without proper mitigation efforts.

### Addressing coastal infrastructure vulnerability and building port resilience

Climate risks on coastal infrastructure notably include disruption of transport due to flooding, damage to assets, such as bridges, inundation of coastal energy and telecommunication infrastructure, and salinisation of water supplies (OECD 2018). Reducing these risks involves complex adaptation decisions and processes. A key pattern is the changing demand for Arctic ports, as sea routes open due to the melting ice. The United States does not currently have a deepwater port in the Arctic. The port of Nome, in Alaska, is undertaking a massive expansion project that will cost 600 million US dollars, although permafrost thaw may compromise the very infrastructure it supports.

More importantly, the submersion of certain agricultural zones, fisheries, industrial areas, and associated ports will disrupt supply chains for food and manufactured products (Izaguirre et al. 2020). Ports serve as critical linkages between global economies, with almost 90% of internationally traded goods transiting through them. Their failure may directly affect food security, human safety, and local economies. Although port infrastructure is generally robust, they are still at risk from hurricanes and sea-level rise. In May 2021, for example, the Port of Pipavav, in India, was hit by Cyclone Tauktae and had to suspend its operations until early June, mostly due to power outages. Similarly, when Hurricane Sandy hit the Port of New York and New Jersey in 2012, power outages due to saltwater intrusion halted commercial activity for ten days, with some terminals out of business even longer (Bachrach 2020).

In order to be resilient, ports need to adapt to changing climate conditions and recover rapidly from disruptions. Climate resilience of port infrastructure typically involves a package of management measures (e.g. modelling climate impacts on assets, changing maintenance schedules, updating operating rules, and creating disaster mitigation plans) and structural measures (e.g. developing traffic diversion and redundancy efforts to overcome disruptions, fortifying coastal and offshore infrastructure, using natural infrastructure such as living shorelines, and locating new facilities outside high-risk zones) (Brandstäter 2021). These solutions reduce repair costs and increase service efficiency. Using natural infrastructure also has the co-benefits of biodiversity conservation and climate mitigation. One example is the DARPA REEFE-NSE (Reef Defence) programme. By building new, living reefs, it attenuates waves and prevents coastal erosion. However, seaport decision makers tend to delay the implementation of climate adaptation strategies, mostly due to governance disconnect, lack of communication or funding, and biased risk perception. To overcome these challenges, decision makers need funding and better guidance, education, and advocacy (Mclean and Becker 2021). Collaboration among the public sector, infrastructure owners and operators, and investors can ensure seaport decision makers have the capacity and incentive to implement climate-resilience measures (OECD 2018).

### Improving Maritime Domain Awareness

Adaptation for maritime environmental security requires improving maritime domain awareness (MDA), defined by the International Maritime Organisation (IMO) as 'the effective understanding of anything associated with the maritime domain that could impact the security, safety, economy, or environment' (IMO 2022). Improved MDA includes, for example, better predictive tools for climatic phenomena as well as real-time monitoring of vessels and illegal activities at sea. To this end, technological innovation is crucial. Satellite solutions like *Copernicus Sentinel-1* have been monitoring sea-level rise and sea-ice evolution as well as identifying risky iceberg areas, which are increasing due to melting ice caps (Vauraste 2021). As sea-ice melting opens alternative – yet unsafe – shipping routes in the Arctic, and as rising tensions in the region lead to more frequent military patrols, such intelligence is vital for early warning and search-and-rescue operations. Robust understanding of this harsh environment will also give significant advantage to national navies (Kuersten 2015).

Enhanced MDA also entails more effective monitoring and policing of 'blue crime'. Illegal activities, such as piracy in Southeast Asia and West Africa, human trafficking in the Mediterranean, narcotics smuggling in the Indian Ocean, and IUU have been expanding – in both reach and scope – due to climate change and biodiversity loss. Combating them effectively depends upon adequate monitoring, control, and surveillance (MCS). Here again, information sharing, open data, and the development of new radar satellite constellations are key. The latter can inform fishery authorities almost in real-time of suspected vessels' exact location, speed, and behaviour, including in challenging weather and at night. Artificial Intelligence (AI) can detect fishing methods, fish species caught, and potential onboard presence of child labour (Withrow 2021). In the future, partnerships between governments, maritime agencies, regional fisheries organisations, and the technology industry will be important to enhance MCS capabilities.

Regional cooperation can also help poor nations with limited MCS capabilities secure their exclusive economic zones (EEZs) and deter blue crime.

### Climate adaptation for military and security forces at sea

Navies operate on the front lines of climate change and are highly aware that the nature and geographical scope of their missions will expand. In particular, the Pentagon recognises that the US Navy has more climate exposure than the other military forces (Eversden 2021). Demands for response – both domestically and abroad – to intensifying natural disasters, forced displacements, resource conflicts, and disputed land and sea areas will continue to increase.

Given the long military procurement timeline, the US Navy needs to plan 20–30 years in advance to make decisions on its future assets today (Reinhardt and Toffel 2017). This may include improving US Arctic capabilities through additional icebreakers (Kuersten 2015), acquiring unmanned aerial vehicles to extend search-and-rescue patrols, or investing in a permanently assigned hospital vessel for the Southern Command. (The latter given the variety of extreme natural disasters that annually hit SOUTHCOM's area of responsibility.) Climate-sensitive technological innovation is also crucial. For example, sonars will need to adapt to changing water salinity, temperature, and pressure (Acoustical Society of America 2016), and boat engines operating in the Caribbean need to be more resistant to surging sargassum seaweed.

Climate change also severely undermines US Navy's readiness, as sea-level rise and severe weather are already damaging its domestic and international bases (Marlatt 2020). A well-known example is the military base in Norfolk, Virginia. The largest military complex in the world – tied to the water by the very nature of its mission – faces increasing flooding from sea-level rise. Adding to land subsidence from tectonic shifting (Phillips 2019), Naval Station Norfolk has recorded over 45.7 cm of sea level in the past 100 years. The projected rise of at least 50.9 cm by 2050 (Malmquist 2020) makes it a living laboratory for climate security adaptation. Pier inundation occurs monthly – undermining training and fleet readiness – and is expected to get worse. In 2020, Virginia unveiled a new resilience master plan, developed by Rear Admiral (Ret.) Ann Phillips, prioritising nature-based solutions beneficial to ecosystems. As for the Norfolk base, the plan involves raising infrastructure, reinforcing floodwalls, and enhancing naval assets' recovery capacities. However, the Norfolk base may still need to relocate some of its assets and residents to safer locations in the future (Turken 2020).

### Climate justice and adapted maritime governance

The range of adaptation efforts detailed here clearly demonstrate how costly adaptation and risk mitigation are, as well as the associated extent of investments and technical know-how. Ensuring security against climate change, including in the maritime domain, will increasingly be a matter of wealth. With sea-level rise of 0.4–2 m expected by 2100, and with 90% of the global coastal population inhabiting 13% of the coastal areas (Stockholm Climate Security Hub 2020), efforts need to be concentrated. Given the divergence of climate impacts on different communities and nations exacerbating existing inequalities, there is a pressing need for cooperation towards climate justice through more equitable distribution of the benefits and burdens of climate change.

Cooperation efforts should focus on addressing the root causes of blue crime by reducing recruitment vulnerabilities amongst coastal communities' (J. Tasse and S. Kabbej, pers. comm.). Criminal networks often recruit from communities where poor governance, poverty, and inequality – each exacerbated by climate change – force people to make difficult coping choices. Combating crime at sea starts on land: raising youth awareness; providing alternative livelihoods; economic security guarantees; and support for local businesses, especially fisheries. There is, however, a mismatch between current programmes and the needs of coastal communities. Designing bespoke strategies for each community requires more intimate knowledge about local dynamics, as well as responses that fit the scope of the problems faced.

Effective mitigation with climate justice also demands a major update of maritime governance. As coastlines change, the delimitation of maritime boundaries under UNCLOS is shifted inland, and with them coastal states' rights to natural marine resources. This shift may lead to conflict from overlapping EEZ claims, as revenues from fish stocks constitute an important part of small insular islands and developing nations' budgets and food systems. This begs the question as to whether small island nations, such as Kiribati or Tonga, will still have legitimate claims to their EEZs when they can no longer inhabit parts of their own land. Moreover, how will they enforce such claims without a navy or coastguard to enforce their rights? The issue should be addressed through legal responses that prioritise fairness and economic justice, and particularly factor in the existential challenges faced by small Pacific island nations as well as larger developing states (Stockholm Climate Security Hub 2020).

Ultimately, the different case studies detailed here make it clear that the key drivers of maritime adaptation to climate change will be innovation, climate science literacy, infrastructure resilience, as well as collaboration and environmental justice.

## Conclusion: setting sail towards a climate-proof future

There are four key conclusions to be drawn from this discussion. The optimal approach to climate change from the maritime military perspective involves multilateral and cross-sectoral cooperation, science diplomacy, the enforcement of security regimes, and concern for human rights and international development to promote social justice.

### Multilateral cooperation and science diplomacy for the high (and rising) seas

Data forms the core of any robust maritime environmental protection and adaptation regime. Climate data collected at the government and academic levels can also inform multilateral cooperation to protect agreed-upon maritime geographies and resources. Governments can spark this process by tabling conservation agreements at existing regional fora, leveraging the '30 × 30 goal' or other politically popular regimes. Subsequent steps include creating and institutionalising data-based joint management of sensitive resources; initiating regular, comprehensive data collection and monitoring regimes run by respected scientists; and cross-training scientists, military, and coastguard personnel on geographically relevant climate security risks. This should provide the intellectual infrastructure for science diplomacy, encompassing transparency and information sharing. The process can be set up to demonstrate early 'wins' by, for example, advising on fishing grounds management practices or climate-resilience strategies for vulnerable coastal areas.

### Cross-sectoral collaboration to batten down the hatches

The multilateral approach described above provides solid foundations for cross-sectoral dialogues and coordinated policy deployment. The current, trusted data streams it proffers can build the science literacy needed for technical and financial assistance, education, and advocacy. With data as a common currency, policymakers can incentivise mitigation, business and non-profit groups can better collaborate on building resilience, and vessel owners can become more informed stewards of the marine environment.

### The military role: getting everyone on the same boat

Nationally sanctioned, internationally recognised naval and coastguard forces are critical to the enforcement of maritime climate security regimes. Efforts can start with systematic coordination across national fleets in order to procure, train, and equip for the full scope of this evolving mission. NATO could be a first mover. Building on the expertise in its existing Centre for Maritime Research and

Experimentation, NATO can hone its conservation and policing capabilities in friendly waters. The 2021 NATO Climate Change and Security Action Plan calls for enhancing outreach with partner countries and international and regional organisations active on climate change and security issues. Center of Excellence on Climate and Security, hosted by Canada, is a promising framework to lead the way in these efforts. At a national level, the US National Maritime Intelligence-Integration Office Arctic programme (NMIO nd) is an example of a maritime intelligence organisation working with private, public, academic, and scientific communities to advance security in the global marine commons. Navies and coastguards should also strive to lead by example in decarbonisation. One obvious collaboration opportunity is the commitment to procure progressively more low- or no-carbon vessel fuels. Others include requiring all equipment to meet increasingly stringent energy efficiency standards and deploying nature-based solutions for climate resilience. Investments in emerging technologies and solutions have the co-benefits of advancing their development, introducing scale economies, and validating broad deployment and use.

### *Making up leeway on climate inequalities*

Climate impacts differ from one region and one community to another and tend to exacerbate inequalities. They also impact more strongly the communities least responsible for carbon emissions. Developing climate justice to achieve more equitable distribution of the benefits and burdens of climate change requires approaches based on cooperation, human rights, and international development. In the maritime domain, local communities must be integrated into the design of coastal resilience strategies, including by institutionalising local knowledge and ensuring the interconnection with the coastguards and militaries operating in their living areas. Climate justice also involves implementing or increasing funding for areas facing an existential threat in order to support critical mitigation and adaptation efforts. Lastly, climate justice cannot be attained without acute attention to protecting the ecosystems which vulnerable communities depend upon. This means prioritising nature-based adaptation solutions allowing for mitigation and conservation co-benefits. Ultimately, the future security of the millions of people living in coastal and island communities around the world depends on these deliberate efforts to prepare for climate risks and protect our future.

### Acknowledgements

For their valuable contributions to this chapter, the authors would like to thank Marisol Maddox, Senior Arctic Analyst at the Polar Institute of the Wilson Center, Julia Tasse, Head of the Climate, Energy, and Security Programme at the French Institute for International and Strategic Affairs

(IRIS) and Co-Director of the French Defence and Climate Observatory, and Sofia Kabbej, PhD student at the University of Queensland's School of Political Science and International Studies and Research Fellow at the French Defence and Climate Observatory.

## Note

1 Figures 10.1–10.3 have been reproduced with the kind permission of SPM/IPCC. The permission requires that the images are not altered in any way. For colour versions of the figures, please see the original reports.

## References

Acoustical Society of America 2016. The future of sonar in semi-heated oceans: naval researchers are studying the effect of climate change on underwater sound propagation and sonar. *ScienceDaily*. www.sciencedaily.com/releases/2016/05/160525111225.htm (accessed 1 May 2023).

Arctic Council 2021. EPPR and Arctic Coast Guard Forum hold joint exercise to improve Arctic maritime emergency response. *Arctic Council*. https://www.arctic-council.org/news/eppr-arctic-coast-guard-forum-joint-exercise-arctic-maritime-emergency-response/ (accessed 1 May 2023).

Asian Development Bank 2022. Asian Development Outlook (ADO) 2022: mobilizing taxes for development. *Asian Development Bank*. 10.22617/FLS220141-3 (accessed 1 May 2023).

Bachrach, T. 2020. Why ports matter for resilience. *The Resilience Shift*. https://www.resilienceshift.org/why-ports-matter-for-resilience/ (accessed 1 May 2023).

Barquet, K., Y. Rylander, and M. Ericksson 2020. Climate, ocean and security: response to ocean-driven security challenges. *Stockholm Climate Security Hub*. https://www.climatesecurityhub.org/wp-content/uploads/2021/01/ClimateOceanSecurity_Pr37pp-Final_WEB.pdf (accessed 1 May 2023).

Beech, H. and M. Suhartono 2020. China chases Indonesia's fishing fleets, staking claim to sea's riches. *The New York Times*. https://www.nytimes.com/2020/03/31/world/asia/Indonesia-south-china-sea-fishing.html (accessed 1 May 2023).

Bengali, S. and V. K.Bao Uyen 2020. Sunken boats, stolen gear: fishermen are prey as China conquers a strategic sea. *Los Angeles Times*. https://www.latimes.com/world-nation/story/2020-11-12/china-attacks-fishing-boats-in-conquest-of-south-china-sea (accessed 1 May 2023).

Beta n.d. The future of flight. *Beta*. https://www.beta.team/ (accessed 1 May 2023).

BCI (The Blue Carbon initiative) 2019. About blue carbon. *The Blue Carbon Initiative*. https://www.thebluecarboninitiative.org/about (accessed 1 May 2023).

Bindoff, N. L., W. W. L. Cheung, J. G. Kairo, J. Arístegui, V. A. Guinder, R. Hallberg, N. Hilmi, N. Jiao, M. S. Karim, L. Levin, S. O'Donoghue, S. R. Purca Cuicapusa, B. Rinkevich, T. Suga, A. Tagliabue, and P. Williamson 2019. Changing ocean, marine ecosystems, and dependent communities. (In) H-O. Pörtner, D. C. Roberts, V. Masson-Delmotte, P. Zhai, M. Tignor, E. Poloczanska, K. Mintenbeck, A. Alegría, M. Nicolai, A. Okem, J. Petzold, B. Rama, and N.M. Weyer (eds) *IPCC Special Report on the Ocean and Cryosphere in a Changing Climate.* Cambridge: Cambridge University Press, pp. 447–587. 10.1017/ 9781009157964.007.

Bolaji, K. 2020. *Climate-Related Security Risks and Violent Crime in Caribbean 'Frontier' Coastal Communities: Issues, Challenges and Policy Options.* New York: UNDP and Folke Bernadotte Academy.

Boulègue, M. 2019. Russia's military posture in the Arctic. *Chatham House.* https:// www.chathamhouse.org/2019/06/russias-military-posture-arctic (accessed 1 May 2023).

Brandstäter, L. 2021. Port resilience: How to make ports immune to future disruptions. *FleetMon.* https://blog.fleetmon.com/2021/08/18/port-resilience-how-to-make-ports-immune-to-future-disruptions/ (accessed 1 May 2023).

Campbell, K. M. 1991. The US-Soviet agreement on the prevention of dangerous military activities. *Security Studies* 1(1): 109–131.

Caribbean Community Climate Change Centre nd. CCCCC mission statement. *Caribbean Community Climate Change Centre.* https://www.caribbeanclimate.bz/ about-us/mission/ (accessed February 2023).

Chavez, L. 2021. Chinese 'fishing fleet' anchored on Philippine reef raises tensions. *Mongabay.* https://news.mongabay.com/2021/03/chinese-fishing-fleet-anchored-on-philippine-reef-raises-tensions/ (accessed 1 May 2023).

Clack, T. and R. Nugee 2022. Climate change is creating security threats around the world – and militaries are responding. *The Conversation* https://theconversation. com/climate-change-is-creating-security-threats-around-the-world-and-militaries-are-responding-173668 (accessed 1 May 2023).

Conley, H. A. and M. Melino 2016. An Arctic redesign: recommendations to rejuvenate the Arctic Council. *Center for Strategic and International Studies.* https:// csis-website-prod.s3.amazonaws.com/s3fs-public/legacy_files/files/publication/ 160302_Conley_ArcticRedesign_Web.pdf (accessed 1 May 2023).

Copernicus nd. Prepare ships: increased safety and efficiency in shipping. *Copernicus Marine Service.* https://marine.copernicus.eu/services/use-cases/prepare-ships-increased-safety-and-efficiency-shipping (accessed 1 May 2023).

Copernicus 2023. 'Record high global sea surface temperatures continue in August'. *Copernicus Climate Change Service.* https://climate.copernicus.eu/record-high-global-sea-surface-temperatures-continue-august (accessed 5 September 2023).

Eversden, A. 2021. 'Climate change is going to cost us': how the US military is preparing for harsher environments. *Defense News.* https://www.defensenews.com/ smr/energy-and-environment/2021/08/09/climate-change-is-going-to-cost-us-how-the-us-military-is-preparing-for-harsher-environments/ (accessed 1 May 2023).

Fuller, C., H. E. Kurnoth, and B. Mosello 2020. Climate-fragility risk brief: the Caribbean. *Climate Security Expert Network/ Adelphi.* https://climate-security-expert-network.org/sites/climate-security-expert-network.org/files/documents/ csen_caribbean_riskbrief.pdf (accessed 1 May 2023).

Gautier, D. L., K. J. Bird, R. R. Charpentier, A. Grantz, D. W. Houseknecht, T. Klett, T. E. Moore, J. Pitman, C. J. Schenk, J. H. Schuenemeyer, K. Sorensen, M. E. Tennyson, Z. Valin, and C. J. Wandrey 2009. Assessment of undiscovered oil and gas in the Arctic. *Science* 324(5931): 1175–1179.

Guy, K., A. Naegele, N. Baillargeon, M. Holland, and C. Schwalm 2021. Temperatures and tensions rise: security and climate risk in the Arctic. *Woodwell Climate Research Center & Council on Strategic Risks.* https://councilonstrategicrisks.org/wp-content/

uploads/2021/06/Security-and-Climate-Risk-in-the-Arctic_Tensions-and-Temperatures-Rise_2021_06_21.pdf (accessed 1 May 2023).

Hayton, B. 2020. China's pressure costs Vietnam $1 Billion in the South China Sea. *The Diplomat*. https://thediplomat.com/2020/07/chinas-pressure-costs-vietnam-1-billion-in-the-south-china-sea/ (accessed 1 May 2023).

Hock, R., G. Rasul, C. Adler, B. Cáceres, S. Gruber, Y. Hirabayashi, M. Jackson, A. Kääb, S. Kang, S. Kutuzov, A. Milner, U. Molau, S. Morin, B. Orlove, and H. Steltzer 2019. High mountain areas. (In) H-O. Pörtner, D. C. Roberts, V. Masson-Delmotte, P. Zhai, M. Tignor, E. Poloczanska, K. Mintenbeck, A. Alegría, M. Nicolai, A. Okem, J. Petzold, B. Rama, and N. M. Weyer (eds) *IPCC Special Report on the Ocean and Cryosphere in a Changing Climate*. Cambridge: Cambridge University Press, pp. 131–202. 10.1017/9781009157964.004.

Hooijer, A. and R. Vernimmen 2021. Global LiDAR land elevation data reveal greatest sea-level rise vulnerability in the tropics. *Nature Communications* 12: 3592. 10.1038/s41467-021-23810-9

IMO (International Maritime Organization) nd. Convention on the prevention of marine pollution by dumping of wastes and other matter. *International Maritime Organization*. https://www.imo.org/en/OurWork/Environment/Pages/London-Convention-Protocol.aspx (accessed 1 May 2023).

IMO (International Maritime Organization) 2017. Milestone for polar protection as comprehensive new ship regulations come into force. *International Maritime Organization*. https://www.imo.org/en/MediaCentre/PressBriefings/Pages/02-Polar-Code.aspx (accessed 1 May 2023).

IMO (International Maritime Organization) 2020. Fourth IMO greenhouse gas study 2020. *International Maritime Organization*. https://wwwcdn.imo.org/localresources/en/OurWork/Environment/Documents/Fourth%20IMO%20GHG%20Study%202020%20Executive-Summary.pdf (accessed 1 May 2023).

IMO (International Maritime Organization) 2022. Introduction to the IMO. *International Maritime Organization*. https://www.imo.org/en/about/pages/default.aspx (accessed 1 May 2023).

IPCC (Intergovernmental Panel on Climate Change) 2014. Climate change 2014: impacts, adaptation and vulnerability: summary for policymakers. https://www.ipcc.ch/site/assets/uploads/2018/02/ar5_wgII_spm_en.pdf (accessed 1 May 2023).

IPCC (Intergovernmental Panel on Climate Change) 2021. Chapter 12. Sixth assessment report, AR6 climate change 2021: the physical science basis. https://www.ipcc.ch/report/ar6/wg1/ (accessed 1 May 2023).

IPCC (Intergovernmental Panel on Climate Change) 2022a. Climate change 2022: mitigation of climate change. https://www.ipcc.ch/report/ar6/wg3/ (accessed 1 May 2023).

IPCC (Intergovernmental Panel on Climate Change) 2022b. Summary for policymakers. (In) P. R. Shukla, J. Skea, R. Slade, A. Al Khourdajie, R. van Diemen, D. McCollum, M. Pathak, S. Some, P. Vyas, R. Fradera, M. Belkacemi, A. Hasija, G. Lisboa, S. Luz, J. Malley (eds.). *Climate Change 2022: Mitigation of Climate Change. Contribution of Working Group III to the Sixth Assessment Report of the Intergovernmental Panel on Climate Change*. Cambridge: Cambridge University Press. 10.1017/9781009157926.001

Izaguirre, C., I. J. Losada, P. Camus, J. L. Vigh, and V. Stenek 2020. Climate change risk to global port operations. *Nature Climate Change* 11: 14–20.

Juhasz, A. 2021. Exxon's oil drilling gamble off Guyana coast 'poses major environmental risk'. *The Guardian.* https://www.theguardian.com/environment/2021/aug/17/exxon-oil-drilling-guyana-disaster-risk (accessed 1 May 2023).

Kuersten, A. 2015. Assessing the US Navy's Arctic roadmap. *Center for International Maritime Security.* https://cimsec.org/assessing-the-u-s-navys-arctic-roadmap/ (accessed 1 May 2023).

Kuo, M. A. 2018. The geopolitics of oil and gas in the South China Sea. *The Diplomat.* https://thediplomat.com/2018/12/the-geopolitics-of-oil-and-gas-in-the-south-china-sea/ (accessed 1 May 2023).

Leiva, M. 2020. What is the blue economy, and is it green? *Investment Monitor.* https://www.investmentmonitor.ai/global/what-is-the-blue-economy/ (accessed 1 May 2023).

Maddox, M. 2021. Climate-fragility risk brief: the Arctic. *Climate Security Expert Network/ Adelphi.* https://5g.wilsoncenter.org/sites/default/files/media/uploads/documents/csen_risk_brief_arctic.pdf (accessed 1 May 2023).

Magill, B. 2020. Military researching ways to suck carbon from air to make fuel. *Bloomberg Law.* https://news.bloomberglaw.com/environment-and-energy/military-researching-ways-to-suck-carbon-from-air-to-make-fuel (accessed 1 May 2023).

Malmquist, D. 2020. Sea-level report cards: 2019 data adds to trend in acceleration. *Virginia Institute of Marine Science.* https://www.vims.edu/newsandevents/topstories/2020/slrc_2019.php (accessed 1 May 2023).

Marlatt, R. 2020. The intersection of U.S. military infrastructure and Alaskan permafrost through the 21st Century. *The Arctic Institute.* https://www.thearcticinstitute.org/intersection-military-infrastructure-alaskan-permafrost-21st-century/ (accessed 1 May 2023).

Maritime Executive 2018. Escaped fishing vessel recaptured in Indonesia. *Maritime Executive.* https://www.maritime-executive.com/article/escaped-fishing-vessel-recaptured-in-indonesia (accessed 1 May 2023).

Mclean, E.L. and A. Becker 2021. Advancing seaport resilience to natural hazards due to climate change: strategies to overcome decision making barriers. *Frontiers in Sustainability* 2. 10.3389/frsus.2021.673630.

Military Emissions Gap 2021. View your government's military emissions data. *Military Emissions Gap.* https://militaryemissions.org/ (accessed 1 May 2023).

National Maritime Intelligence-Integration Office nd. Advance global maritime community of interest priorities. *National Maritime Intelligence-Integration Office.* https://nmio.ise.gov/NMIOs-Missions/Advance-GMCOI/ (accessed 1 May 2023).

National Research Council 2011. *National Security Implications of Climate Change for U.S. Naval Forces.* Washington, DC: The National Academies Press.

NATO 2021a. Climate change and security action plan NATO. https://www.nato.int/cps/en/natohq/official_texts_185174.htm (accessed 1 May 2023).

NATO 2021b. NATO allies Denmark and the UK are on the way to becoming carbon neutral. *NATO.* https://ac.nato.int/archive/2021/nato-allies-denmark-and-the-uk-are-on-the-way-to-becoming-carbon-neutral- (accessed 1 May 2023).

NOAA Fisheries nd. Understanding illegal, unreported, and unregulated fishing. *NOAA.* https://www.fisheries.noaa.gov/insight/understanding-illegal-unreported-and-unregulated-fishing#what-are-some-examples-of-iuu-fishing-activities (accessed 1 May 2023).

NOAA Arctic Program 2021. Arctic report card. *NOAA.* https://www.arctic.noaa.gov/Report-Card/Report-Card-2021 (accessed 1 May 2023).

OECD 2018. Climate-resilient infrastructure: policy perspectives. Environment Policy Paper No. 14. *OECD.* https://www.oecd.org/environment/cc/policy-perspectives-climate-resilient-infrastructure.pdf (accessed 1 May 2023).

OECD 2023. Economic outlook for Southeast Asia, China and India 2023. *OECD Development Centre.* https://www.oecd.org/dev/asia-pacific/economic-outlook/Overview-Economic-Outlook-Southeast-Asia-China-India.pdf (accessed 1 September 2023).

Operational Endurance from Environmental Carbon nd. *Office of Naval Research.* https://www.onr.navy.mil/en/Science-Technology/Departments/Code-33/All-Programs/331-advanced-naval-platforms/operational-endurance-from-environmental-carbon (accessed 1 May 2023).

O'Shea, H., Z. Smith, and K. Poole 2021. Biden administration lays out 30x30 vision to conserve nature. *NRDC.* Available from: https://www.nrdc.org/experts/helen-oshea/biden-administration-lays-out-30x30-vision-conserve-nature (accessed 1 May 2023).

Østhagen, A. 2020. The good, the bad and the ugly: three levels of Arctic geopolitics. *Balsillie Papers.* https://balsilliepapers.ca/bsia-paper/the-good-the-bad-and-the-ugly-three-levels-of-arctic-geopolitics/ (accessed 1 May 2023).

Our Ocean 2016. Sustainable fisheries. *Our Ocean.* http://ourocean2016.org/sustainable-fisheries (accessed 1 May 2023).

Penn, J. L. and C. Deutsch 2022. Avoiding ocean mass extinction from climate warming. *Science* 376(6592): 524–526.

Phillips, A. C. 2019. Statement to the United States House of Representatives Subcommittee on Water Resources and Environment.

Pinsky, M. L., A. M. Eikeset, D. J. McCauley, J. Payne, and J. M. Sunday 2019. Greater vulnerability to warming of marine versus terrestrial ectotherms. *Nature* 569: 108–111.

Reinhardt, F.L. and M. W. Toffel 2017. Managing climate change: lessons from the U.S. Navy. *Harvard Business Review.* https://hbr.org/2017/07/managing-climate-change (accessed 1 May 2023).

Schreiber, M. 2022. A key Arctic Alaska port expansion gets $250 million in federal funding. *Arctic Today.* Available from: https://www.arctictoday.com/a-key-arctic-alaska-port-expansion-get-250-million-in-federal-funding/ (accessed 1 May 2023).

Seminario, M. R., L. Sandin, and I. Parham 2021. Development solutions to address illegal, unreported, and unregulated fishing in Latin America and the Caribbean. *Center for Strategic and International Studies.* https://www.csis.org/analysis/development-solutions-address-illegal-unreported-and-unregulated-fishing-latin-america-and (accessed 1 May 2023).

Simmons, D. 2020. What is 'climate justice'? *Yale Climate Connections.* https://yaleclimateconnections.org/2020/07/what-is-climate-justice/ (accessed 1 May 2023).

Simpson, M., D. Scott, and U. Trotz 2011. *Climate Change's Impact on the Caribbean's Ability to Sustain Tourism, Natural Assets, and Livelihoods.* Washington, DC: Inter-American Development Bank.

Singh, A. 2021. Soil salinity: a global threat to sustainable development. *Soil Use and Management* 38(1): 39–67.

Turken, S. 2020. Virginia unveils new framework to deal with sea level rise. *WHRO.* https://whro.org/news/local-news/14390-virginia-unveils-new-framework-to-deal-with-sea-level-rise (accessed 1 May 2023).

UN FAO 2022. Sargassum. *Food and Agriculture Organization of the United Nations.* https://www.fao.org/in-action/climate-change-adaptation-eastern-caribbean-fisheries/topics/sargassum/en/ (accessed 1 May 2023).

US Department of State 1972. *Agreement Between the Government of The United States of America and the Government of The Union of Soviet Socialist Republics on the Prevention of Incidents On and Over the High Seas.* Washington DC: US Department of State.

Vauraste, T. 2021. The last frontier from space. Polar Perspectives No.6. *Wilson Center.* https://www.wilsoncenter.org/publication/polar-perspectives-no-6-last-frontier-space (accessed 1 May 2023).

Vergun, D. 2022. DOD turns to industry to meet carbon pollution-free energy targets. *DOD News.* https://www.defense.gov/News/News-Stories/Article/Article/2922149/dod-turns-to-industry-to-meet-carbon-pollution-free-energy-targets/ (accessed 1 May 2023).

Withrow, A. 2021. Near real-time monitoring solutions for maritime security. *American Security Project.* https://www.americansecurityproject.org/near-real-time-monitoring-solutions-for-maritime-security/?mc_cid=e5bf94315f&mc_eid=f9b7bbfba0 (accessed 1 May 2023).

# 11

# CLIMATE DISRUPTION TO HIDDEN NETWORKS

Understanding human–animal–ecological relationships for conflict and security

*Alex Tasker*

## Introduction

This book provides a wealth of examples of how climate change is (re)shaping the nature, location, and intensity of present and future conflicts. From pan-global atmospheric alterations to micro-local tensions over water, shifts in climatic conditions complicate many of the most significant challenges we currently face as a species. Despite huge international efforts, the majority of our mitigation and response measures still fail to operate at a sufficient scale and impact to change the current direction of travel. These observations have prompted researchers to look more closely at the shared foundations of climate and conflict in an attempt to unpack the interplay of complex drivers and consider not the 'what', 'where', or 'who', but the fundamentals of 'how' and 'why'. Why does decreasing rainfall lead to violent social change? Why would rising sea level result in economic downturns inland? Answering these questions requires us to radically rethink not only how we plan our responses, but how we define the problem.

To understand the magnitude and existential nature of the climate change threat requires us to consider effects and drivers across a huge number of dimensions. In response, we most often curate and create reductive, linear cause-and-effect models which set out steps from A to B (sometimes via C or D); these stories are useful as they provide us with a reassuring mechanistic sense of control and act as political rallying points to mobilise others to our way of thinking. The reality is that climate change is far more complex; a multidimensional evolving web of interconnected relationships which continually shape, and are shaped by, one another. In order to engage with climate change, we must set aside these attractive linear models and embrace

DOI: 10.4324/9781003377641-14

a relational view of climate drivers and effects that stimulate and sustain new conflicts in unexpected ways. This chapter shows how relational thinking can help uncover the hidden contributions of climate to conflict in order to develop our preparation and responses. Drawing on current scholarship, I will show how a closer focus on interrelationships between the 'big three' – human, animal, and ecological dimensions – can answer some of the questions that continue to trouble those grappling with operational or strategic decision making. The chapter concludes by offering challenges to current thinking in order to help practitioners and researchers engage more effectively with the hidden dynamics of climate-related conflict.

## Climate, conflict, and connections

There is little need to restate the origins of the climate crisis here – other chapters in this volume provide an excellent introduction of our journey to date. My fellow authors introduce specific dimensions of climate impacts, for example, detailing economic and political aspects supported by work on regional and national outcomes. Taken as a whole this volume demonstrates that climate issues transcend geographical borders and disciplinary boundaries, and, as such, require a multidimensional approach to gain greater insight. At the heart of this complexity are relationships between environments and those that inhabit them, and vital issues, such as soil degradation, water access, and utilisation, and micro-level temperature changes. Each provides clear examples of connections between climate, conflict, and the environment.

Despite considerable research, there remains no evidence of a clear causal link between these environmental impacts and direct conflict, instead these factors are mediated through complex nonlinear pathways such as the water–food–energy nexus. Researchers have demonstrated links between co-factors such as terrorism and $CO_2$ rises (Bildirici and Gokmenoglu 2020), yet the casual mechanisms remain less well understood. Other researchers (e.g. Rüttinger et al. 2015) have moved beyond the contextual to propose meta-level climate-fragility risks linking climate change with social, economic, and environmental dimensions to include local resource competition, livelihood insecurity, migration, volatile food prices and provision, and transboundary water management issues. These studies provide multiple perspectives on how environmental factors can generate and sustain insecurity – although they stop short of providing a universal explanation - and underline how we ignore these dynamics at our peril.

Despite accelerating in frequency and severity, connections between environments, human and animal factors, and climate are not new. Ecological conflict mitigation strategies have been central to many cultures for millennia. What is new, however, is our understanding of how existing mitigation techniques are being overcome at increasing rates as climate change impacts are combined with emerging technological and political landscapes to intensify conflicts and increase the scale, frequency, and lethality of traditionally less violent actions.

Speaking to humanity's turbulent history, researchers have suggested that the bodies left by a 10,000-year-old resource conflict in Nataruk (in present-day Kenya) represent the earliest evidence of inter-group warfare stimulated by competition for high-value, productive food-baring ecosystems (Lahr et al. 2016). What can such historical examples tell us about modern aggressions? The clearest messages are those that show how the environmental and eco-logical impacts of climate change must be considered as being layered upon complex socio-political and economic landscapes, which can lead to a daz-zling array of conflicts ranging from civil war and state repression to localised warfare and insurgencies (Shemyakina 2022). The following sections examine examples of human–animal–climate relationships to show how these interac-tions are central to the generation, sustainment, and intensification of conflict.

## Agriculture and conflict

The origins of contemporary nation-states can in many cases be traced back to natural resource acquisition and defence that would not be entirely unfamiliar to our Nataruk ancestors. Agriculture represents one of the oldest practices of natural resource management, with a clear connection between human, animal, and ecological domains. In many rural communities, local ecologies and socio-economic systems remain linked through diverse, com-plex, and often labour-intensive practices. Many of these systems centre on the use of technological and human-centred practices to stabilise often complex ecological conditions, for example, the construction of tiered rice paddies, cultivation of agricultural irrigation systems, or cropping of seasonal fruits. Disruptions to human–animal–ecological systems can have multiple impacts on the health, wellbeing, and livelihoods of marginalised peoples – conditions which can promote disruption and violence.

Even in geographical areas which do not favour traditional settled agri-culture, such as the African Sahel, exposure to extreme temperature and climatic variation can disrupt mobile production systems accelerating conflict (Link et al. 2015). Recent work in the Sahel has shown how this region disproportionately suffers from an unequal distribution of global tempera-ture rises, accelerating at three times the global mean in the last 40 years. These threats are magnified by socio-political decisions around land use, which disrupt relationships between traditional land users and practices, and further accelerate ecological degradation and impact human and animal health. These can be seen profoundly affecting relationships between Sahelian farmers reliant upon livestock for production, where increasing frequency of droughts directly impact forage cover and reduce the number and productivity of livestock.

These disruptions impact established networks between settled farmers and mobile herders which, in turn, increase social, cultural, and political tensions within indigenous groups. This destabilises relationships and provokes and intensifies additional violence in regions which have long histories of conflict. In many cases, violence has been shown to worsen around growing seasons. When these are disrupted, we see the eruption of conflict along the lines of previous grievances (Von Uexkull et al. 2016). Despite this attractively simple narrative, researchers recognise increasingly that these agricultural conflicts occur with greater frequency and severity in more ethnically divided regions than in less diverse areas, underlining the importance of inter-group histories and relationships in determining the likelihood of insecurity following climatic shocks (Schleussner et al. 2016).

Agriculturally-linked violence is not limited to inter-community conflicts. Culturally embedded human–ecological relationships and agriculture also generate tensions between indigenous and state actors. The opium poppy (*Papaver somniferum*) has been cultivated by cultures for over 7,000 years (Grey-Wilson 1993) to the extent that no wild populations exist in the modern world (Chouvy 2010). Opium poppies are capable of growth in relatively harsh conditions. However, advanced ecological knowledge and expertise are required to maximise the food and psychoactive potential of the crop. As Western empires expanded, opium, opium farmers, and opium-producing environments became useful political objects. Britain effectively weaponised the growing opium addiction of China for strategic advantage, enabled and supplied by poppy production from peasant farmers in British colonial India (Richards 1981) – these tensions contributed to the First Opium War in 1839 (Lack 2016). Nearly two centuries later, states and international groups continue to use opium production as a political tool, trading on the imagery of damaging addiction to control international markets (Berger 2014). The impact of these policies on small-scale producers ranges from criminalisation to full-scale militarised responses depending on international objectives.

The majority of the global supply of illicit opium is now produced in Afghanistan, where producers remain caught up in global struggles for influence (Parenti 2015). The ecological knowledge and relationships of Afghan farmers to their environments continue to play a key geopolitical role. Up to the 1970s, Afghanistan had a relatively small capacity to produce opium as the country continued to fight to expel the forces of the USSR (Kreutzmann 2007) supported secretly by billions of dollars of US funding for mujahadeen guerrillas (Parenti 2015). Against this backdrop, first the mujahadeen, and then Taliban, encouraged – or forced – local farmers to grow poppies to fund campaigns against the Soviets, and then assorted internal tribal conflicts to which the United States turned a blind eye. After the 9/11 terrorist attacks, the US-led (and UK backed) invasion against Afghanistan included poppy eradication in its military strategy (Parenti 2015). However, resistance from farmers meant that despite a collective spending of over a trillion US dollars, poppy eradication was not successful, and Afghan farmers continue to work their fields under the shadow of instability.

The examples above demonstrate how agriculture can provide a useful lens to explore how human–ecological relationships can initiate conflict; it is also possible to suggest how agricultural relationships can be weaponised to perpetuate instability and commit socio-ecological harm. For many actors involved in conflicts, agriculture and human–environmental relationships represent key tactical and strategic domains. Alongside the destruction of civil infrastructures and political upheaval, we now recognise armed conflicts as major contributors to food insecurity (Benton et al., this volume). It is no great leap to suggest how assaults on fragile food systems are compounded by climate events such as droughts, floods, and hurricanes. Despite these intuitive connections, however, no simple pathway from climate change to food insecurity to conflict has been shown empirically (Link et al. 2015). The 'black box' leading from food insecurity to conflict suggests that social, cultural, and political variables are more likely to be responsible for the nonlinear emergence of violence, an observation borne out across the globe.

Control of agricultural production now features as a key strategic and tactical objective for many belligerents in war. International and state participants, insurgents, paramilitaries, criminals, and drug gangs have all pursued agricultural objectives for political and logistical campaign success in multiple ways (Adelaja and George 2019). Groups intent on forming parallel states such as ISIS in Iraq and Syria, Boko Haram in Nigeria, and al-Shabaab in Somalia have all worked to establish firm agricultural bases. The degradation of human–ecological relationships and agricultural capacity through climate change and/ or deliberate acts enable these groups to garner public support by engineering or exploiting new and existing scarcity. This scarcity can then be reframed as a failure of the state to provide for their populations whilst simultaneously positioning and legitimising aggressors as a viable alternative to resolve shortages.

Conflicts may also accidentally fracture human–ecological relationships. Attempts to eradicate sub-national criminal and violent activity can shift destructive agricultural production practices into areas of pristine forest such as clampdowns on *coca* production in the Andes (Fjeldså et al. 2005). In these areas, political and economic power is held by armed groups who use violence to assume control of land for other powerbrokers (Álvarez 2020), including conducting illicit drug cultivation and extraction from regions considered under ongoing state control (Alvarez 2002). Increasing demands for illicit crops such as marijuana (*Cannabis sativa*) and opium poppies have radically shifted human relationships with these environments. Coca (*Erythroxylum coca* and *Erythroxylum novogranatense*) were traditionally grown in Andean valleys for local consumption, with ecologically sparse impact compared to areas cleared for cash crops, such as citrus and coffee. Nonetheless, as international demand grew, so did the destruction of forest areas for cultivation in Bolivia and Peru, followed by violent crackdowns which shifted cultivation into Colombia in the 1990s (UN Office on Drugs and Crime 2000).

Economic pressures have forced illicit cultivation into more inaccessible and pristine areas of major forests, destroying biodiversity at greater rates than ever before (Cavelier and Etter 1995). To combat illegal farms, state and military actors have historically used chemical agents to retake and remove contested agricultural ground; the environmental impacts of these approaches are illustrated by the use of glyphosphate in South America (McSweeney 2015) and Agent Orange in the US-Vietnam conflict (Bencko and Foong 2013). The sheer scale and impact of these approaches have informed the development of recent ideas of 'ecocide' (Zierler 2011), and the operational, strategic, and tactical success of these approaches remains debated, especially as belligerents continue to relocate and clear pristine land further from state oversight (Young 1996). In the face of these challenges it is somewhat ironic that whilst many paramilitary groups often seek to support the interests of traffickers, some left-leaning guerrillas have had significant successes in developing and enforcing their own environmental protection agendas (Davalos 2001).

## Conflict and health

The previous section examined agriculture to demonstrate the role of human–animal–ecological relationships in conflict. The exact role of agriculture remains unclear in pathways to insecurity, but within these networks, food security, and by extension, human and animal health, are central considerations. Human, animal, and ecological health are closely bound; just as climate change has been suggested as the greatest global environmental threat of the century, it represents a direct threat to global health. As with agriculture, the specific mechanisms through which climate affects health to generate and sustain conflict remain disputed. However, despite interest in direct impacts such as heat and cold waves, and modified infectious disease behaviours, these mechanisms are likely to be dwarfed by the impacts of second-order effects from reduced agricultural production and increased violent conflict (Bowles et al. 2015). Researchers have been advocating for closer investigation of links between climate-related conflict and health, yet the academic community has been slow to respond with notable exceptions around infectious disease and natural disaster research.

During conflict, population and individual health is often assaulted in multiple ways. Violence may be committed directly on civilians, or health infrastructures (such as clinics and hospitals). Environmental harms such as overcrowding, migration, and reduced nutrition create further impacts, compounded by the reduction of preventative health services and the degradation of human capacities to respond. Many of these impacts, including psychological damages, can extend far beyond the initial insults and may significantly hamper peacebuilding efforts, particularly if marginalised groups remain in controlled zones, or where terrorist activity continues (Miller and Rasmussen 2010). It is vital to understand that these impacts are felt unequally around the globe, with marginalised urban and peri-urban inhabitants of low-income countries baring the brunt of the impacts of climate-related conflict health inequalities (Friel et al. 2008) – those same populations that are most vulnerable to violent instability.

Some researchers have suggested that economic poverty, rather than poor health should be considered as the fundamental driver of violent resistance, proposing Liberia, Haiti, Sierra Leone, and/or the Democratic Republic of Congo as evidence of this. In these countries, we see populations suffering from long-term poverty which undermines state and community capacities to distribute wealth in the face of economic grievances and enable elites to capture the assets of the poor (Nafziger and Auvinen 2003). In these situations, profound inequalities ultimately result in violent confrontation. However, this remains a rather simplistic and possibly unhelpful view of a complex situation. This is countered by Kett and Rowson (2007, p. 404) who suggest, 'many poor countries are not at war; shared poverty may not be a destabilising influence. Indeed, economic growth itself can destabilise, as the wars in countries afflicted by an abundance of particular natural resources appear to show'. What is clear is that climate variability is exacerbating existing links between urban social and health inequalities (Friel et al. 2011).

Despite the complexity of health impacts from human–ecological relationships, recent work by various militaries has made inroads into establishing collaborative civil society projects to offset the risks (Jarvis et al. 2011; Morisetti 2012). The majority of this work is underpinned by ideas that climate change acts not as a primary insult, but as a 'risk multiplier' to human–ecological relationships which increases the probability and severity of violent conflict through reduced resource access. These are confounded through economic impacts and generalised insecurities which further rupture social fabrics and degrade resilience by upsetting strained social and ecological relationships. These reductions of state capacity to ensure the provision of public goods and services, in combination with social disruption, can increase the risk of civil disturbance, terrorism, and ultimately revolution (Kahl 2018; Mazo 2010). Tracing these disruptions further, we find alarmingly similar echoes with agriculture and food security; in times of uncertainty, populations may look to other actors to ensure the provision of services, including health.

## Syria

The previous sections detail how human, animal, and ecological relationships can generate and sustain conflicts compounded by accelerating climate change. These linkages lay at the heart of many historical and contemporary conflicts, including the ongoing violence seen in the Middle East. Syria provides a case in point; a conflict with a long and complex history under-pinned by multiple social, political, and ecological drivers. Here we will rethink the trajectory of the Syrian conflict in terms of the human, animal, and ecological relationship pathways previously discussed to provide new insights into past events, current circumstances, and future trajectories of the conflict.

Before hostilities began, Syria had seen a steady move towards urban migration, mirroring wider global predictions that, by 2030, the numbers living in urban centres will have risen from 5% to nearly 45% (Howson et al. 1998). In Syria, this migration was haphazard and unplanned, driving asymmetries in resource access and inhibiting service planning and provision. This hampers our search for evidence of a discrete cause of conflict, as a limited amount of data on health and the environmental situation is available from before the conflict (Maziak et al. 2005). Despite these blind spots, human–ecological links emerged between the instability we now see and Syria's high population growth rate and limited access to natural resources (Müller et al. 2016). Examples of failing relationships include the rapid expansion of irrigation projects in the 1980s which undermined traditional human–animal–ecological practices and depleted groundwater reserves, placing increased stress on urban water supplies – a trend mirrored in other Middle Eastern locations (Jaafar et al. 2020). Many researchers have sug-gested how the conflict escalated following the severe drought from 2006 to 2010. Reduced rainfall and ecological damage significantly degraded the already fragile agricultural sector, resulting in mass unemployment in rural areas, worsening food insecurity, and intensified existing mass migrations away from farmland towards urban centres. This marked upturn in popu-lation displacement intensified Syria's already high growth rate and worsened ongoing drought-linked water scarcity in urban areas, damaging existing human–animal–ecological relationships, overwhelming coping strategies, and establishing the conditions for social unrest.

These challenges occurred within a worsening ecological landscape that had degraded human–ecological practices over decades. Immediately prior to the drought, for example, vehicle and industrial air pollution had risen to unprecedented levels. These levels intersected with other environmental harms including the Syrian government's enabling of unacceptably high levels of mining pollution and mismanagement of other natural resources. Prior to the conflict, in these urban areas, nearly three-quarters of Syria's population were exposed to high levels of particulate matter from industrial, domestic, and seasonal pollution. Ironically, the percentage of the population exposed to pollution dropped in real terms shortly after the onset of violence in 2011, before rapidly rising to a new high in 2015 – a rise attributed to the bombardment of oil facilities, bushfires, dust storms, and chemical attacks. It is perhaps of little surprise that in the traditionally lush regions of Syria – dependent on rainfall agriculture and with the highest population densities – the combination of increased pollution, ecological damage, drought, and in-migration led these areas to become the most violent sites of protest (Houghstow 2018).

These impacts were not limited to urban areas. The close human–animal–ecological ties found in rural communities were subject to different disruptions to high-density settings. In 2020, wildfires destroyed over 9,000 hectares of agricultural and forested land and resulted in massive disruptions of power and basic services. These losses stimulated further degradation, including the targeting of power lines. Following attacks on 30 power stations, timber became an essential resource for power generation and heating. Timber collection extended far beyond the replaceable, rapidly growing cash crops to include centuries-old heritage trees in Hasakah and Al-Belas reserves. Wildfire also destroyed fruit-producing trees in Heffeh, Lattakia Markaz, and Jableh, and dust storms damaged food production across the region.

The destruction of economic crops such as fruit trees – and potentially more significantly for Syria, olive oil – has significant implications for long-term stability and conflict resolution. Prior to the conflict, Syria had over 79 m trees producing over a million tons of olives for consumption and oil production combined, accounting for between 1.5% and 3.5% of GDP. This sector employed nearly 10% of Syria's pre-conflict population; Turkish-backed Syrian opposition groups then targeted and destroyed over half a million olive trees in Afrin (northwest Syria) as part of offensive operations, decimating the industry. Continuing hostilities in these same areas continue to degrade and contaminate previously fertile areas. This is compounded by marked displacement which has increased demands on already fragile eco-systems, especially in the western coastal region which contains over 90% of the vegetation. These have been significantly degraded during the 2011–2017 conflict and seen greatly increased soil erosion rates, compounded by fluctuant use by displaced communities (Abdo 2018).

These conditions set the scene for the ongoing violent conflict. As noted, many researchers have asserted that the multi-year drought that began in 2006 led to the 2011 uprisings, an argument commonly cited to support the idea of climate change as a cause of conflict. By rethinking climate not in terms of cause and effect, but of unbalancing existing human–animal–ecological relationships and the subsequent erosion of livelihoods, security, and health, we can see how events, such as the drought, impact established webs of relationships that stretch back into history. In her work examining the role of climate in the Syrian uprising, De Châtel (2014, p. 532) notes that 'by focusing on external factors like drought and climate change in the context of the Syrian uprising is counterproductive as it diverts attention from more fundamental political and economic motives behind the protests and shifts responsibility away from the Syrian government'. Through examining the complex human, animal, and ecological factors that underpin these 'political and economic motives', we can extend our understanding beyond these dimensions to offer a deeper and more nuanced picture of the present and future Syrian conflict.

## Lessons for the future

Under all reasonably predicted trajectories, climate change will have profound impacts on the availability of natural resources including water, arable land, and food. Impacts on these systems will not be distributed equally, with much of the burden falling on low and middle-income countries in the tropics (Bowles et al. 2014). In the majority of these settings, agriculture remains central to local cultures, social practices, and economies. Communities that retain strong animal, human, and ecological connections are profoundly vulnerable to disruptions in these relationships.

These relationships are vulnerable to multidimensional threats. Pakistan, Iraq, and Afghanistan are subject to increased state and climate vulnerability, and 30 of the most weak, fragile, and failing states can be found in Sub-Saharan Africa (Mazo 2010). Macro-climatic threats, such as variations in Monsoon rains, have the potential to destroy crops in South Asia or reduce the flows of major rivers. The stakes are high. Should the ecological coping mechanisms and harvests of Pakistani populations living near the banks of the Indus fail, might Pakistan, for example, place (nuclear) pressure on India for increased water sharing?

All too often climate change appears to be a zero-sum game in which states must secure their interests against those of other competitor nations. By considering the human–animal–ecological relationships at the core of our societies, we may be able to offer alternative pathways through these current challenges. Instead of investing in business-as-usual, we could look to combine mitigation efforts with searches for new strategies and new relationships between our societies and ecologies. By centring cultural and economic connections between groups and their environments, we become more aware of the ways in which impacts are felt and addressed. Moreover, by prioritising these contextual, relational dynamics we provide new perspectives on those pathways that lead towards and away from conflict. Defence ignores these at their peril.

## References

Abdo, H. G. 2018. Impacts of war in Syria on vegetation dynamics and erosion risks in Safita area, Tartous, Syria. *Regional Environmental Change* 18: 1707–1719.

Adelaja, A. and J. George 2019. Effects of conflict on agriculture: evidence from the Boko Haram insurgency. *World Development* 117: 184–195.

Alvarez, M. D. 2002. Illicit crops and bird conservation priorities in Colombia. *Conservation Biology* 16: 1086–1096.

Álvarez, M. D. 2020. Forests in the time of violence: conservation implications of the Colombian war. *Journal of Sustainable Forestry* 16(3–4): 47–68.

Bencko, V. and F. Y. Foong 2013. The history, toxicity and adverse human health and environmental effects related to the use of agent orange. (In) B. G. Simeonova, F. Macaev, B. Simeonova (eds) *Environmental Security*. Dordrecht: Springer. pp. 119–130.

Berger, L. 2014. Heroin use and harm reduction in Afghanistan: an interview with Helen Redmond, LCSW. *Journal of Social Work Practice in the Addictions* 14: 425–434.

Bildirici, M. and S. M. Gokmenoglu 2020. The impact of terrorism and FDI on environmental pollution: evidence from Afghanistan, Iraq, Nigeria, Pakistan, Philippines, Syria, Somalia, Thailand and Yemen. *Environmental Impact Assessment Review* 81: 106340.

Bowles, D., C. Butler, and S. Friel 2014. Climate change and health in Earth's future. *Earth's Future* 2: 60–67.

Bowles, D. C., C. D. Butler, and N. Morisetti 2015. Climate change, conflict and health. *Journal of the Royal Society of Medicine* 108: 390–395.

Cavelier, J. and A. Etter 1995. Deforestation of montane forests in Colombia as a result of illegal plantations of opium (Papaver somniferum). (In) S. P. Churchill, H. Balslev, E. Forero, and J. L. Luteyn (eds) *Proceedings of the Neotropical Montane Forest Conservation Symposium, The New York Botanical Gardens, 21–26 June 1993*. New York: NYBG, pp. 541–550.

Chouvy, P-A. 2010. *Opium: Uncovering the Politics of the Poppy*. Cambridge, MA: Harvard: University Press.

Davalos, L. M. 2001. The San Lucas mountain range in Colombia: how much conservation is owed to the violence? *Biodiversity & Conservation* 10: 69–78.

De Châtel, F. 2014. The role of drought and climate change in the Syrian uprising: untangling the triggers of the revolution. *Middle Eastern Studies* 50: 521–535.

Fjeldså, J., M. D. Alvarez, J. M. Lazcano, and B. Leon 2005. Illicit crops and armed conflict as constraints on biodiversity conservation in the Andes region. *AMBIO: A Journal of the Human Environment* 34: 205–211.

Friel S., T. Hancock, T. Kjellstorm, G. McGranahan, P. Monge, and J. Roy 2011. Urban health inequities and the added pressure of climate change: an action-oriented research agenda. *Journal of Urban Health* 88: 886–895.

Friel S., M. Marmot, A. J. McMichael, T. Kjellstrom, and D. Vågero 2008. Global health equity and climate stabilisation: a common agenda. *The Lancet* 372: 1677–1683.

Grey-Wilson, C. 1993. *Poppies: A Guide to the Poppy Family in the Wild and in Cultivation*. London: BT Batsford Ltd.

Houghstow, A. B. 2018. *Learning from Syria: Applying Environmental Modeling Toward Strategic Peacebuilding Interventions*. Cambridge, MA: Harvard University.

Howson, C. P., H. V. Fineberg, and B. R. Bloom 1998. The pursuit of global health: the relevance of engagement for developed countries. *The Lancet* 351: 586–590.

Jaafar, H., F. Ahmad, L. Holtmeier, and C. King-Okumu 2020. Refugees, water balance, and water stress: lessons learned from Lebanon. *Ambio* 49: 1179–1193.

Jarvis, L., H. Montgomery, N. Morisetti, and I. Gilmore 2011. Climate change, ill health, and conflict. *British Medical Journal* 342: 10.1136/bmj.d1819

Kahl, C. H. 2018. *States, Scarcity, and Civil Strife in the Developing World*. Princeton, NJ: Princeton University Press.

Kett, M. and M. Rowson 2007. Drivers of violent conflict. *Journal of the Royal Society of Medicine* 100: 403–406.

Kreutzmann, H. 2007. Afghanistan and the opium world market: poppy production and trade. *Iranian Studies* 40: 605–621.

Lack, A. 2016. *Poppy*. London: Reaktion Books.

Lahr, M. M., F. Rivera, R. Power, A. Mounier, B. Copsey, F. Crivellaro, J. Edung, J. Fernandez, C. Kiarie, and J. Lawrence 2016. Inter-group violence among early Holocene hunter-gatherers of West Turkana, Kenya. *Nature* 529: 394–398.

Link, P. M., T. Brucher, M. Claussen, J. S. Link, and J. Scheffran 2015. The nexus of climate change, land use, and conflict: complex human–environment interactions in Northern Africa. *Bulletin of the American Meteorological Society* 96: 1561–1564.

Maziak, W., K. Ward, F. Mzayek, S. Rastam, M. Bachir, M. Fouad, F. Hammal, T. Asfar, J. Mock, and I. Nuwayhid 2005. Mapping the health and environmental situation in informal zones in Aleppo, Syria: report from the Aleppo household survey. *International Archives of Occupational and Environmental Health* 78: 547–558.

Mazo, J. 2010. *Climate Conflict: How Global Warming Threatens Security and What to Do About It*. London: Routledge.

McSweeney, K. 2015. *The Impact of Drug Policy on the Environment*. New York: Open Society Foundations.

Miller, K. E. and A. Rasmussen 2010. War exposure, daily stressors, and mental health in conflict and post-conflict settings: bridging the divide between trauma-focused and psychosocial frameworks. *Social Science & Medicine* 70: 7–16.

Morisetti, N. 2012. Climate change and resource security. *British Medical Journal* 344: 34–35.

Müller, M. F., J. Yoon, S. M. Gorelick, N. Avisse, and A. Tilmant 2016. Impact of the Syrian refugee crisis on land use and transboundary freshwater resources. *Proceedings of the National Academy of Sciences of the United States of America* 113: 14932–14937.

Nafziger, E. W. and J. Auvinen 2003. *Economic Development, Inequality, and War: Humanitarian Emergencies in Developing Countries*. Dordrecht: Springer.

Parenti, C. 2015. Flower of war. *The SAIS Review of International Affairs* 35: 183–200.

Richards, J. F. 1981. The Indian empire and peasant production of opium in the nineteenth century. *Modern Asian Studies* 15: 59–82.

Rüttinger, L., G. Stang, D. Smith, D. Taenzler, and J. E. Vivekananda 2015. *A New Climate for Peace: Taking Action on Climate and Fragility Risks*. London: The Wilson Center.

Schleussner, C-F., J. F. Donges, R. V. Donner, H. J. Schellnhuber 2016. Armed-conflict risks enhanced by climate-related disasters in ethnically fractionalized countries. *Proceedings of the National Academy of Sciences of the United States of America* 113: 9216–9221.

Shemyakina, O. 2022. War, conflict, and food insecurity. *Annual Review of Resource Economics* 14: 313–332.

UN Office on Drugs and Crime 2000. *World Drug Report*. Oxford: United Nations Office on Drugs and Crime.

Von Uexkull, N., M. Croicu, H. Fjelde, and H. Buhaug 2016. Civil conflict sensitivity to growing-season drought. *Proceedings of the National Academy of Sciences of the United States of America* 113: 12391–12396.

Young, K. R. 1996. Threats to biological diversity caused by coca/ cocaine deforestation in Peru. *Environmental Conservation* 23: 7–15.

Zierler, D. 2011. *The Invention of Ecocide: Agent Orange, Vietnam, and the Scientists Who Changed the Way We Think About the Environment*. Athens, GA: University of Georgia Press.

# 12

# CLIMATE INTELLIGENCE IN THEORY AND PRACTICE

*Louise Selisny, Timothy Clack, Tristan Burwell, and Richard Nugee*

## Introduction

In *The Art of War*, Sun Tzu asserted that foreknowledge enables, 'the wise sovereign and the good general to strike and conquer, and achieve things beyond the reach of ordinary men' (Sun Tzu 1910, p. 59; see also McNeilly 2015). Intelligence must be timely if it is to be of use. And so, for at least 2,500 years, it has been recognised that intelligence sets the foundation for successful warfighting and stabilisation operations. Intelligence, of course, is not limited in its use to the military domain and also offers advantage in statecraft, politics, and economics. It also holds the potential to help constructively shape policies and operations to mitigate climate change threats and their consequences.

This chapter argues that hyper-localised climate change and insecurity data is required to inform appropriate state responses and resource allocations across contexts of regional, national, and subnational variability. The chapter outlines potential mechanisms for the collection and collation of climate intelligence (CLINT). It argues that the leveraging of deployed military force elements for data collection, particularly those with reconnaissance and intelligence capabilities, offers a package of multi-scalar benefits, without the addition of a significant specialist training burden. Drawing on interview data collected in 2021–2022 from soldiers serving in the United Nations Multidimensional Integrated Stabilisation Mission in Mali (MINUSMA), the chapter assesses the operational value of CLINT in a context where climate change has, and is, exacerbating security deterioration.

The operational reach of deployed military assets – underpinned by logistical support, force protection, and surveillance capabilities – is greater

DOI: 10.4324/9781003377641-15

than that of civilian actors and ensures data can be collected linked to local climatic conditions, conflict dynamics, community adaptations, and intervention effects. With appropriate sanitation, this data can be used by military forces and/or distributed freely to civilian and scientific actors for independent interpretation and analysis, in 'real time' where possible. As well as providing a collective vernacular and methodology for collection, such granular data can be used to inform local climate forecasts and, in turn, predictive analyses of the operational and tactical conflict environment.

## Climate security context

Climate security concerns unite military, academic, and civil organisations. From the High North, across the Sahel, and through Europe, climate security has become a prominent issue. At COP26, the international climate change debate was focused on mitigation and the lack of robust emission reduction targets to keep planetary warming under 1.5°C. This had implications for military posturing (see Nugee and Clack, this volume). In turn, the focus of COP27 was on action and implementation, particularly on financing initiatives relating to loss and damage (EC 2022). The role of the military in the sphere of response – adaptations, operations, and programming – has yet to be contextualised.

When deploying in stabilisation roles, particularly in upstream environments, Western militaries tend to be reactive. The threat from insecurity, insurgency, and terrorism – to both Western interests and the human security of locals – is usually acute. Indeed, since the underperformance in, and conclusion of, the early 21st century low-intensity wars in Iraq and Afghanistan, the capabilities of Western militaries and appetites of their political executives for intervention have diminished (Osinga 2021; Self 2022; Wintour 2021). Moreover, the West's current strategic calculus affords primacy to great power competition and preparation for peer or near-peer contest. However, there is an erroneous internal logic to this position, which sees preparedness for peer competition and counter-insurgency/ counter-terrorism as mutually exclusive (see Jones 2021). Irregular and surrogate warfare remain core characteristics of great power competition in the so-called 'grey zone' (Johnson 2021). As such, the West's capacity for operations in low-intensity environments should not be deprioritised.

In light of the Russia–Ukraine War, the case for Western military adaptation, if not transformation, has been made repeatedly (Atlantic Council 2023; Binnendijk and Hamilton 2022; Ryan 2022). NATO (2022) has itself reported its intent to, 'continue to adapt and develop, politically and militarily, to meet the challenges of a more unpredictable and competitive world'. The necessity of adaptation to the forecasted operating environment, of course, remains central to mainstream military planning. Maintaining operational reach is vital.

Sustainability, operating in harsher conditions, and responding to the impacts of climate change through activities such as Humanitarian Assistance and Disaster Relief (HADR), Military Aid to the Civilian Authorities (MACA) (see Evans and Lewis, this volume), and the augmentation of military estates are commonly cited as ways in which Western defence will have to adapt.

Providing the foundation for this pre-emptive action, The UK Ministry of Defence's 'Climate Change and Sustainability Strategic Approach' (MOD 2021a) highlighted that the UK defence sector accounts for 50% of the central government's emissions and prescribed the identification of products, practices, and behaviours that are more climate aware, environmentally sound, and work towards reducing emissions. This vision was shared by the 2021 UK Defence Command Paper (MOD 2021b) which expressed an intent for the United Kingdom to become a world leader in responding to threats exacerbated by climate change, building resilience to more extreme weather conditions, mitigating the impact of military's carbon footprint, and seizing opportunities to improve sustainability. CLINT holds the prospect to be an additional resource in this battle.

## Climate intelligence

### *CLINT and C-CLINT*

The role of intelligence is to give policymakers a decision advantage. In practice, this can, for example, inform assessments as to threats and priorities and provide early warning and longer-range forecasts. Recent conflicts have demonstrated the value of open-source intelligence (OSINT). In Ukraine, commercial satellite imagery indicated Russian preparation and stockpiling in the pre-invasion phase, traffic data on Google Maps exposed troop activity on either side of the border, and the public in Ukraine provided 'crowd-sourced' intelligence linked to Russian troop movements (Karalis 2022). OSINT can be triangulated with other intelligence to enhance reliability, illumination, and assessments.

Militaries can also make intelligence available for public consumption – make it open-source – in order to deter, counter disinformation, inform and shape attitudes, and facilitate analyses. The Russian–Ukraine War, for example, has been fought 'in the open' (The Economist 2023). Indeed, since early 2022, UK defence intelligence has provided 'intelligence updates' via Twitter. As Dylan (2022) has pointed out, the deployment of intelligence in the public arena 'denied Russia the advantage of surprise, and undercut its claims … in the information space'. The point is that using OSINT and/or making intelligence available has clear value in certain contexts.

There are various areas where intelligence can add to tackling climate insecurity. Covert influence might have an impact in shaping the appetites of political and corporate leaders for transformational change, facilitating access to resources, and deescalating frictions over energy and water. Moreover, the collection of secret intelligence might be used to inform multilateral climate change negotiations. It is reported, for example, that signals intelligence (SIGINT) has been used by Western states to confer advantage in international negotiations linked to political and trade agreements and conflict resolution (Mainwaring 2020; Poteat 2000; Risen and Poitras 2014). There is also the prospect of climate counterintelligence (C-CLINT). If states have climate treaty obligations, such as to reduce carbon emissions, it may be that their claims in relation to these can and should be verified. The intent and capability to collect such intelligence could have an incentivising effect as it facilitates the exposure of those not meeting their commitments, with implications for ongoing cooperation in this and other spheres.

Nugee et al. (2022) outlined the potential for an enhanced UK military offering in relation to CLINT, with a focus on widening the breadth of indicators reported on, as well as increasing collaboration with academic institutions in order to enhance climate security forecasting and predictive analysis. Moreover, Sikorsky (2023, pp. 11–15) has outlined other areas where intelligence on climate could be usefully collected and incorporated into existing NATO intelligence programmes. These include integrating climate and environmental change data into analysis; evaluating and developing climate security risk assessment frameworks; and refining climate security intelligence education for member nations and their intelligencers.

Further to these areas, this chapter proposes that CLINT can be collected by national militaries and employed as an instrument to inform their strategic calculus and operational activities. As Barry (2022, p. 6) notes, deployed forces will need to, 'better understand how climate change is impacting their areas of operations and the local population'. Beyond the traditional intelligence process,[1] the CLINT cycle informs understanding, advantage, capacity, and resilience (Figure 12.1).

### Collection and analysis

A variety of active and passive remote sensing techniques are employed to observe climate data. These include assorted satellite-fitted sensors and techniques. For example, infra-red (IR) sensors can measure land and oceanic temperature and evapotranspiration rate and imaging spectroscopy can examine the chemical composition of the atmosphere, water vapour, and levels of precipitation. Various non-orbital remote sensing platforms, including aircraft, boats, and Argo floats, are used, for example, to measure the levels of atmospheric pollutants, such as greenhouse gases and particulate matter, and oceanic circulations, surface, and depth temperatures. Ground-based instruments are also used. To measure solar radiation, for example, sun

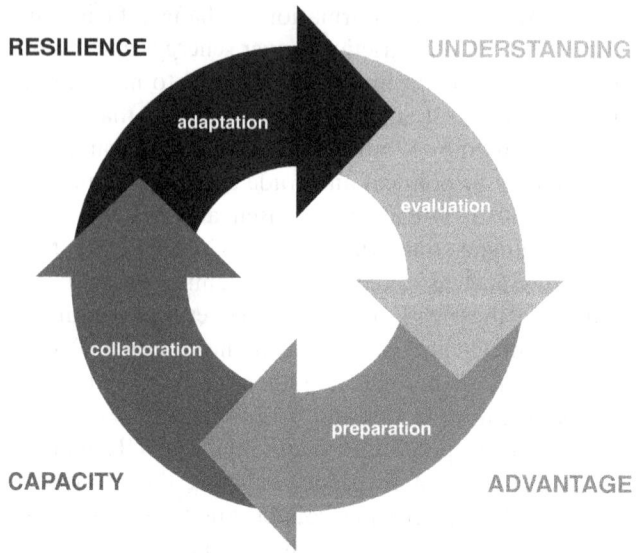

**FIGURE 12.1** The CLINT cycle.

*Source:* Authors' creation.

spectral radiometers and pyranometers are used (Argo 2023; Yang et al. 2013). In addition to rain gauges and lightning detectors, hygrometers, anemometers, and barometers are used to measure humidity, wind speed, and air pressure, respectively.

Given the primacy of remote sensing data from orbital platforms in climate change modelling, it is unsurprising that macro-level analyses have become dominant.[2] As Yang et al. (2013, p. 880) note, coarse-resolution sensors 'cannot capture climate processes occurring at finer spatial scales'. Local-level climate modelling requires greater precision and volume of sensor data. These sensor technologies are available – deployable in both ground and orbital contexts – and hold much potential. However, it is also the case that, as noted by the editors of a special issue of the journal *Sensors*, 'new tools and concepts are needed to combine long-term (coarse resolution) data sets with these novel high-resolution observations' (Ludwig and Sanchez-Azofeifa 2019). Advances in data mining, data fusion, and integration of ground- and satellite-based sensors should yield significant understanding (e.g. see Gallant 2018).

In certain localities, such as Sub-Saharan Africa, it has been shown that even ensembles of global climate models perform poorly (Charles et al. 2021, pp. 7–8). The collection of specific environmental data is central to robust forecasting. Innovative research into rainfall patterns in East Africa used sub-seasonal diagnostics with high-resolution topography to better understand the role of low-level jets – fast-moving ribbons of air in the low levels of the atmosphere – in order to improve drought and flood forecasting (Charles et al. 2021, p. 7). Similarly, recent improvements to sub-basin scale seasonal forecasting in Ethiopia have relied on the analysis of global climate tele-connections – relationships between weather phenomena separated by great distance – and sea-surface temperatures (2021, pp. 7–8). A localised and fusion approach to CLINT is thus optimal. Data from multiple sensors can be collected to provide granular and hyper-localised insights.

In addition, CLINT collection is also likely to focus on climate-related effects in an area of operations, including local community access to water (e.g. mapping of perennial watercourses, number as well as levels and use of different types of wells, etc.); environmental indicators (e.g. rate of bush encroachment, detail on wildfires, salinity in groundwater, etc.); and human responses to shifting conditions (e.g. levels of in and out-migration, rangeland mobility of nomadic cattle-herders, etc.). In specific areas, local communities could also be trained to record certain environmental and response data, including information on water source use, consumption levels, and fluctu-ations in price in water economies. As such, CLINT provides the necessary nuance for understanding how climate impacts vary between groups and individuals, by gender, ethnicity, poverty, unequal social and political power, and other processes of exclusion and marginalisation.

### Algorithmic processing

CLINT can provide historic, current, and predictive insight into natural (primary) and human (secondary) systems, thereby facilitating informed decision making for climate security mitigation and adaptation. Advances in artificial intelligence (AI), machine learning (ML), and acceleration in the processing time for climate data could be harnessed.[3] Haupt et al. (2022, p. e1354) recognise that, in general, modellers have turned to 'automation infused with AI and ML to derive new knowledge and provide real-time actionable predictions, augmenting the capabilities of researchers and fore-casters'. It is clear that the longer time scales required in climate forecasting as opposed to weather forecasting have meant there have been fewer appli-cations developed for the former. Regrettably, a detailed assessment of the merits and precision of different forms of algorithmic forecasting is beyond the scope of this chapter (see Hoch et al. 2022; Strobach and Bel 2020; Thomas et al. 2021).

The potential of algorithmic climate forecasting is indicated in recent studies. Haupt et al. (2022), for example, have noted that 'transfer learning' has proved a fruitful approach. Pasini et al. (2017) utilised both anthropo-genic and natural environment data to construct a neural network to model climate change over one and a half centuries and then used sensitivity analysis by fixing certain variables and permitting others to change to determine shifts and their associated drivers. Moreover, Ham et al. (2019) employed a convolutional neural network (CNN) to train El Niño–Southern Oscillation from coupled dynamic models, which was then further trained with reanalysis data. A key lesson from developments in this field is that forecast accuracy depends less on the quality of the forecast algorithms and more on the quality of the input data – both training data and, in real-time forecasts, current meteorological conditions (Haupt et al. 2022, p. e1359). Access to quality data is crucial to realising the opportunities afforded by technologies.

### Other prospects

There are other possible opportunities afforded by CLINT. It could, for example, be used to establish a baseline in long-duration campaigns so as to offer insight into trends over time and impacts of mitigation or other trends in theatre. In addition to supporting military operations, CLINT might also provide data that can be used by academics, civil society, and policymakers to examine the interactive drivers of conflict within the context of climate change. If employed correctly, CLINT holds the potential to mitigate tactical to strategic level climate risks and enhance the security of some of the world's most vulnerable communities. It can also confer greater fidelity to global and regional climate analyses. Here we should note that the scale of technical resources required to deliver climate forecasting in and around conflict zones is often beyond the means, logistics, and risk appetite of civilian research teams and charities. Militaries are better positioned to engage in forecasting activities, particularly if this is done cooperatively with the research and technological sectors.

The character of CLINT is such that, in many contexts, it can be shared. Whilst at times collection techniques may need to be protected, sensor data and assessments could be made openly available. In many situations, it may prove possible and desirable to do this in real time. Such sharing of granular data would inform scientific actors and wider publics. Scientific communities lament the lack of localised climate data available for their analyses, particularly relating to the most at risk, fragile, and remote environments (Bell-Pasht and Krechowicz 2015; Reyes-Garcia et al. 2016). Given the growing public awareness of climate issues, such activity is likely to also confer a reputational dividend for militaries. Being seen to be an active participant in the understanding and mitigating of climate change threats is a significant engagement opportunity at home and aboard.

There is also demand from both the policy world and energy industry for forecasts and other insights into the climate effects of scalar adjustments to the energy mix, e.g. in the transition to renewable energies, such as wind and solar, in certain localities (see Ham et al. 2019; Haupt et al. 2016; Stengel et al. 2020). CLINT could help meet these requirements, which are present in and beyond conflict environments, and inform more context-sensitive decision making linked to infrastructural investment.

## Capacity and integration

This chapter suggests that military-generated CLINT could greatly enhance climate security forecasting, predictive analysis, and military decision making, particularly when it is produced in collaboration with civilian research competencies. Militaries have the capacity to maintain footprints and operational effectiveness in the fragile environments where the impacts of climate change are most acute. In short, the military's reach, force protection capabilities, and risk thresholds facilitate data collection in areas inaccessible to civilian counterparts. To demonstrate the applicability of the approach, the viability of incorporating CLINT collection will be described through the lens of the UK context.

The UK MOD's security mission extends to 'global reach and influence' (MOD 2023). The footprint of the UK armed forces extends across multiple continents and a diversity of human and natural environments. By exploiting this characteristic reach and flexibility to deliver CLINT, the UK military is well-placed to position itself at the frontier of a more considered approach to climate security. Integrating CLINT into intelligence collection plans enhances predictive analysis and advance warning for climate risks. Given the increasing awareness of the climate-security nexus, the collection of CLINT would facilitate a greater understanding of fragile theatres and climate-impacted communities. CLINT could also be used to identify target areas for engagements and resources, thus enabling earlier mitigation of the security risks posed by climate variation and change. By default, CLINT collection complements the UK's 2021 Command Paper's vision of a persistent and proactive global engagement that, as mentioned, increases understanding of and pre-empts threats to the UK's national security and economic interests (MOD 2021b). The UK's 2023 Command Paper refresh, drawn up in response to the significant security shifts brought about by the Russia-Ukraine War, continued to recognise the realities of climate change on the UK's global strategy and operations and the requirements of adaptation (UK MOD 2023b, p. 6, 35).

The collection of CLINT is possible without significant adjustments to operational activities or the burden of many additional duties. Certain parts of the military have experience in deploying and working with sensors and are otherwise well-positioned to conduct data and intelligence collection. Force elements, such as special operations forces (SOF/SF), for example, are deployed routinely alongside partner forces around the world, often have intelligence experience – as collectors and customers – and access to, and regularly learn to use, cutting-edge technologies in high threat environments.

In the United Kingdom, given their role to undermine adversaries and contribute to collective deterrence by training, advising, and, if necessary, fighting alongside allies and partners, the Special Operations Brigade[4] (formerly the Specialised Infantry Group) could also collect CLINT. Given this is a new formation and intended to operate alongside specialised forces in complex threat environments, where understandings of climate-induced insecurity would confer opportunities for operational impact, this would perhaps be optimal. The 11th Security Force Assistance Brigade (SFAB) is also well-positioned to collect and use CLINT, with its mission to contribute to early stage conflict prevention and resilience through the commission, design, delivery, and assessment of security force assistance activity around the world.

The capabilities in UK specialist reserve forces, for example in the 77th Brigade, combine military training with in-depth subject matter expertise to support regular units. The contribution of reserves covers both augmentation and support to operations. Specialist skills are necessary to help meet the challenges of modern conflict, including the threats from climate change. Recruitments into the specialist reserve and wider collaboration would build organic resilience, for example, in enhancing understanding. Moreover, the opportunity to train deployable climate security experts within the UK's 6th Division – the formation that coordinates assorted intelligence, information, and partner operations – could be explored.

Making them especially suitable to CLINT work, commonalities across these units include their reach (often into fragile geographical areas), adaptive and flexible operators, innovative use of technologies, experience of working with partner and liaison forces, and delivery of upstream effects (in terms of both location and phase of the conflict cycle). Intelligence collection in operational theatres is routine – e.g. the intelligence preparation of the battlefield – but often fails to achieve a comprehensive understanding. This is to be expected due to the complexities of different environments, scale of and whether capabilities are available in theatre, episodic character of sources, and adversarial countermeasures. Given the amplification of climate security impacts (see Introduction, this volume), CLINT is increasingly likely to enhance the breadth of indicators available for assessment and illuminate key dynamics in select areas of operations.

The next section considers a limited number of ways that CLINT can be integrated into mainstream military tactics, techniques, and procedures (TTPs). This is discussed in relation to the UK military but similar integrative opportunities are available for exploitable in other military contexts.

### Integration

The Intergovernmental Panel on Climate Change (IPCC) provides a repository of socioeconomic information that offers a consistent framework of factors that can be applied in climate change impact assessments. The collection of this data was conducted at the macro/aggregate regional level. There is, however, also the need for micro/local-level data collection to improve understandings of, for example, climatic conditions, demographic change, economic shifts, land use, water access, agricultural activity, human–animal co-dependencies, and biodiversity.

Engagement with locals and other stakeholders, and the observations of deployed forces and data from their assorted sensors, would generate a useful bank of information. Furthermore, investigation into loss of livelihoods due to environmental degradation and extreme weather, as well as food insecurity and its secondary impact on social tensions, would increase and inform contextual understanding. Similarly, understanding of local political dynamics and the quality of governance retains great significance for managing climate-related security risks. In the context of horizon scanning, knowledge of local capacity to adapt to wider climatic stresses becomes a core part of conflict forecasting and mitigation.

To help achieve the necessary level of granularity across socio-political themes that are atypical to certain traditional forms of intelligence collection, the augmentation of forces with specialist knowledge and skills would be optimal. Just as 'Cultural Advisors' and 'Human Security Advisors' have been embedded within units deployed on operations to enhance situational awareness (Clack and Dunkley 2022), attaching 'Climate Security Advisors' who could analyse local climate-impacted economic and political dynamics would increase operational effectiveness and permit militaries to offer a valuable contribution to other upstream conflict prevention capabilities and options (see Clack and Johnson 2019). Furthermore, minor adjustments could have a major impact. A Q1 brief, for example, could be tailored to include climate impacts. Moreover, the collection of CLINT and assessment of climate effects could be encompassed within operational Human Terrain Reconnaissance (HTR), Human Terrain Mapping (HTM), and Target Audience Analysis (TAA) activities.

In order to develop a military understanding of the complex and nonlinear relationships between climate change and conflict, CLINT should not be siloed from wider government and scientific expertise. A fusion approach would be optimal, in particular one which integrates and interfaces national security, diplomatic, and development capabilities. In the UK's case, this would involve integrative working between the deployed units, single services, MOD, Foreign, Commonwealth and Development Office (FCDO), and various others.

NATO is particularly relevant here as it is set to continue its ambitious programme of climate security deliverables, with commitments on enhanced climate impact assessments, forecasting, and exercises (see Kertysova, this volume). The United Kingdom has much to offer the Alliance with direct contributions to the NATO work strands of Allied Command Transformation (ACT), Emerging Security Challenges Division (ESCD), and the Climate Change and Security Centre of Excellence (CCASCOE).[5] With one of the primary aims of the CCASCOE to be, a 'platform through which both military actors and civilians will develop, enhance, and share knowledge on climate change security impacts' (Government of Canada 2022), the contribution from CLINT could be significant. At the strategic level, collaborative working linked to the sharing of intelligence would serve to strengthen (inter)national policy decisions and bolster collective defence against climate insecurities.[6]

## CLINT in the Sahel

This chapter takes the position that hyper-localised climate intelligence is required to inform not only military operations but also appropriate state responses and resource allocations across contexts of regional, national, and subnational variability. As such, this section describes the value of CLINT in practice through consideration of the situation in parts of Mali, a locality that could be described as a canary in the coal mine for climate change insecurity. In this theatre, ongoing conflict, political instability, and weak government institutions – each exacerbated by climate change – degrade the security situation.[7] The situation is compounded by rapid population growth – in excess of 3% per annum – and pervasive poverty. The ongoing armed conflict has also weakened food production and health systems and undermined the functioning of markets and institutions, as well as profoundly impacting health outcomes, including mortality, morbidity, and malnutrition (Tranchant et al. 2018). USAID (2020) have projected a 3–6°C temperature increase by the end of the century and, in the forthcoming decades, high occurrence of flooding and droughts due to increased inter-annual variability in rainfall, with sudden oscillations between extremely wet and extremely dry years.

The West African Sahel is a semi-arid region extending from Chad westwards to the Atlantic Ocean. It is sandwiched between the Sahara to the north and savannah lands to the south. The region is amongst the poorest and most environmentally degraded in the world. Indeed, across Sahelian West Africa, temperatures are increasing at a rate 1.5 times higher than the global average, the frequency of droughts has doubled between 2015 and 2020, and, in parts, flooding is common (Red Cross 2021). In Mali, 20% of the country's population (3.6 m) is also currently food insecure (UN 2021). Moreover, the combination of decreased seasonal regularity of vegetation regrowth with high drought frequency and rainfall variability has the potential to amplify food insecurity for the 80% of the population that depend on agropastoral forms of subsistence (ICRC 2021). Climate change has also been reported to hamper the capacity of civilian humanitarian workers to reach vulnerable groups (ICRC 2021; UN 2021). Militaries offer an alternative and, of course, have experience in delivering HADR operations, including in sub-Saharan Africa (see Jaff et al. 2022; Newsome and Herron 2012). Indeed, there have been calls for the strengthening of HADR operations in the context of climate change (Retter et al. 2021).

A lack of, or delay in, humanitarian response to extreme weather events and environmental crises risks human security, drives displacement, and reconfigures transhumant migratory patterns. Without access to grass and water, for example, cattle-herding communities are unable to sustain themselves. In drought conditions, those that practise rain-fed cropping (Traore 2013) or those that have a subsistence regime of semi-nomadic seasonal fishing are imperilled (WPS 2023). Under such circumstances, such people must move or die (USAID 2020). In 2021, the UN's High Commissioner for Refugees estimated the number of Malians displaced from their homes to be in excess of 300,000 (ICRC 2021). This is within a wider regional context of 1.1 m refugees and internally displaced persons in Burkina Faso, Chad, Mauritania, and Niger cumulatively (USAID 2020). Enhancing the prospect of conflict, migration increases stress on infrastructure in urban areas and food and water resources in rural areas. A senior UN official reported that the influx of migrants into Mali's urban centres, for example, Timbuktu, Bamako, and Mopti has resulted in episodes of resistance and violence (see also UNHCR 2023).

It is within this context that, in December 2020, the British Army's Long Range Reconnaissance Group (LRRG) was deployed on Operation SEKA to support the United Nations Multidimensional Integrated Stabilisation Mission in Mali (MINUSMA).[8] From the outset, the MINUSMA mandate recognised explicitly the impact of climate change on the security and stability of Mali (UNSC 2020). Deploying to Gao initially, the basis of the first rotation of the British Army task group comprised approximately 300 soldiers from The Light Dragoons and Royal Anglian Regiment (British Army 2021a). The LRRG deployed with orders to recce across the area of operations, to protect the local population, ensure the freedom of movement of friendly forces, and deter hostile activity (British Army 2021b). A key factor in the mission was to 'enable understanding of the needs of local communities' (British Army 2021b). Moreover, part of the publicly stated intent of the LRRG was to provide 'persistent presence' (British Army 2021b) in remote parts of the country in order to anticipate and react to emerging crises.[9]

During the course of their first deployment, the LRRG exploited the off-road manoeuvrability of Jackal 2 and Coyote vehicles to conduct their long-range reconnaissance patrols and, in doing so, covered more than 1,500 km and conducted local engagement at over 60 villages (Figure 12.2). The kinetic nature of the LRRG's area of operations had seen the deployment described as the 'most dangerous peacekeeping mission' (BBC 2021), with threat conditions comparable to those faced during the insurgency in Afghanistan.

**FIGURE 12.2** British Army's Long Range Reconnaissance Group (LRRG) in central Mali in 2021.

*Credit:* UK PJHQ; Crown copyright.

Amidst their patrolling, the LRRG was reported, for example, to have engaged and neutralised terrorist fighters (British Army 2021c). To date, MINUSMA has also suffered almost 300 fatal casualties from disease, small arms, and IED attacks (Al Jazeera 2023).

To inform the analysis in this chapter, during 2021–2022, the authors interviewed soldiers from the LRRG in Mali and on their return to the United Kingdom. In total, 12 individuals were interviewed. With a focus on CLINT, the interviews were designed to collate responses and perspectives on climate change security broadly and explore the potential for new methodologies for data collection specifically. The interviewees were representative of the entire task group and included almost the full rank profile from the Commanding Officer to troopers. As discussed in the next section, a range of pertinent observations were recorded from these interviews. An additional output of these interviews was climate security lessons, which are reproduced in the appendix.

### 'No need for mortars': climate security observations

All the interviewees recognised the central role played by climate on the security situation situation in their area of operations. Local disputes over water, food, and grassland, in particular, were noted to interconnect with national and regional conflict dynamics. This, in turn, amplified community tensions and violence, recruitment into and facilitation of armed groups, and formation and activities of self-defence militias. These trends have been identified in various reports and analyses (see BBC 2019; Benjaminsen and Ba 2018; Bodian 2020; Mbaye 2020; SIPRI 2021; see also Cabot 2017). Moreover, the UN has indicated explicitly that the 'cycles of farmer-herder violence and reprisals have become increasingly lethal since 2015' (UN 2021).

The importance of connections and movement in the region was emphasised (see Scheele 2012; McDougall and Scheele, 2012). Again 'ground-truthing' a range of other analyses (Beeler 2006; SIPRI 2021), the LRRG interviewees described how environmental stresses were clearly changing traditional migration and mobility patterns. This exacerbated the scale and risk of violence between herders and agriculturists, in particular increasing friction between Dogon and Fulani communities. Whilst such responses were noted to lead to a greater exposure to human security violations for all, as seen elsewhere (Fordham 1999; Le Masson et al. 2019; Pelling 2011), the burden was recognised to fall disproportionately on women as they more often lacked the resources, access to services, and networks to navigate stress or implement alternative lifestyle strategies.

As noted, the MINUSMA mandate recognised climate change as impacting the conflict in Mali. However, the level of prioritisation, which affects the focus and allocation of resources by the UN and troop-contributing countries, meant that it was never seen as anything more than a complicating environmental factor. Climate change was neither treated as a threat multiplier, threat shaper, and threat driver nor a target that should be understood in detail and have resource expended in its mitigation. Many of the interviewees noted that the collation, processing, and dissemination of tactical-level climate and environmental reporting – CLINT – could have informed operational decisions and, potentially, the posture of MINUSMA. There were conflicting views expressed as to whether additional manpower would have been required to capture granular data. One of the respondents summed up the suboptimal alignment of approach and equipment with the words, 'no need for mortars'. In order to increase flexibility across the task group, it was considered advisable for all members to have undertaken a level of engagement training. It was also noted that the Force HQ should have had the infrastructure in place to curate and process data and, where appropriate, share it with the UN and other agencies.

Lots of evidence for clashes between herders and sedentary communities was highlighted. It was accepted that, given the harsh conditions, even small changes to the environment and climate could have a significant effect, including the disruption of livelihoods and cross-border migration. Also recognised was a paucity of data relating to multi-scalar migratory and mobility patterns. The evidence for climate insecurity driving recruitment into violent and extremist organisations was unclear but most respondents assessed that – in order for locals to navigate increasing resource scarcity and be on the right side of power dynamics – this was likely to be happening.[10]

A number of interviewees articulated that a shift in mission focus would have improved operational outcomes. The key metrics of operational success were related to the scale of acts of violence – including theft, robbery, and extortion – against civilians. However, it was observed that if the mission had seen a greater focus on reducing the freedom of movement of extremist groups, this would have facilitated additional attention on the root causes of radicalisation and malign power leverage. With such a recalibration, climate change dynamics would have been much more visible and relevant in theatre. This would have seen a pivot from 'effect' to 'cause', with a concomitant licence to protect or degrade the means by which the environment and natural resources were being used by malign actors. Importantly, this would also have had positive implications for human security.

The value of embedded advisors in the task group was consistently made clear. The suggestion that at least one of these be a specialist (subject matter expert) on climate and its impacts was thought at once worthwhile and workable. A range of ways that the task group could have been better prepared pre-deployment were described and a number of these were suggested in order to offer insight into local climate and effects. These included input from local and specialist advisors (in and beyond Mission Specific Training); short courses by academics and experts, specifically on how climate change affects the lives of people; wargaming of climate dynamics to explore implications for the Human Terrain and Target Audience Analysis; and exercising collaborations with the UN, NGOs, and other agencies. It was also suggested that the latter could become part of the battlecraft syllabus. Given the lack of baseline data, it was also considered critical that climate and environment data was stored in electronic repositories and made available to relief task groups. The detailed debriefing of key personnel in the task group by intelligence debriefing teams to include climate factors in the area of operations was also proposed. This was on the basis of ensuring mission continuity through the transfer of high-fidelity resolution and understanding.

Given the above observations from the LRRG, it is clear that CLINT has the potential to confer significant dividend. There are other examples of non-military collection and analysis from the Malian theatre which not only further demonstrate value but could be incorporated into military tools and assessments. To give one example, the competition over diminishing arable and pastoral lands and how this relates to conflict in Mali has been identified as an intelligence gap. Various analytical enterprises offer insight. Analysis conducted by Mercy Corps in Mali, for example, used geo-coded Armed Conflict Location and Event Data Project (ACLED) data that mapped the distribution of conflict across the country and then overlaid it with several different environmentally linked factors, including transhumance patterns.

Additionally, the International Organisation for Migration's Transhumance Tracking Tool (TTT) monitors transhumance routes and provides early warning of potential conflicts that may arise due to food and water insecurity or the movement of rival groups into contested territories.

### Campaigning for climate security

The potential of CLINT exists in a potential feedback loop whereby the sharing of data and insight catalyses analytical effort from the research community, NGOs, and other actors. Over a decade of widespread insecurity in Mali and across the wider region has constrained the range and reach of feasible scientific research, including at the intersection of climate and conflict. With its training and equipment, the UK task group, for example, could have used its unique manoeuvrability and force protection capabilities to reach even the remotest and conflict-affected areas in order to generate CLINT.

The LRRG could have collected intelligence to illuminate shifting environmental conditions and food stress and, in turn, developed indicators and warnings to identify communities most at risk of escalating food scarcity, displacement, and conflict. Such identifications could have been used to support the work of the UN World Food Programme (WFP) and other NGOs. To be clear, this is not a case of competing for a space already occupied by other actors, but rather of supplementing efforts linked to the former's ability to project a presence into the most remote and 'at risk' areas. This offers prospects for an integrated approach to human security operations.

There is much other data that could be collected in order to inform climate security assessments. For example, food insecurity indicators include crop loss; livestock and herd numbers; drought and flood effects; soil erosion; and presence of pests. Water insecurity indicators include water levels; source access; salinity levels; and seasonal variation. Whereas livestock numbers and productivity, deterioration of grasslands and water sources, loss of rangeland access, and scale of migration and displacement are potential indicators of conflict between and within farming and herding subsistence groups.[11]

In Mali, force elements could also have been used to verify climate-related imagery intelligence (IMINT) and geo-spatial intelligence (GEOINT). Satellite data, for example, is often of low resolution, with indistinct features. As part of their routine patrol activities, reconnaissance units could have been utilised to provide real-time information on vegetation health, land use, and water locations and levels. Patrols also had the ability to conduct human terrain mapping. With a focus on baselining and adaptations, such techniques and capabilities could have been deployed to identify specific transhumance routes and displacement. As became apparent from the interviews with

LRRG personnel, deployed forces became aware of conflict incidents that go unreported on open-source databases, such as the International Peace Information Service (IPIS) and ACLED, but are logged on their own systems, such as the Tactical Ground Reporting System (TIGR). Where appropriate, these data could be shared and overlaid to enrich climate security analysis and the development of prevention, resilience, and preparedness measures.

Mindful of human security considerations, CLINT provides opportunities for insight into a battery of demographic factors. In the context of gender, for example, resource scarcity, conflicts, and displacement caused by environmental degradation are deepening existing inequalities and, at times, amplifying rates of gender-based violence (GBV) in Mali (UN 2021). Climate change increases the distances women and girls travel to access water. This makes them more vulnerable to sexual assault and also results in girls missing days from school or being withdrawn completely in order to assist with increasingly protracted household chores (UN 2021, 2022). Studies have also highlighted examples of girls being traded for resources during times of peak scarcity, thereby increasing their vulnerability to sexual exploitation and trafficking. In the Sahel and elsewhere, girls have also been reportedly married off at increasingly younger ages so as to pass the burden of provision on to the husband's family from the parents' family. On this basis, it has been noted that in climate-changed environments, 'women and girls face double victimization as human beings as well as because of their gender' (Desai and Mandal 2021, p. 138).

The relevance of gender to the LRRG's mission was emphasised in the interviews. It was noted that effective data collection and collation, for example, linked to the location and use of potable water, cost and regulation of local food and water economies, and incidence of GBV, would have informed trend analyses. This, in turn, could have protected vulnerable communities and demographics by informing operational decisions as to resource allocations. Moreover, it was suggested that data and insight could have been shared with MINUSMA and other relevant agencies in order to inform coordinated action. For example, deconflicted effort could have resulted in engineering projects to construct wells targeted at areas where human security risks were greatest and where the winning of 'hearts and minds' would most undermine the attraction of malign actors.

Other opportunities can also be identified from the context of Mali. In 2020/1, approximately 6% of the UK task group were women (17 of 283). Across the wider MINUSMA force, women made up only 3.7%.[12] Given such ratios, each uniformed woman represents a significant critical enabler and engagement asset. Balancing security concerns with cultural sensitives, with appropriate training, these personnel could conduct activities such as community relationship building, communication, searches, and intelligence collection. This could take the model of Female Engagement Teams (see US Army Command 2014; cf Greenburg 2023), but gendered engagement could be more informal and agile. The prospects for the collection of CLINT through such female engagement efforts are worth exploring in future operational theatres.

## Conclusion

Climate is increasingly an actor in conflict. Climate change is a shaping threat, threat multiplier, and direct threat. In recognition, this chapter has proposed a new military intelligence trade of CLINT and shown how it informs a multifaceted understanding of the drivers and catalysts of climate insecurities, as well as changes to operating environments. In the main, CLINT is not a covert capability but it does, nonetheless, provide policy-makers and theatre commanders with decision advantage in order to influence events. It also confers wider benefits and, as well as informing pre-emptive mitigation and adaptation, responses and support to human security protections, and reputational enhancements, it generates data for scientific and infrastructural analyses.

With minor adjustments to kit and TTPs, the UK military has the opportunity to take the lead in developing this offering. CLINT would enhance both civilian and military capabilities, thereby strengthening collective climate defence and security. Moreover, the United Kingdom announced in 2019 that it was to set up a Centre of Excellence for Human Security (MOD 2019). The development of CLINT could enhance the mitigation of GBV and intersectional risks in operational theatres.

Drawing on the experiences and insights of the UK's LRRG in Mali, the chapter proposes that the leveraging of deployed military assets, particularly those with reconnaissance and intelligence capabilities, for CLINT offers a package of multi-scalar benefits, without the addition of a significant specialist training burden. CLINT can also draw upon existing competencies, such as HTR, GEOINT, and OSINT, to generate timely and impactful multi-source intelligence. Other scaling competencies, such as AI, can amplify scope and scale, and reduce processing time. As such, CLINT is also a unifying approach for drawing together existing and new competencies for operationalising climate security within military intelligence.

## Acknowledgements

Sincere thanks are owed to the personnel of the LRRG for participating in the interviews and sharing their insights and experiences. We also gratefully acknowledge permissions from the UK's Permanent Joint Headquarters (PJHQ) to reproduce Figure 12.2.

## Notes

1 The intelligence process/cycle is a conceptual pathway used extensively by military and civilian intelligence agencies. Existing in a closed loop, the pathway consists of five stages: direction; collection; processing; analysis/production; and dissemination. The approach is not without its critics or alternatives (see Phythian 2013).

2 Other limitations are also evident for common satellite-fitted sensors, e.g. recalibration of satellite sensors is often impossible after launch and sensors are known to lose radiometric sensitivity and stability during their operational deployment.

3 In line with Anning et al. (2022), we highlight three cautions: the risks of 'AI-hype'; ethical challenges and the need for a 'human in the loop'; and issues around the availability of obtaining quality data from fragile environments (see also O'Hara 2020; Slota et al. 2020; Wieltschnig et al. 2021).

4 The Special Operations Brigade is composed of four Ranger Battalions and the Joint Counter-Terrorism Training and Advisory Team (JCTTAT).

5 In May 2022, NATO officially accepted Canada's formal offer of a Climate Change and Security Centre of Excellence (CCASCOE). Canada announced in June 2022, that Montréal would be the host city. In July 2023, NATO representatives from 12 sponsoring nations signed the CCASCOE into existence.

6 Other collaborations would also be possible at the international level, including, for example, with the United Nations Environment Programme's Strata platform that seeks to, 'democratize environmental and climate security intelligence by making analytical capacity available to practitioners and policymakers' (UNEP 2022). Strata works by overlaying environmental and climate indicators, socio-economic indicators, and conflict and humanitarian indicators.

7 It is not possible in the confines of this chapter to discuss the origins and history of the current conflicts in Mali. Readers are directed to the following works for insight linked to various themes: violent and extremist organisations (Thurston 2020); French intervention and counter-terrorism (Guichaoua 2020; Wing 2016); failed peace accords (Pezard et al. 2015); political transitions (Haidara 2021; The Economist 2020); and governance issues (Bleck and Logvinenko 2018).

8 Following a coup d'état and subsequent increase in violence, including the so-called Tuareg Rebellion, MINUSMA was established by UN Security Council Resolution 2100 in April 2013. The intervention maintains nearly 13,500 military and 2,000 police personnel from 65 countries, with the strategic priorities of implementing and supporting stabilisation and reconciliation, reducing inter-communal conflict, protecting civilians, and strengthening state presence (UN 2014).

9 This posture was, of course, undercut with the termination of operations in late 2022. The British withdrawal was, however, inevitable once the French announced that their much larger military footprint (via Operation BARKHANE) was leaving in response to the military coup, Mali's leadership partnering with Russia's Wagner Group on security and commercial matters, and growing local sentiment that the French (and West) should depart (BBC 2022).

10 A number of violent and extremist organisations present a significant threat in central and northern Mali, including Islamic State West Africa (ISWA), Jamaat Nusratul Islam wal-Muslimin (JNIM) and Katiba Macina (Clack and Nugee 2022). These groups have the intent and capability to mount attacks against government, military, and civilian targets (Redaction 2022; Risemberg 2022). The Malian military and, since 2022, Wagner Group mercenaries have also been involved in the extra-judicial killing of civilians (France24 2022).

11 Such data could inform studies on the causal relationship between conflict and resource scarcity (Percival and Homer-Dixon 1996; cf Martin et al. 2006). The

collation of granular data would reduce the chance of misdiagnoses of association and enable less prominent variables to be incorporated into analyses.

12 This situation has improved and, as of May 2023, the proportion of uniformed women in MINUSMA was 5.2% (UN 2023).

## References

Al Jazeera 2023. Mali court sentences man to death over UN peacekeeping deaths. *Al Jazeera*. https://www.aljazeera.com/news/2023/1/26/mali-court-sentences-man-to-death-over-u-n-peacekeeper-deaths (accessed 1 May 2023).

Anning, S., T. Fenton, J. Muraszkiewicz, and H. Watson 2022. Operationalising human security in the contemporary operating environment: proposing population intelligence (POPINT). *Journal of Intelligence, Conflict, and Warfare* 4(3): 30–61.

Argo 2023. Argo and climate change: Argo and the warming sea. *Argo Programme*. https://argo.ucsd.edu/science/argo-and-climate-change/ (accessed 1 May 2023).

Atlantic Council 2023. In 2022, the war in Ukraine awakened Europe: here's how it must adapt in 2023. *Atlantic Council*. https://www.atlanticcouncil.org/blogs/new-atlanticist/in-2022-the-war-in-ukraine-awakened-europe-heres-how-it-must-adapt-in-2023/ (accessed 1 May 2023).

Barry, B. 2022. *Green Defence: The Defence and Military Implications of Climate Change for Europe*. London: The International Institute for Strategic Studies.

BBC 2019. Mali attack: more than 130 Fulani villagers killed. *BBC News*. https://www.bbc.co.uk/news/world-africa-47680836 (accessed 1 May 2023).

BBC 2021. Mali: the world's 'most dangerous peacekeeping mission'. *BBC News*. https://www.bbc.co.uk/news/av/world-africa-56949408 (accessed 1 May 2023).

BBC 2022. UK withdraws troops from Lai early blaming political instability. *BBC News*. https://www.bbc.co.uk/news/uk-politics-63627489 (accessed 1 May 2023)

Beeler, S. 2006. Conflicts between farmers and herders in north-western Mali. *Sida*. https://www.iied.org/sites/default/files/pdfs/migrate/12533IIED.pdf (accessed 1 May 2023).

Bell-Pasht, K. and D. Krechowicz 2015. Why does access to good climate data matter? *World Meteorological Organization Bulletin* 64(2). https://public.wmo.int/en/resources/bulletin/why-does-access-good-climate-data-matter (accessed 1 May 2023).

Benjaminsen, T. A. and B. Ba 2018. Why do pastoralists in Mali join jihadist groups? A political ecological explanation. *The Journal of Peasant Studies* 46(1): 1–20.

Binnendijk, H. and D. S. Hamilton 2022. Lessons for NATO from the war in Ukraine. *Politico*. https://www.politico.eu/article/nato-summit-europe-brussels-war-in-ukraine/ (accessed 1 May 2023).

Bleck, J. and I. Logvinenko 2018. Weak states and uneven pluralism: lessons from Mali and Kyrgyzstan. *Democratization* 25(5): 804–823.

Bodian, M., A. Tobie, and M. Marending 2020. The challenges of governance, development and security in the central regions of Mali. *SIPRI Insights on Peace and Security*. https://www.sipri.org/sites/default/files/2020-04/sipriinsight2004.pdf (accessed 1 May 2023).

British Army 2021a. Mali – Operation Newcombe. *UK MOD*. https://www.army.mod.uk/deployments/mali/ (accessed 1 May 2023).

British Army 2021b. Long Range Reconnaissance Group completes deterrence operation in Mali. *UK MOD*. https://www.army.mod.uk/news-and-events/news/2021/04/op-seka-mali/ (accessed 1 May 2023).

British Army 2021c. British soldiers engage with terrorist fighters in Mali. *UK MOD.* https://www.army.mod.uk/news-and-events/news/2021/10/british-soldiers-engage-with-terrorist-fighters-in-mali/ (accessed 1 May 2023).

Cabot, C. 2017. *Climate Change, Security Risks and Conflict Reduction in Africa: A Case Study of Farmer-Herder Conflicts over Natural Resources in Cote d'Ivoire, Ghana and Burkina Faso, 1960–2000.* Dordrecht: Springer.

Charles, K., A. Murgatroyd, D. Yeo, and A. Nileshwar 2021. *Water Security for Climate Resilience Report: A Synthesis of Research from the Oxford University REACH Programme.* Oxford: REACH.

Clack, T. and M. Dunkley 2022. Culture, heritage, security: an interview with Colonel Rosie Stone, Captain Mark Waring, Major Anne Seton-Sykes, and Major Luke Wattam. (In) T. Clack and M. Dunkley (eds) *Cultural Heritage in Modern Conflict: Past, Propaganda, Parade.* London: Routledge, pp. 289–308.

Clack, T. and R. Johnson (eds) 2019. *Before Military Intervention: Upstream Stabilisation in Theory and Practice.* Basingstoke: Palgrave.

Clack, T. and R. Nugee 2021. Climate change is creating security threats around the world – and militaries are responding. *The Conversation.* https://theconversation.com/climate-change-is-creating-security-threats-around-the-world-and-militaries-are-responding-173668 (accessed 1 May 2023).

Desai, B. H. and M. Mandal 2021. Role of climate change in exacerbating sexual and gender-based violence against women: a new challenge for international law. *Environmental Policy and Law* 51: 137–157.

Dylan, H. 2022. How has public intelligence transformed the way this war has been reported? *KCL News.* https://www.kcl.ac.uk/how-has-public-intelligence-transformed-the-way-this-war-has-been-reported (accessed 1 May 2023).

EC (European Commission) 2022. COP27: team Europe steps up support for climate change adaptation and resilience in Africa under Global Gateway. *European Commission.* https://ec.europa.eu/commission/presscorner/detail/en/ip_22_6888 (accessed 1 May 2023).

Fordham, M. 1999. The intersection of gender and social class in disaster: balancing resilience and vulnerability. *International Journal of Mass Emergencies Disasters* 17(1): 15– 37.

France 24 2022. Malian and 'white' soldiers involved in 33 civilian deaths, UN experts say. *France24 News.* https://www.france24.com/en/africa/20220805-malian-and-white-soldiers-involved-in-33-civilian-deaths-un-experts-say (accessed 1 May 2023).

Gallant, A. L., W. Sadinski, J. F. Brown, G. B. Senay, and M. F. Roth 2018. Challenges in complementing data from ground-based sensors with satellite-derived products to measure ecological changes in relation to climate. *Sensors* 18(3): 880. 10.3390/s18030880

Government of Canada 2022. NATO Climate Change and Security Centre of Excellence. *Government of Canada.* https://www.international.gc.ca/world-monde/international_relations-relations_internationales/nato-otan/centre-excellence.aspx?lang=eng (accessed 1 May 2023).

Greenburg, J. H. 2023. *At War with Women: Military Humanitarianism and Imperial Feminism in an Era of Permanent War.* Ithaca, NY: Cornell University Press.

Guichaoua, Y 2020. The bitter harvest of French interventionism in the Sahel. *International Affairs* 96(4): 895–911.

Haidara, B. 2021. Inside Mali's coup within a coup. *The Conversation.* https://
theconversation.com/inside-malis-coup-within-a-coup-161621 (accessed 1 May 2023).

Ham, Y-G., J-H. Kim, and J-J. Luo 2019. Deep learning for multi-year ENSO
forecasts. *Nature* 573: 568–572.

Haupt, S. E., J. Copeland, W. Y. Cheng, C. Amman, Y. Zhang, and P. Sullivan 2016.
Quantifying the wind and solar power resource and their inter-annual variability
over the United States under current and projected future climate. *Journal of
Applied Meteorology and Climatology* 55 345–363.

Haupt, S. E., D. J. Gagne, J. David, W. W. Hsieh, V. Krasnopolsky, A. McGovern,
C. Marzban, W. Moninger, V. Lakshmanan, P. Tissot, and J. K. Williams 2022.
The history and practice of AI in the environmental sciences. *Bulletin of the
American Meteorological Society* 103(5): e1351–e1370

Hoch, J., N. Wanders, and S. de Bruin 2022. We built an algorithm to predict how
climate change will affect future conflict in the Horn of Africa: here's what we
found. *The Conversation.* https://theconversation.com/we-built-an-algorithm-to-
predict-how-climate-change-will-affect-future-conflict-in-the-horn-of-africa-heres-
what-we-found-185627 (accessed 1 May 2023).

ICRC 2021. Mali's invisible front line: climate change in a conflict zone. *International
Committee of the Red Cross.* https://www.icrc.org/en/document/mali-invisible-
front-line-climate-change-conflict-zone (accessed 1 May 2023).

Jaff, D., L. Margolis, and E. Reeder 2022. Civil-military interactions during non-
conflict humanitarian crises: a time to reassess the relationship. *Defence Studies*
22(3): 398–413.

Johnson, R. 2021. Hybrid warfare and counter-coercion. (In) R. Johnson, M. Kitzen,
and T. Sweijs (eds) *The Conduct of War in the 21st Century: Kinetic, Connected and
Synthetic.* London: Routledge, pp.45–57.

Jones, S. G. 2021. *Three Dangerous Men: Russia, China, Iran and the Rise of Irregular
Warfare.* New York, NY: W. Norton & Co.

Karalis, M. 2022. Open-source intelligence in Ukraine: asset or liability? *Chatham
House.* https://www.chathamhouse.org/2022/12/open-source-intelligence-ukraine-
asset-or-liability (accessed 1 May 2023).

Le Masson, V., C. Benoudji, S. Sotelo Reyes, and G. Bernard 2019. How violence
against women and girls undermines resilience to climate risks in Chad. *Diasters* 43:
S245–S270.

Ludwig, R. and A. Sanchez-Azofeifa 2019. Special issue information. *Sensors* 19(21):
4770.

Mainwaring, S. 2020. Division D: Operation Rubicon and the CIA's secret SIGINT
empire. *Intelligence and National Security* 35(5): 623–640.

Martin, A., A. Blowers, and J. Boersema 2006. Is environmental scarcity a cause of
civil wars? *Environmental Sciences* 3(1): 1–4.

Mbaye, A. A. 2020. Climate change, livelihoods, and conflict in the Sahel. *Georgetown
Journal of International Affairs* 21: 12–20.

McDougall, J. and J. Scheele (eds) 2012. *Saharan Frontiers: Space and Mobility in
Northwest Africa.* Bloomington, IN: Indiana University Press.

MOD 2019. MOD to establish Centre of Excellence for Human Security. *MOD News.*
https://www.gov.uk/government/news/mod-to-establish-centre-of-excellence-for-
human-security (accessed 1 May 2023).

MOD 2021a. Climate change and sustainability strategic approach. *UK MOD.* https://www.gov.uk/government/publications/ministry-of-defence-climate-change-and-sustainability-strategic-approach (accessed 1 May 2023).

MOD 2021b. Defence command paper. *UK MOD.* https://www.gov.uk/government/news/the-defence-command-paper-sets-out-the-future-for-our-armed-forces (accessed 1 May 2023).

MOD 2023. About us. *GOV.UK.* https://www.gov.uk/government/organisations/ministry-of-defence/about (accessed 1 May 2023).

McNeilly, M. R. 2015. *Sun Tzu and the Art of Modern Warfare.* Oxford: Oxford University Press.

NATO 2022. The strategic concept. *NATO.* https://www.nato.int/strategic-concept/ (accessed 1 May 2023).

Newsome, C. and H. Herron 2012. *Humanitarian Crisis and Response in the Horn of Africa.* New York: Nova.

Nugee, R., L. Selisny, T. Burwell, and T. Clack 2022. Defence evolution: climate intelligence and modern warfare. *Climate Change & (In)Security Project.* https://cciproject.uk/resourcesblog/defence-evolution-climate-intelligence-amp-modern-warfare (accessed 1 May 2023).

Osinga, F. 2021. Strategic underperformance: the West and three decades of war. (In) R. Johnson, M. Kitzen, and T. Sweijs (eds) *The Conduct of War in the 21st Century: Kinetic, Connected and Synthetic.* London: Routledge, pp. 17–41.

O'Hara, K. 2020. Explainable AI and the philosophy and practice of explanation. *Computer Law and Security Review* 39: 105474. 10.1016/j.clsr.2020.105474

Pasini, A., P. Racca, S. Amendola, G. Cartocci, and C. Cassardo 2017. Attribution of recent temperature behavior reassessed by a neural-network method. *Scientific Reports* 7: 17681. 10.1038/s41598-017-18011-8.

Pelling, M. 2011. *Adaptation to Climate Change: From Resilience to Transformation.* London: Routledge.

Percival, V. and T. Homer-Dixon 1996. Environmental scarcity and violent conflict: the case of Rwanda. *The Journal of Environment and Development* 5(3): 270–291.

Pezard, S., M. Shurkin, and S. Paezard 2015. *Achieving Peace in Northern Mali: Past Agreements, Local Conflicts, and the Prospects for a Durable Settlement.* London: RAND.

Phythian, M. (ed.) 2013. *Understanding the Intelligence Cycle.* London: Routledge.

Poteat, E. 2000. The use and abuse of intelligence: an intelligence provider's perspective. Diplomacy & Statecraft 11(2): 1–16.

Redaction 2022. Peacekeepers and soldiers killed in attacks in Mali. *AfricaNews.* https://www.africanews.com/2022/03/08/peacekeepers-and-soldiers-killed-in-attacks-in-mali// (accessed 1 May 2023).

Red Cross 2021. UN: Sahel region one of the most vulnerable to climate change. *Red Cross.* https://www.climatecentre.org/981/un-sahel-region-one-of-the-most-vulnerable-to-climate-change/ (accessed 1 May 2023).

Retter, L., A. Knack, Z. Hernandez, R. Harris, B. Caves, M. Robson, and N. Adger. 2021. Crisis response in a changing climate: implications of climate change for UK defence logistics in humanitarian assistance and disaster relief (HADR) and military aid to the civil authorities (MACA) operations. *RAND Corporation.* https://www.rand.org/pubs/research_reports/RRA1024-1.html (accessed 1 May 2023).

Reyes-Garcia, V., A. Fernandez-Llamazares, M. Gueze, A. Garces, M. Mallo, M. Villa-Gomez, and M. Vilaseca 2016. Local indicators of climate change: the potential contribution of local knowledge to climate research. *WIREs Climate Change* 7(1): 109–124.

Risemberg, A. 2022. Suspected Islamist militants kill 132 civilians in Central Mali. *VOA News.* https://www.voanews.com/a/6626488.html (accessed 1 May 2023).

Risen, J. and L. Poitras 2014. Spying by NSA: Ally entangled US law firm. *The New York Times.* https://www.mcslaw.com/wp-content/uploads/2014/02/Spying-by-N.S. A.-Ally-Entangled-U.S.-Law-Firm-NYTimes.pdf (accessed 1 May 2023).

Ryan, M. 2022. How Ukraine is winning in the adaptation battle against Russia. *Engelsberg Ideas.* https://engelsbergideas.com/essays/how-ukraine-is-winning-in-the-adaptation-battle-against-russia/ (accessed 1 May 2023).

Scheele, J. 2012. *Smugglers and Saints of the Sahara: Regional Connectivity in the Twentieth Century.* Cambridge: Cambridge University Press.

Self, R. 2022. *Making British Defence Policy: Continuity and Change in an Uncertain World.* London: Routledge.

SIPRI 2021. Climate, peace and security factsheet: Mali. *SIPRI.* https://sipri.org/sites/ default/files/210526%20Final%20Mali%20Climate%20Peace%20Security%20Fact %20Sheet_EN.pdf (accessed 1 May 2023).

Sikorsky, E. 2023. Warning on a warming planet: integrating climate change into NATO's intelligence programs. (In) C. Maternowski (ed) *Navigating a Global Crisis: Climate Change and NATO.* Toronto: NATO Association of Canada, pp. 11–15.

Slota, S. C., K. R. Fleischmann, S. Greenberg, N. Verma, B. Cummings, L. Li, and C. Shenefiel 2020. Good systems, bad data? Interpretations of AI hype and failures. *Proceedings of the Association for Information Science and Technology* 57(1): e275. https://doi-org.ezproxy-prd.bodleian.ox.ac.uk/10.1002/pra2.275

Stengel, K., A. Glaws, D. Hettinger, and R. N. King 2020. Adversarial super-resolution of climatological wind and solar data. *Proceedings of the National Academy of Sciences of the United States of America* 117: 16805–16815.

Strobach, E. and G. Bel 2020. Learning algorithms allow for improved reliability and accuracy of global mean surface temperature projections. *Nature Communications* 11: 451. 10.1038/s41467-020-14342-9

Sun Tzu 1910. *The Art of War: The Oldest Military Treatise in the World* (trans. L. Giles). London: Allendale.

The Economist 2020. Past coups predict future coups, and Mali has had a series of them: what happens after this one? *Economist Intelligence Unit.*

The Economist 2023. Open-source intelligence is piercing the fog of war in Ukraine. *The Economist.* https://www.economist.com/interactive/international/2023/01/13/ open-source-intelligence-is-piercing-the-fog-of-war-in-ukraine (accessed 1 May 2023).

Thomas, M., R. Bury, I. Sujith, M. Sccheffer, T. Lenton, M. Anand, and C. Bauch 2021. Deep learning for early warning signals of tipping points. *Proceedings of the National Academy of Sciences of the United States of America* 118(39): e2106140118

Thurston, A. 2020. *Jihadists of North Africa and the Sahel: Local Politics and Rebel Groups.* Cambridge: Cambridge University Press.

Tranchant, J-P., A. Gelli, L. Bliznashka, A. Sekou Diallo, M. Sacko, A. Assima, E. Siegel, E. Aurino, and E. Masset 2018. The impact of food assistance on food insecure populations during conflict: evidence from a quasi-experiment in Mali. *World Development* 119: 185–202.

Traore, B., M. Corbeels, M. T. van Wijk, M. C. Rufino, and K. E. Giller 2013. Effects of climate variability and climate change on crop production in southern Mali. *European Journal of Agronomy* 49: 115–125.

UK MOD 2023. *Defence's Response to a More Contested and Volatile World.* London: Ministry of Defence.

UN 2014. Agreement for peace and reconciliation in Mali resulting from the Algiers process. *United Nations.* https://www.un.org/en/pdfs/EN-ML_150620_Accord-pour-la-paix-et-la-reconciliation-au-Mali_Issu-du-Processus-d%27Alger.pdf (accessed 1 May 2023)

UN 2021. Five things you need to know about the climate emergency in West and Central Africa. *UN Office for the Coordination of Humanitarian Affairs.* https://unocha.exposure.co/five-things-you-need-to-know-about-the-climate-emergency-in-west-and-central-africa (accessed 1 May 2023).

UN 2022. Climate change exacerbates violence against women and girls. *United Nations.* https://www.ohchr.org/en/stories/2022/07/climate-change-exacerbates-violence-against-women-and-girls (accessed 1 May 2023).

UN 2023. Uniformed personnel deployed by gender. *United Nations.* https://peacekeeping.un.org/en/gender (accessed 1 May 2023).

UNEP 2022. Custom climate security analytics. *UN Environment Programme.* https://unepstrata.org (accessed 1 May 2023).

UNSC 2020. Resolution 2351. *United Nations Security Council.* https://minusma.unmissions.org/sites/default/files/s_res_25312020_e.pdf (accessed 1 May 2023).

UNHCR 2023. Violence and threats by armed groups continue to displace refugees and civilians in Mali. *UNHCR.* https://www.unhcr.org/uk/news/briefing/2023/1/63cf9c524/violence-threats-armed-groups-continue-displace-refugees-civilians-mali.html (accessed 1 May 2023).

USAID 2020. Climate change risk profile: West Africa Sahel. *USAID.* https://www.climatelinks.org/sites/default/files/asset/document/2017%20April_USAID%20ATLAS_Climate%20Change%20Risk%20Profile%20-%20Sahel.pdf (accessed 1 May 2023).

US Army Command 2014. *Female Engagement Teams: Making the Case for Institutionalization Based on US Security Objectives in Africa.* Washington, DC: Createspace.

Wieltschnig, P., J. Muraszkiewicz, and T. Fenton 2021. Without data we are fighting blind: the need for human security data in defence sector responses to human trafficking. *Journal of Modern Slavery* 6(3): 64–86.

Wing, S. 2016. French intervention in mali: strategic alliances, long-term regional presence? *Small Wars & Insurgencies* 27(1): 59–80.

Wintour, P. 2021. Anniversary of 9/11 and fall of Kabul trigger questions over US interventionism. *The Guardian.* https://www.theguardian.com/world/2021/sep/08/what-next-for-american-foreign-policy-9-11-afghanistan-military-intervention (accessed 1 May 2023).

WPS 2023. Mopti's fishermen and boatmen: resisting a receding river. *Water, Peace and Security.* https://waterpeacesecurity.org/info/mali-impact-story-02-14-2023 (accessed 1 May 2023).

Yang, J., P. Gong, R. Fu, M. Zhang, J. Chen, S. Liang, B. Xu, J. Shi, and R. Dickinson 2013. The role of satellite remote sensing in climate change studies. *Nature Climate Change* 3: 875–883.

## Appendix: list of climate security lessons

Neither the UN nor the United Kingdom currently focus enough resources on climate and security. This constrains the potential for pertinent data collection and analysis, thereby risking operational outcomes, and preparedness and resilience operations, in relation to localised climate conflict.

The collection, collation, storage, and analysis of climate data would establish a baseline that would facilitate coherent and ongoing analysis and the targeting of relevant resources.

Climate intelligence (CLINT) supports the vital link between threat analysis and decision making. It maximises prospects to support stabilisation, mitigate threats, and strengthen connections between in-field activity and headquarters' deliberation.

The lack of local climate data compromises effective operational and tactical decision making. CLINT should develop understandings of seasonal dynamics and water and food scarcity.

CLINT should involve assessments of how climate-induced scarcity escalates sectarian violence, shapes sentiment towards government, drives civil unrest, and exacerbates displacement (internal and cross-border).

CLINT should assess the economic implications of climate change on an area of operations, including issues of capital flight, re-configuration of supply and demand, health shocks, out-migration of skills and labour, and foreign food aid undermining local markets.

The more extreme the environment is, the more significant even small changes are in terms of implications for food and water security.

A lack of accessible climate data prevents robust evaluation of indigenous claims about increasingly extreme weather patterns and climate hazards.

Climate change increases the requirement to integrate planning across in-theatre military, police, and civilian actors.

Large areas of operations with complex human terrain increase the need for additional linguists and interpreters. A more effective distribution of interpreters would enhance the ability of soldiers to collect CLINT from local sources.

Operations in climate change-impacted environments amplify the requirement for specialist kit that is designed specifically for temperature and weather extremes.

Increased temperatures exhaust soldiers on patrols more quickly requiring a re-evaluation of expectations for tactical-level engagement in terms of distance, duration, and tasking.

Human terrain reconnaissance (HTR) requires specialist training to develop skillsets (rapport building) and mindsets (desire to understand) to enhance effectiveness. This has implications for pre-deployment training (PDT) and decisions around the selection of deployees.

Pre-deployment training (PDT) should include coverage of climate change threats, including how impacts become drivers of conflict at the local level. This should include a focus on second and third-order effects.

Military presence can both enable and constrain the ability of government and civilian humanitarian actors. The military can offer a security envelope affording protection, but can also inadvertently increase kinetic threats, including from the seeding of IEDs by malign actors.

The extended reach and manoeuvrability offered by Jackal mounted battlegroups increase prospects for CLINT collection from otherwise inaccessible areas.

Female engagement offers significant potential to understand dimensions of the battlespace, human terrain, and the impacts of climate change on security. The training and deployment of female engagement personnel should be prioritised in climate change-impacted theatres.

When areas become uninhabitable, out-migration becomes the most effective mode of adaptation. Enhancing upstream capacity and resilience in terms of maintaining habitable areas is key to reducing displacement.

Failure to deliver on commitments to deliver on climate change mitigation, such as the distribution of food aid and completion of water access projects, significantly reduces positive sentiment towards military and government actors.

# 13

## OPERATIONAL RISKS AND OPPORTUNITIES FROM CLIMATE CHANGE ON WESTERN MILITARIES' CONSERVATION ACTIVITIES

*Richard Milburn*

### Introduction

Scepticism. The most common response this author receives when discussing the benefits and importance of military involvement in biodiversity conservation activity. Those in the military think it is not their job and is irrelevant to warfighting, those in conservation think it leads to death and destruction, not wildlife protection. This chapter seeks to explain why both sides are wrong, outlining the value of biodiversity to support human prosperity and absorb carbon emissions in the global fight against climate change, and showcasing the value that militaries can provide by supporting conservation operations and the benefits they, in turn, receive from doing so through enhancing their warfighting capabilities.

The chapter first examines the impact of armed conflict on biodiversity, finding that it is overwhelmingly negative, but that sometimes the presence of armed actors discourages environmental exploitation, leading to short-term conservation. It then discusses the value of conservation to the global fight against climate change, explaining the role that ecosystems play in reducing disaster risk and sequestering carbon. The central assertion is that military commanders and planners need an understanding of operating in biodiverse areas, an awareness many such areas may be, or become, sites of armed conflict, and to understand the value of protecting and restoring biodiversity to support efforts to fight climate change.

The chapter then examines international military conservation activities, including India's Ecological Battalions and South American military involvement in Amazon protection to showcase current practise, before going into greater depth specifically on the illegal wildlife trade (IWT) and the

DOI: 10.4324/9781003377641-16

British Army's countering IWT (CIWT) Operation CORDED. The chapter outlines the key drivers of IWT and showcases how a counter-insurgency (COIN) framework can be applied to the IWT environment. Analysis of Op CORDED is used to show the positive role that the military can play in conservation, as well as the limitations of law enforcement and the need for interventions to win the human terrain, which, in turn, provides important lessons for COIN operations. The chapter finishes by considering what other operations Western militaries could engage with to support conservation activity and highlights the benefits this would deliver for climate change, prosperity, and enhancing warfighting capability by providing low-risk, permissive operating environments to teach, test, and innovate conventional warfare and COIN tactics, equipment, and capabilities.

The central argument of the chapter is twofold. Firstly, that militaries have a unique set of skills which can have a significant impact to improve nature conservation activities, if utilised correctly, and that those activities can be a useful component of efforts to achieve militaries' internal commitments to net zero as well as supporting global efforts to fight climate change. And, secondly, that engaging in conservation activity provides a unique opportunity for militaries to enhance their warfighting capabilities, particularly with regards to COIN. As such, it contends that, even if commanders have no interest in the environment or climate change, the opportunities to improve the core capabilities critical to COIN and conventional warfighting that are provided by conservation activities mean Western militaries should be much more actively involved in those operations, and at greater scale than is currently the case.

## Biodiversity and armed conflict

For a long time, biodiversity has not been seen as an important consideration for the military, but with the 2022 UK Integrated Review's inclusion of the importance of biodiversity and climate change as well as a number of outputs from the UK and US military on a similar theme (e.g. MOD 2021; DOD 2021), that is starting to change. While biodiversity protection is unlikely to ever be the primary concern for militaries and war is often highly destructive of the environment, an understanding of the value and importance of protecting biodiversity, both for its human value and also for its contribution to averting climate change, is important for effective military command. The Romans understood this well when they salted the fields of Carthage – if you degrade the environment, human life becomes impossible. On the other hand, effective protection and restoration of biodiversity can improve human security and prosperity.

Research in the early millennium found that 90% of armed conflicts took place in countries containing biodiversity hotspots, and 80% took place within the biodiversity hotspots themselves (Hanson et al. 2009). While the presence of biodiversity was not the cause of these conflicts, the geographic correlation between war and biodiversity hotspots increased interest in the effect of those wars on biodiversity (Plumptree et al. 2001; Price 2003; Brauer 2009). Much of that research came from a 'nature-first' perspective – from environmentally focused research – as opposed to a 'war-first' approach which considered the wider security implications of biodiversity loss, but the spatial correlation between biodiversity hotspots and armed conflict raises important considerations for military operations (Milburn 2018). Many more recent conflicts have also occurred within biodiversity hotspots, such as the conflict in the Democratic Republic of Congo (DRC), in which sits the world's second-largest tropical rainforest, described as one of the two 'lungs of the earth' alongside the Amazon, as well as large peatlands, another key store of carbon (Dargie et al. 2017). The DRC also provides a useful case study of the diverse direct and indirect impacts that armed conflict has on biodiversity under different conditions, offering lessons for a range of armed conflicts around the world.

The effects of the 1994 Rwandan genocide swept across the borders into eastern DRC, with millions of refugees housed in camps followed by a series of intra- and inter-state conflicts in the DRC, including what has been described as 'Africa's World War', and which is estimated to have led to the deaths of around 5 m people (Deibert 2013; Prunier 2009). The conflict and the refugee flows that resulted from that conflict heavily impacted biodiversity in the country. A refugee camp established close to the Virunga National Park, home to various species, including the mountain gorillas, led to a significant increase in timber exploitation in the park, with an estimated 89 hectares of forest lost each day (UNEP 2011, p. 26). The scale of people in the camps and their dependency on the natural environment for fuel created unprecedented pressures on the park's forest resources (Brauer 2009).

A contrasting example of the indirect effects of war is offered from Kahuzi-Biega National Park in a neighbouring province. Hundreds of thousands of people fled through the park's forest, living off the land as they moved, leading to the decimation of wildlife species within the park (Shalukoma 2000). The forest elephant population, which plays an important role in the ecosystem by clearing trails through the forest and encouraging new tree growth, was driven to extinction, and the endemic Grauer's gorillas population was reduced by around 90%, from 20,000 down to less than 3,000 (Yamagiwa 2003). Described as the 'gardeners' of the forest, the gorillas are responsible for 80% of seed dispersal and play a key role in the maintenance of the forest ecosystem (Goldenburg 2009).

In addition to the subsistence consumption described above, commercial exploitation of timber within the DRC context varied significantly. The annual logging rate was 0.33%, significantly below other countries in West and Central Africa, despite having much larger forests. This low rate can be attributed to the conflict in the country as it prevented much commercial exploitation although this varied through different parts of the country (Seyler et al. 2010; Baker et al. 2004). In the west of the country, which was mostly under central government control, timber licenses were issued as a method of mobilising revenues for the government, leading to increased deforestation rates (Baker et al. 2004). The presence of the port city of Matadi made export of timber more feasible than in eastern DRC, where road infrastructure is poor meaning there are limited routes to export timber. Deforestation rates in the east of the country were lower due to the lack of infrastructure as well as the presence of more valuable alternative minerals such as gold, diamonds, and coltan (Baker et al. 2004). The value density of such minerals is much higher, making them far more attractive to exploit than large, heavy timber logs that are difficult to transport (Blundell 2010).

The DRC case illustrates some of the damage that armed conflict can cause to biodiversity and the reasons for that damage. While war is generally assumed to be damaging to the environment, such as with the use of Agent Orange in Vietnam, or the burning of oil wells in Iraq (Brauer 2009), it is not always so. The 'warzone refuge' effect has been identified in some conflicts, whereby the presence of armed groups dissuades human presence and commercial exploitation of environmental resources, thereby leading to areas being protected (Dudley et al. 2002). In such instances, for example in a forest environment that an insurgent group uses as its base of operations, the trees may remain well-protected even while many animal species are killed to provide food for guerrilla fighters (Oglethorpe et al. 2002). In other locations, such as the Korean De-Militarised Zone (DMZ), the risk of death from unexploded ordnance has kept people out and allowed the area to become a *de-facto* nature reserve (Kim 2007). At the time of writing, the impact of the Ukraine conflict on biodiversity seems to follow a similar pattern, with a combination of environmental damage from war, but also some areas of Russian occupation where forest is *de facto* protected but where animal populations are likely to reduce due to illegal hunting by soldiers for meat (Humphreys 2022).

Despite these examples of *de-facto* conservation as a result of armed conflict, war is usually bad for nature, as may be expected. However, an important consideration is not just conflict, but the post-conflict period, during which time demands on natural resources for reconstruction and exploitation for revenue may be highly acute, leading to increased exploitation and destruction of biodiversity. McNeely (2002, p. 45) notes, '[w]hile war is bad for biodiversity, peace can be even worse... Market forces may be more destructive than military forces'. Once security improves and natural resource areas open up to commercial exploitation, and while demand for resources for rebuilding areas affected by armed conflict is high, the damage done to biodiversity can be severe, destroying key ecosystems and resources that the populations and nations will need for medium- and long-term prosperity (McNeely 2002).

The exploitation of the environment in the postwar phase can also perpetuate insecurity, as former armed groups utilise their structures, weapon systems, and knowledge of the terrain to engage in illegal exploitation (Swatuk 2007). This undermines stabilisation efforts, increases the risk of a return to armed conflict, and degrades environmental resources which are important to a country's long-term prosperity. However, if these risks are addressed, it also offers an opportunity to help with disarmament, demobilisation, and reintegration (DDR) programmes, with former soldiers being given specific conservation roles as part of the DDR process. This happened in Afghanistan, with the creation of the Afghanistan Conservation Corps (UNEP 2012), as well as in Gorongosa National Park in Mozambique, where conservation patrols were organised with members of both sides of the conflict, providing a specific, albeit small scale, tool to help bring peace to the country (Hatton et al. 2001). If the idea was adopted more formally, integrating DDR into conservation offers an opportunity to provide hundreds of jobs to ex-combatants to support stabilisation, and that, in turn, provides the large volume of personnel needed for effective conservation and restoration activities. Such a conservation DDR programme could be managed by outside partners, such as the UN or other military organisations, to create a 'Yellow Berets' unit of former combatants turned *en masse* into eco-guards (Milburn 2012).

For those involved in warfighting, it is important to have an awareness of the above impacts of war to then consider how they may be reduced and redressed in order to protect biodiversity. The next section demonstrates that such activities can provide important services to support human life and reduce emissions to help prevent climate change. In much the same way as Western militaries give consideration to the protection of cultural heritage and critical national infrastructure during conflict, so too should greater consideration be given to the protection of biodiversity, what might be described as a country's 'critical natural infrastructure'.

## Biodiversity and climate change

The protection of biodiversity is an important consideration with regard to climate change for three core reasons. Firstly, it provides a large store and sequestering capacity of carbon into the world's terrestrial and marine ecosystems, helping to control the rate of carbon in the atmosphere. Secondly, biodiversity provides 'natural capital' that supports the prosperity of developed and developing countries alike. Thirdly, protecting and restoring critical ecosystems helps reduce the intensity of, and negative effects from, natural disasters.

Efforts to prevent global temperatures from rising above 1.5–2°C rely on rapidly reducing emissions by 2030 and bringing net emissions down to zero by 2050 (IPCC 2022). Much of the work to achieve that goal will focus on reducing the volume of emissions into the atmosphere as a result of human activity. However, increasing the earth's capacity to sequester carbon will also play a vital role in bringing net emissions down to zero and in the longer term helping to reduce the volume of carbon in the atmosphere to provide long-term control to prevent temperature rises (Hawken 2021). While there are technological solutions being designed to store and sequester carbon, these remain expensive and small scale, such as the Orca plant owned by Climeworks in Iceland, which charged US$1,200 per tonne of carbon sequestered (Economist 2021a). As such, biodiversity remains a key component in global efforts to reduce net emissions. Furthermore, many gigatons of carbon are stored in the world's ecosystems, such as the Artic permafrost, Amazon rainforest, and the Congo Basin's rainforest and peatland. If these ecosystems are lost, emissions will rise drastically and rapidly, leading to concurrent increases in temperatures detrimental to human habitation (Dargie et al. 2017). In contrast, when biodiversity is protected and restored, emissions can be sequestered in greater quantity, leading to an overall reduction in a highly cost-effective way. While there remain some debates about the extent to which natural solutions will be able to reduce emissions to net zero (Seddon et al. 2020), they are likely to play an important part in carbon sequestration. Protection and restoration of biodiversity should therefore be a key concern of all those interested in achieving net zero.

In addition to the carbon sequestration benefits of biodiversity, the concept of natural capital and associated Green Growth has been advanced in recent decades to try to bring the economic discipline to nature protection and provide a more specific and useful measure of the value of biodiversity to policymakers (MEA 2005; World Bank 2012). Some of the valuations are too large to be credible or meaningfully useful, however, the broad thrust of the research is useful in identifying the value that biodiversity provides (Holzman 2012). It also helps to frame exploitation not solely as a profit-making endeavour, but as one that, if conducted unsustainably, ultimately degrades the capital value of assets, much like selling off the family silver. If, for example, a forest is felled, cash is raised in the short-term, increasing the cash wealth of those involved, but the planet and the human populations that relied on the forest are made poorer.

Where ecosystems are under stress, especially where there are few technological solutions to adapt to such stress, such as among poor communities, the effects of climate change are increased. As Mrema et al. (2009, p. 4) have pointed out, natural resources are the 'wealth of the poor'. Intact, flourishing ecosystems enable people to survive and thrive, but when those ecosystems are degraded, they expose populations to greater risk and reduce their 'adaptive capacity' to cope with the effects of climate change (Brown et al. 2007). With climate change now recognised as a security threat multiplier, exacerbating existing weaknesses, keeping underlying ecosystems intact to reduce a population's overall vulnerabilities will help to reduce the risk of conflict breaking out.

Intact ecosystems not only help to sequester carbon, but also provide a proven, simple, and cost-effective method to reduce the risk and intensity of natural disasters (Estrella et al. 2016), leading to calls for conservation to support disaster risk reduction (Renaud et al. 2013; Sudmeier-Rieux et al. 2006). Around 600 m people are estimated to live on land vulnerable to environmental disruptions, relying on ecosystem services to support agricultural livelihoods. Where natural disaster risk is increased, especially in areas that have recently experienced conflict or are prone to conflict due to existing societal divisions, it can exacerbate those underlying issues and contribute to the outbreak of armed conflict (Austin and Bruch 2007; Bergholt and Lujala 2012; Nel and Righarts 2008).

Intact and restored ecosystems can provide a range of benefits, for example, evidence suggests mangroves reduced the impact of the 2004 Indian Ocean as well as more generally helping to improve fishermen's catches and incomes (Chevallier 2013). Coral reefs also provide important sea protection for coastal communities, with an estimated 150,000 km of shoreline in over 100 countries protected (The Economist 2016). Intact forests help to reduce landslides by stabilising the soil and also provide a wealth of other benefits, such as the regulation of rainfall patterns through transpiration and maintenance of soil quality and fertility. Yet, despite the value of these 'shock-absorbing' effects provided by mangroves and corals, many have been degraded for short-term gain, undermining the long-term protection they offer and increasing the risk of natural disaster (Girot 2002).

Climate change is likely to lead to increases in migration and intra-state – and possibly inter-state – conflict and the need for militaries to maintain readiness to deploy into those environments. Equally, however, the ability to identify areas at risk and proactively intervene not with a kinetic military operation but instead with support to enhance the natural environment and resilience against climate change will allow military resources to be deployed more effectively and help to prevent the risk of large numbers of conflicts breaking out in different parts of the world. As noted, helping to protect and restore biodiversity is a simple, low-cost method to help enhance human resilience to the effects of climate change and reduce the risk of armed conflict. It is thus something Western militaries could, and arguably should, be involved in. The following section examines international military involvement in biodiversity protection to date in order to understand approaches and opportunities for Western militaries' involvement in conservation activity today.

## Global military involvement in biodiversity protection

Due to the recognised importance of protecting the environment in some countries, militaries have already been involved in conservation and restoration activities in different parts of the world. One of the longest-running roles has been the Indian Army's Ecological Battalions. These battalions were originally formed from within the Territorial Army to provide jobs for veterans and have continued to provide a role in tree-planting and other restoration activity within India, with 10 battalions operational (Balwan 2021). Recognising that the environment is important to the Indian population, for example, by reducing disaster risk and increasing natural capital as discussed in the previous section, and that the Indian Armed Forces are the second-largest landholder and therefore have the potential for significant carbon sequestration, the units provide both an ecological benefit as well as activity and employment for veterans (Dhanasree 2017).

The Brazilian military has also been involved in environmental projects for decades (The Economist 2020). As of 2019, they took on a more specific terrestrial and marine conservation mandate under Operation Green Brazil, with a stated intent to deploy the military to reduce illegal activity in the Amazon. The deployment has been unsuccessful, however, failing to significantly reduce illegal activity while costing US$71 m, more than some other environmental agencies' annual budgets (Paes 2021). Although the Brazilian military has claimed success, pointing to the thousands of arrests that have been made and over US$300 m of fines issued, there is widespread scepticism about the operation's outcome, and criticisms that the military was ill-suited to their role, and poorly planned and executed the operation (Paes 2021; McCoy 2021).

The Colombian military has also been tasked with supporting conservation operations in the Amazon, under Operation Artemis, deploying 22,000 service personnel to prevent illegal deforestation (Barrett 2019). This was in response to the increasing deforestation levels, which illustrate the quote earlier in the chapter that peace is often worse for biodiversity. The Fuerzas Armades Revolucionarias de Colombia (FARC) were known to prevent deforestation to maintain concealment in the forest, but since the peace agreement more exploitation has taken place (Noriega 2022). However, whether this operation will prove successful also remains in doubt; in theory, the Colombian military could replicate the FARC's activities to restrict deforestation, but in practise this is more challenging due to the difficulties of operating within the law. Unless the underlying causes of deforestation are addressed, it is unlikely military action will achieve much conservation effect, as will be discussed below.

The Botswana Defence Force (BDF) has provided support to the country's Department for Wildlife and National Parks (DWNP) in their counter-poaching operations for several decades. The use of soldiers in counter-poaching roles provides for a well-trained and equipped counter-poaching capability, which is better able to counteract increasingly heavily armed poachers than in many cases where poorly armed rangers are the only source of defence. This provides Botswana with a strong conservation record and useful lessons for other countries seeking to counteract wildlife poaching (Henk 2006).

In terms of Western military involvement in conservation activities, to date there has been ongoing involvement in conservation of the training estate and wider military-owned land and the UK MOD has released a plan to plant two million trees on the UK training estate to help sequester the carbon produced by the organisation as a whole (Kalkowski 2020). Militaries must comply with changing domestic legislation while still maintaining their capability, something which is also true of climate change, especially the UK MOD's legally binding commitment to achieve net zero by 2050 as a result of UK government legislation (MOD 2021; see also Nugee and Clack, this volume).

In terms of international efforts to actively support conservation activity, the US Corps of Engineers in Iraq helped to restore the Mesopotamian Marshlands back to their wetland status after Saddam Hussein had drained them (Stevens 2007). The marshes are the traditional and cultural home of the Marsh Arabs, a group persecuted by Saddam's regime, as well as a key migration site for a host of bird species. Their restoration played an important cultural and conservation role in Iraq (Malfatto 2021). Of all Western militaries, the British Army is perhaps most directly involved in international conservation activity through its CIWT Operation CORDED. This has provided support to counter-poaching operations in Sub-Saharan Africa, most recently in Kafue National Park, Zambia (Brewer 2021).

Operation CORDED has shown the benefits of military involvement in conservation activity in terms of enhancing the effectiveness of conservation, as well as the benefits to Western militaries from the lessons learned for wider operations (Vinall and Milburn 2021). CIWT also helps to counter the global security risks of zoonotic disease transfer, organised crime networks, and biodiversity and ecosystem loss contributing to climate change. The following sections therefore delve deeper into the IWT and British Army's response to draw out the wider lessons for Western militaries involvement in conservation.

## The illegal wildlife trade

Calculating the size of the IWT is fraught with difficulty due to its black-market nature, but the World Wildlife Fund estimates it to be worth over £15bn per year (WWF 2022). The estimates of the size of the trade have not changed significantly for at least a decade, suggesting that current approaches to countering IWT are ineffective in global terms, even if some success is achieved at a smaller scale in certain localities. While iconic species, such as rhinoceros, elephants, and tigers, capture most of the headlines related to the trade, much of the multi-billion market consists of reptiles and birds (Warchol 2004). In addition to the global high-value trade in IWT products, such as rhinoceros' horn, localised, domestic demand for bushmeat and wildlife products for traditional medicine also drive the illegal killing and capture of wildlife. Alongside the IWT, the illegal timber and charcoal trades are a significant contributor to the loss of biodiversity, with estimates placing the value of the illegal timber trade at up to US$150bn annually (Wallen 2018). International demand for high-value timber products such as Rosewoods, as well as the dependency of many poorer populations on timber and charcoal for their building materials and fuels, places a significant burden on natural resources, driving illegal deforestation (Ong and Carver 2019).

The security threat from IWT is threefold. Firstly, organised criminal groups are involved in the trade, often using the smuggling of IWT products through their existing criminal network to provide extra revenue for their operations or to fill revenue shortfalls when their involvement in other illicit trades, such as narcotics, is compromised by enhanced law enforcement (South and Wyatt 2011). Additionally, some insurgent groups have been linked to IWT, purportedly using sales of wildlife products, such as ivory, to generate revenue, although the extent to which these links exist and the amount of revenue generated remain contested (Maguire and Haenlein 2015). The second threat comes from the loss of biodiversity and associated natural capital, ecosystem services, and revenue from enterprises such as eco-tourism, which undermines the ability of communities and states to develop, and which also degrades ecosystems that provide vital carbon sinks to help address climate change, as discussed earlier in the chapter. Thirdly, as the recent COVID-19 pandemic has brought into focus, the proximity of people to wild animals and the movement of different animals and species around the globe, without any regulation or oversight, increases the risk of zoonotic disease transfer to human beings and future pandemics. For example, 89% of recognised RNA viruses known to harm humans have zoonotic origins and 60% of emerging infectious disease outbreaks are zoonotic, of which 70% originate in wildlife (Watsa et al. 2020).

Understanding the drivers of IWT at both the strategic and tactical levels enables responses to be developed to address the underlying causes of the trade and also provides useful insights into how to address the issue of climate change. At the strategic, global level, IWT is driven by the 'economic equation'. Put simply, poaching is cheap and lucrative, whereas conservation is expensive and poorly funded (Vinall and Milburn 2021). For example, rhinoceros' horn is estimated to be worth over US$60,000 per kg, making it more valuable than gold. This value creates a significant incentive for organised criminal groups to employ people living close to rhinoceros' habitats as poachers (Ellis-Peterson 2019). Such poachers often cost little and are considered expendable assets by organised criminal groups. In contrast, for those seeking to protect the rhinoceros, teams of rangers, fully trained, paid, and equipped, need to be employed every day of the year to keep poachers at bay, and the only source of revenue to finance those deployments is a small percentage of tourism income combined with limited charitable grant funding. Asymmetric economics drives continued exploitation of wildlife and undermines CIWT efforts.

At the tactical level, the 'poacher's equation' provides the framework to understand why poaching takes place (Vinall and Milburn 2021). The equation considers what an individual or community has to gain from poaching, subtracts what they have to lose if they are caught, and sets this against the risk of getting caught and prosecuted or killed. Though not a mathematical equation *per se*, it provides a framework to understand the drivers of poaching at a local level and how to effectively intervene. For example, potential gain is not only the revenue from selling IWT products, but may also be social status drawn from being part of a poaching gang, revenge for wildlife destroying crops or livestock, or retaliation against national park authorities for arresting and possibly mistreating members of the community or denying access to, and benefit from, the conservation areas. Similarly, the potential loss is not only a fine or prison term if caught, but also the current benefits received from conservation as well as general quality of life. Studies have shown that someone with a stable, even if modest, income who sees conservation as beneficial is unlikely to poach. Finally, the risk of getting caught is key to the whole framework, since even if someone does not have much to gain from poaching, if they know they are unlikely to get caught they are more easily persuaded into doing so (Figure 13.1).

For those earning the most from poaching, with links to organised crime networks to sell high-value products, the law-enforcement piece also becomes crucial as their potential gain is significantly higher than people simply poaching just to survive. Effective counter-poaching teams combined with visibly high rates of prosecution of poachers are key to creating a high risk of being caught. It is important to note within the poacher's equation framework that if the potential loss is zero or close to zero, for example, where someone is poor and unemployed and trying to feed their family, the risk of poaching is very high, regardless of the other two elements in the equation.

$$\text{Propensity to Poach} = \frac{\text{Potential Gain} - \text{Potential Loss}}{\text{(Perceived) Risk of Getting Caught}}$$

**FIGURE 13.1** The poacher's equation.

*Source:* Author's creation.

As Earthshot 2021 Finalist John Kahekwa has said, 'empty stomachs have no ears'; if people are struggling to survive, they will poach, no matter what, highlighting the limitations of a law-enforcement-only approach, and the need for CIWT activity to win the support of local people (Kahekwa 2022).

For those familiar with the counter-insurgency (COIN) environment, the above analysis of IWT is likely to sound familiar. Indeed, IWT should be seen as a low-intensity COIN environment; the organised crime groups are the insurgent organisation recruiting members of the population to their cause, only in this case focused on revenue from wildlife poaching as opposed to political objectives such as the overthrow of government (Vinall and Milburn 2021). It is important to note the term 'low-intensity'. CIWT environments are rarely full-scale COIN environments, with large numbers of heavily armed groups hunting wildlife and attacking rangers, but rather should be seen as smaller-scale COIN environments where a law-enforcement as opposed to full-scale, kinetic military deployment is most appropriate. This would be more comparable to the insurgency in Malaya than the war in Afghanistan, for example.

The central purpose of framing CIWT in COIN terms is to utilise the body of doctrine and experience in this field by applying it to the CIWT space, not only to deliver counter-poaching operations in national parks, but to understand and address the root causes of poaching and win the support of local populations through understanding and winning the human terrain. As discussed with regard to the poachers' equation, it is not simply the financial reward that incentivises poaching, nor solely law enforcement which prevents it, so a more holistic COIN doctrine approach, such as that employed by the British Army, is highly applicable to deliver the most effective solution to stop the trade.

There has been much criticism of the so-called 'militarisation' of conservation and CIWT operations, as more private military companies and military veterans have moved into the space to organise operations, train rangers, and provide enhanced military equipment to counter-poaching (Duffy et al. 2019). Some of this criticism has been valid, especially in incidences where rangers have abused local populations, because it is neither an effective nor cost-efficient way to solve the problem (Schiffman 2020). However, most of the criticism misses the fundamental issue that heavily armed poachers pose a threat to wildlife and rangers, so an enhanced militarised response is required to be effective. Two rangers are, for example, killed in action on average each week (Neme 2014). What is missed in the debate is identifying that this militarised response is only a part of the appropriate and most effective response and must be delivered alongside efforts to provide support to local communities from which the poachers originate. It is for that reason that COIN doctrine is so applicable since, when conducted effectively, it focuses on addressing the root causes of the issue, winning the human terrain, establishing effective governance, and using military action as only a small, targeted component of the overall approach (Vinall and Milburn 2021). To see how this works in practise, the next section will provide a case study of the British Army's CIWT work on Op CORDED and demonstrate the effective use of COIN doctrine in a CIWT environment, which provides not only lessons for how to effectively protect biodiversity, but also demonstrates how militaries can learn from CIWT operations to enhance their own COIN capabilities and effectiveness.

### Wildlife warriors – the British Army's CIWT operations

Op CORDED has provided support to counter-poaching operations in Malawi and Zambia (Brewer 2021). The aim of the operation has been to provide training in key infantry skills and military tracking procedures, as well as medical and legal training, to enhance rangers' skills and capabilities to operate in national parks and track down and arrest poachers, whilst also ensuring their own safety on operations. The UK government has made CIWT a core priority, hosting international conferences and launching the IWT Challenge Fund to support activity which is helping to protect wildlife and counter serious organised crime involvement in the trade. Op CORDED is funded by the IWT Challenge Fund, which is, in turn, managed by the Department for the Environment, Farming, and Rural Affairs, involving some inter-departmental co-operation within government (DEFRA 2022).

Although originally focused on providing infantry skills training to rangers, since 2020 Op CORDED has evolved to provide a more holistic approach to CIWT, integrating principles of military intelligence to better understand the problem in the areas of operation and teach data and pattern analysis techniques to rangers to enable them to predict poacher activity and proactively deploy patrols into the most high-risk areas for poaching. Each Op CORDED deployment now conducts a full analysis of the situation in and around the national park in a similar process to deployment into a COIN environment, having adapted the Army's 'Question 1' analytical framework into a new Strategic Wildlife Protection (SWIPRO) framework relevant to a CIWT environment (Vinall and Milburn 2021). The intent of the operation is to provide the host nation with the tools and training to understand the poaching situation and proactively respond to it, deploying well-trained and well-motivated rangers into high-risk areas to make arrests and protect wildlife, as well as developing an understanding of the underlying drivers of poaching, and working to develop plans to reduce those drivers through community support.

A recent article sets out the CORDED approach as well as a proposed 4-pronged cross-government strategy to address IWT at the tactical (National Park) and strategic (global) levels, although that approach has yet to be adopted by the UK government (see Vinall and Milburn 2021). The strategy involves: (1) disrupting the poacher through an effective counter-poaching force; (2) disrupting IWT by deterring through the disruption of networks and effective prosecution; (3) developing economic alternatives through alternative revenue generation for communities and the monetisation of wildlife, and; (4) behaviour change to reduce demand for illegal wildlife products.

The production of that strategy highlights the limitations of the current CORDED operation, in that it only effects one small element of the needed CIWT response, by focusing its effects on providing support to improve law enforcement to disrupt the poacher. As seen in the poacher's equation and wider economic circumstances, there is a lot of other work required to effectively deliver CIWT that can have a significant and long-lasting impact in reducing global and local illegal trade and consumption of wildlife products. As such, although the current operation provides unique and valuable support to counter-poaching operations in Zambia, it is not actively *solving* IWT because it is failing to engage in cross-government co-operation to deliver economic effects to address the root causes of poaching. In short, Op CORDED addresses the symptoms rather than the cause of IWT. Those same failings have been identified in other COIN deployments, such as those in Mali (The Economist 2021b).

Given the similarities between the COIN and CIWT environments (above), Op CORDED offers an opportunity for the British Army and wider government to test and enhance their COIN capabilities, developing, maintaining, and improving the civil-military co-operation that is required to address the root causes of violence in both a CIWT and COIN environment (Vinall and Milburn 2021). It also offers the opportunity to develop wider lessons relevant to the likely environmental security threats that will emerge as a result of climate change. For example, climate-induced migration is seen as a growing threat over the coming decades, and it is not one that can be prevented through force alone. Unless those involved in human trafficking can be provided with better alternative economic opportunities, they will continue to engage in illegal trafficking (McCullough et al. 2019). Similarly, to disincentivise those who are smuggled requires a similar approach to CIWT, by increasing the potential loss if caught. If people have reliable livelihoods where they live and a reasonable quality of life, they are less inclined to migrate. If law enforcement against migration is then also improved, the 'migration equation' is also likely to shift, just like the poacher's equation.

Furthermore, the analytical process involved in Op CORDED is applicable to more locations suffering environmental stress where armed conflict is breaking out. Indeed, it is not enough simply to deploy military force into the space, it must be combined with development and engineering efforts to address the root causes of violence. The UK MOD's sustainability strategy describes the likely increase in deployments to conflicts caused by climate change (MOD 2021); if these deployments do not address the root causes of conflict, they will likely prove both costly and ineffective. Op CORDED is therefore a useful microcosm or 'petri-dish' to develop the skills, under-standing, and capabilities to respond effectively to COIN conflicts and to future environmental insecurity. It has the additional benefit of being undertaken in a permissive environment; the main threat to soldiers is from wildlife, not people, enabling more work to be carried out and new approaches tested in a relatively low-risk environment.

The British Army's CIWT operations could therefore deliver more than they are doing currently. A gold-standard Op CORDED deployment would address the root causes of IWT set out in this chapter, as well as utilising the protection and restoration of biodiversity in and around national parks in the areas of operation to help absorb carbon emissions to support global efforts to achieve net zero. In addition, it would adopt a specific innovation mission. It would develop, test, and innovate COIN responses, develop and maintain civil-military co-operation methods and relationships, and seek to maximise ecological sustainability, reducing negative operational environmental impacts and increasing positive impacts, and learning lessons that would then be rolled out to the wider military. The operation would therefore become a petri-dish for COIN, sustainability, and emerging operations to address climate change and provide both ecological and military value. The next section considers the further opportunities for Western militaries to be involved in conservation, on both the land and in the sea, and the benefits of doing so.

## Future opportunities for military involvement in conservation

Specifically with regard to climate change, the biggest benefit of involvement in conservation activities for Western militaries is the carbon sequestration effect from conservation and restoration. The UK MOD's commitment to net zero by 2050, for example, is likely to require some carbon off-setting as a result of latent emissions by that time, in particular covering emissions from live operations, which is better achieved through internal activity than pur-chasing credits (Barry et al. 2021). By engaging in conservation activity, militaries not only gain the useful skills and experience discussed in this chapter, but they can also deliver projects that sequester carbon and help to achieve the net-zero target. Although this chapter has focused primarily on land, there are even bigger opportunities in marine conservation.

The world's oceans sequester more carbon than terrestrial ecosystems, but also face limitations on how much more carbon can be absorbed without negative warming and acidification effects occurring, leading to damage to marine ecosystems (Hawken 2021). Restoring marine life to its pre-industrial state would help to avert such damage, by restoring the ecosystem dynamics that enable the oceans to continue to sequester carbon. For example, massively increasing whale populations to eat more carbon-sequestering phytoplankton and take them to the bottom of the sea, with a great whale estimated to absorb 33 tons of $CO_2$ in its lifetime, significantly more than many trees (Chami et al. 2019). Just as Op CORDED offers the British Army's support to terrestrial conservation, so an equivalent nautical operation could support the protection of marine protected areas (MPAs). These MPAs would be established over key breeding grounds both in coastal waters and in the deep ocean, providing protection to allow marine life to breed into abundance. If successful, the seas would not only increase their carbon sequestration capability, but the increase in marine life would provide greater food security from fish protein. Such protein is relied on by billions of people globally already, and an uplift in scale would provide greater economic opportunities for national fishing fleets (Hawken 2021). Given the size of the world's oceans and proposed MPAs, as well as the challenges of maritime operating environments, militaries' unique skills and capabilities are vital to success. This provides an extra role for militaries for decades to come, helping to justify continued, and even increased, central government spending on armed forces and winning wider domestic public support.

In order for militaries to become more involved in conservation activities and achieve success, there needs to be a greater focus on non-kinetic capabilities to address the root causes of poaching or over-fishing as well as supporting law enforcement. Continued and expanding involvement in conservation activity by both Western navies and armies, and likely supported by air force assets in some instances, can help to develop those capabilities. Those capabilities are likely to grow in importance if predicted intra-state conflict increases as a result of climate change. An interesting example is Somali piracy; foreign ships came into Somali waters and over-fished, denying local fishermen a living and creating a pool of potential recruits to engage in piracy (Farquhar 2017). In response, a large naval deployment combined with a significant terrestrial presence focused around Mogadishu has helped bring the piracy problem under control, but at great financial cost, and Somalia remains a violent country. If Western militaries will deploy *en masse* to fight piracy, there is a case to be made that they should similarly deploy to prevent climate change and secure global fish stocks. The urgency of the Somali problem, with ships captured and held to ransom and sailors killed, is always likely to trump longer-term considerations, such as climate change, but it nevertheless shows that large naval deployments can take place to secure marine corridors.

Similarly, the United States has military bases and deployments all over the globe and spends an estimated US\$81bn a year protecting oil supplies (Crawford 2019). As climate change is recognised as a greater economic and security threat and the importance of marine and terrestrial biodiversity conservation is realised, the logical next step is to deploy military capabilities in support of conservation activities. As set out in this chapter, such deployments need to not only be about law enforcement, but also addressing the root causes of the problem and winning the support of the people who would otherwise be the cause of the problem, solving the 'fishermen's equation' as well as the poacher's equation. In so doing, the relationships established and the capabilities developed will enhance Western militaries' ability to address the root causes of COIN conflicts more effectively in future, as well as providing an extra justification for funding to maintain and increase budgets.

## Conclusion

Biodiversity protection has not been a key consideration for Western militaries despite its contribution to human security and prosperity, but with the growing awareness of the threat of climate change this is starting to change. Given the largely negative impacts of armed conflict on biodiversity, working to reduce those impacts, where possible, in order to protect the critical natural capital that populations and states rely on can support postwar recovery and prosperity, and can also help to protect key carbon sinks that are vital to reducing global carbon emissions. Militaries have unique equipment and capabilities that are crucial for effective terrestrial and maritime conservation and using those capabilities for conservation operations can increase public support for the military and mobilise greater budgets for operations from central governments.

Perhaps most important of all, the benefits to Western militaries from involvement in conservation go far beyond its ecological and climate change impacts. While those are significant, offering the opportunity to help reduce natural disaster risk, prevent the outbreak of armed conflict, and sequester carbon to enable militaries to meet net-zero commitments, the benefits to operational capability make engaging in conservation operations worthwhile from a purely conventional military perspective. Having low-risk but real-life operations in permissive environments to test, teach, and innovate COIN and conventional conflict approaches, capabilities, and equipment is a valuable opportunity that should be seized energetically; the benefits of fighting climate change are the icing on the cake.

As the Romans salted the fields of Carthage to prevent human life, so now Western militaries should do the opposite, working to protect and restore the natural environment. Not only will that help address the threat of climate change and support the prosperity and safety of global populations, but it will also enhance Western warfighting capabilities for future conventional and counter-insurgency operations alike.

## References

Austin, J. E. and C. E. Bruch (eds) 2007. *The Environmental Consequences of War: Legal, Environmental and Scientific Perspectives*. Cambridge: Cambridge University Press.

Baker, M., R. Clausen, R. Kanaan, M. N'Goma, T. Roule, and J. Thomson 2004. *Conflict Timber: Dimensions of the Problem in Asia and Africa: Volume III, African Cases*. Vermont: ARD, Inc.

Balwan, N. 2021. Contribution of Ecological Task Force of Indian Army. *Times of India*. https://timesofindia.indiatimes.com/readersblog/col-nagial/contribution-of-ecological-task-force-of-indian-army-33169/ (accessed 1 May 2023).

Barrett, O. 2019. Colombian military combats unprecedented deforestation. https://dialogo-americas.com/articles/colombian-military-combats-unprecedented-deforestation/#.Yl2Hsy8w1PM (accessed 1 May 2023).

Barry, B. D. Barrie, and N. Childs 2021. Dealing with hot air: UK defence and climate change. *IISS Military Balance Blog*. https://www.iiss.org/blogs/military-balance/2021/04/uk-defence-climate-change (accessed 1 May 2023).

Bergholt, D and P. Lujala 2012. Climate-related natural disasters, economic growth, and armed civil conflict. *Journal of Peace Research* 49: 147–162.

Blundell, A. G. 2010. *Forests and Conflict: The Financial Flows That Fuel War*. Washington DC: Program on Forests (PROFOR).

Brauer, J. 2009. *War and Nature: The Environmental Consequences of War in a Globalized World*. Plymouth: AltaMira Press.

Brewer, H. 2021. Op CORDED: 1st Battalion Irish Guards tackle poaching in Zambia. *Sanctuary Magazine* 50: 24–25.

Brown, O., A. Hammill, and R. McLeman 2007. Climate change as the 'new' security threat: implications for Africa. *International Affairs* 83: 1141–1154.

Chami, R., S. Oztosun, T. Cosimano, and C. Fullenkamp 2019. Nature's solution to climate change. *Finance & Development* 56(4): 34–38.

Chevallier, R. 2013. *Balancing Development and Coastal Conservation: Mangroves in Mozambique*. Johannesburg: South African Institute of International Affairs.

Crawford, N. C. 2019. *Pentagon Fuel Use, Climate Change, and the Costs of War*. Providence, NJ: Brown University Press.

Dargie, G. C., S. L. Lewis, I. T. Lawson, E. T. A. Mitchard, Y. E. Bocko, and S. A. Ifo 2017. Age, extent and carbon storage of the central Congo Basin peatland complex. *Nature* 542(7639): 86–90.

DEFRA 2022. Illegal wildlife trade challenge fund. https://iwt.challengefund.org.uk (accessed 1 May 2023).

Deibert, M. 2013. *The Democratic Republic of Congo: Between Hope and Despair*. London: Zed Books.

Dhanasree, J. 2017. Climate diplomacy and India's Ecological Task Force. *Climate Diplomacy*. https://climate-diplomacy.org/magazine/environment/climate-diplomacy-and-indias-ecological-task-force (accessed 1 May 2023).

DOD 2021. *Department of Defense Draft Climate Adaptation Plan*. Washington DC: Department of Defense.

Dudley, J. P., J. R. Ginsberg, A. J. Plumptre, J. A. Hart, and L. C. Campos 2002. Effects of war and civil strife on wildlife and wildlife habitats. *Conservation Biology* 16: 319–329.

Duffy, R., F. Massé, E. Smidt, E. Marijnen, B. Büscher, J. Verweijen, M. Ramutsindela, T. Simlai, L. Joanny, and E. Lunstrum 2019. Why we must question the militarisation of conservation. *Biological Conservation* 232: 66–73.

Ellis-Peterson, H. 2019. Vietnam seizes 125kg of smuggled rhino horns worth $7.5m. *The Guardian.* https://www.theguardian.com/world/2019/jul/29/vietnam-seizes-125kg-of-smuggled-rhino-tusks-worth-75m (accessed 1 May 2023).

Estrella, M., F. G. Renaud, ₖ. Sudmeier-Rieux, and U. Nehren 2016. Defining new pathways for ecosystem-based disaster risk reduction and adaptation in the post-2015 sustainable development agenda. (In) F. G. Renaud, ₖ. Sudmeier-Rieux, M. Estrella, and U. Nehren (eds) *Ecosystem-Based Disaster Risk Reduction and Adaptation in Practice.* Dordrecht: Springer, pp. 553–591.

Farquhar, S. D. 2017. When overfishing leads to terrorism: the case of Somalia. *World Affairs: The Journal of International Issues* 21(2): 68–77.

Girot, P. O. 2002. Overview B: environmental degradation and regional vulnerability: lessons from Hurricane Mitch. (In) R. Matthew, M. Halle, and J. Switzer (eds) *Conserving the Peace: Resources, Livelihoods and Security.* Winnipeg: International Institute for Sustainable Development, pp. 273–324.

Goldenburg, S. 2009. Wildlife expert claims gorilla dung is critical to containing climate change. *The Guardian.* https://www.theguardian.com/environment/2009/oct/13/gorilla-forests-climate-change-redmond (accessed 1 May 2023).

Hanson, T. 2009. Warfare in biodiversity hotspots. *Conservation Biology* 23: 578–587.

Hatton, J., M. Couto, and J. Oglethorpe 2001. *Biodiversity and War: A Case Study of Mozambique.* Washington, DC: Biodiversity Support Program.

Hawken, P. 2021. *Regeneration: Ending the Climate Crisis in One Generation.* London: Penguin Books.

Henk, D. 2006. Biodiversity and the military in Botswana. *Armed Forces & Society* 32(2): 273–291.

Holzman, D. C. 2012. Accounting for nature's benefits: the dollar value of ecosystem services. *Environmental Health Perspectives* 120: 152–157.

Humphreys, J. 2022. War-extinction in Ukraine. *The Marjan Study Group Blog.* https://themarjancentre.wordpress.com/2022/04/17/war-extinction-in-ukraine/ (accessed 1 May 2023).

IPCC 2022. *Mitigation of climate change: summary for policymakers.* Geneva: Intergovernmental Panel on Climate Change.

Kahekwa, J. 2022. Caring about conservation is not enough: we need to make it pay. *African Arguments.* https://africanarguments.org/2022/03/caring-about-conservation-is-not-enough-we-need-to-make-it-pay/ (accessed 1 May 2023).

Kalkowski, J. 2020. DIO's commitment to planting trees. *GOV.UK.* https://insidedio.blog.gov.uk/2020/08/14/dios-commitment-to-planting-trees/ (accessed 1 May 2023).

Kim, K. C. 2007. Preserving Korea's demilitarized corridor for conservation: a green approach to conflict resolution. (In) S. H. Ali (ed) *Peace Parks: Conservation and Conflict Resolution: Global Environmental Accord: Strategies for Institutional Innovation.* Cambridge, MA: MIT Press, pp. 39–60.

Maguire, T. and C. Haenlein 2015. *An Illusion of Complicity: Terrorism and the Illegal Ivory Trade in East Africa.* London: Royal United Services Institute.

Malfatto, E. 2021. Drought and abundance in the Mesopotamian marshes. *New York Times.* https://www.nytimes.com/2021/04/12/travel/iraq-mesopotamian-marshes.html (accessed 1 May 2023).

McCoy, T. 2021. Bolsonaro sent soldiers to the Amazon to curb deforestation: here's how the effort failed. *Washington Post*. https://www.washingtonpost.com/world/the_americas/brazil-bolsonaro-military-amazon-deforestation/2021/01/03/cde4d342-3fc9-11eb-9453-fc36ba051781_story.html (accessed 1 May 2023).

McCullough, A., L. Mayhew, and S. Opitz-Stapleton 2019, When rising temperatures don't lead to rising tempers: climate and insecurity in Niger. *Overseas Development Institute*. https://odi.org/en/publications/when-rising-temperatures-dont-lead-to-rising-tempers-climate-and-insecurity-in-niger/ (accessed 1 May 2023).

McNeely, J. A. 2002. Overview A: biodiversity, conflict and tropical forests. (In) R. Matthew, M. Halle, and J. Switzer (eds) *Conserving the Peace: Resources, Livelihoods and Security*. Winnipeg: International Institute for Sustainable Development, pp. 364–396.

Milburn, R. 2012. Mainstreaming the environment into post-war recovery: the case for 'ecological development'. *International Affairs* 88: 1083–1100.

Milburn, R. 2018. The forgotten pillar of post-war recovery: does biodiversity loss contribute to insecurity and can biodiversity conservation support peacebuilding and development in the DRC, and beyond? Unpublished PhD thesis, King's College London.

Millennium Ecosystem Assessment 2005. *Ecosystems and Human Well-Being: Biodiversity Synthesis*. Washington, DC: World Resources Institute.

MOD 2021. *Climate Change and Sustainability Strategic Approach*. London: Ministry of Defence.

Mrema, E. M., C. Bruch, and J. Diamond 2009. *Protecting the Environment During Armed Conflict: An Inventory and Analysis of International Law*. Nairobi: United Nations Environment Programme.

Nel. P and M. Righarts 2008. Natural disasters and the risk of violent civil conflict. *International Studies Quarterly* 52: 159–185.

Neme, L. 2014. For rangers on the front lines of anti-poaching wars: daily trauma. *National Geographic*. https://www.nationalgeographic.com/animals/article/140627-congo-virunga-wildlife-rangers-elephants-rhinos-poaching (accessed 1 May 2023).

Noriega, C. 2022. Colombia's new anti-deforestation law provokes concern for small-scale farmers. *Mongabay*. https://news.mongabay.com/2022/01/colombias-new-anti-deforestation-law-provokes-concern-for-small-scale-farmers/ (accessed 1 May 2023).

Oglethorpe, J., R. Ham, J. Shambaugh, and H. van der Linde 2002. Overview C: conservation in times of war. (In) R. Matthew, M. Halle, and J. Switzer (eds) *Conserving the Peace: Resources, Livelihoods and Security*. Winnipeg: International Institute for Sustainable Development, pp. 361–384.

Ong, S. and E. Carver 2019. The rosewood trade: an illicit trail from forest to furniture. *Yale 360*. https://e360.yale.edu/features/the-rosewood-trade-the-illicit-trail-from-forest-to-furniture (accessed 1 May 2023).

Paes, C. F. 2021. As Brazil's military pulls out of the Amazon, its legacy is in question. *Mongabay*. https://news.mongabay.com/2021/04/as-brazils-military-pulls-out-of-the-amazon-its-legacy-is-in-question/ (accessed 1 May 2023).

Plumptre, A. J., M. Masozera, and A. Vedder 2001. *The Impact of Civil War on the Conservation of Protected Areas in Rwanda*. Washington, D.C.: Biodiversity Support Program.

Price, S. V. 2003. Preface. *Journal of Sustainable Forestry* 16: xvii–xxii.

Prunier G. 2009. *Africa's World War: Congo, the Rwandan Genocide, and the Making of a Continental Catastrophe.* Oxford: Oxford University Press.

Renaud, F. G., K. Sudmeier-Rieux, and M. Estrella (eds) 2013. *The Role of Ecosystems in Disaster Risk Reduction.* Tokyo: United Nations University Press.

Schiffman, R. 2020. Green violence: 'eco-guards' are abusing indigenous groups in Africa. *Yale 360.* https://e360.yale.edu/features/green-violence-eco-guards-are-abusing-indigenous-groups-in-africa (accessed 1 May 2023).

Seddon, N., A. Chausson, P. Berry, C. A. Girardin, A. Smith, and B. Turner 2020. Understanding the value and limits of nature-based solutions to climate change and other global challenges. *Philosophical Transactions of the Royal Society B: Biological Sciences* 375(1794): 1–12.

Seyler, J. R., D. Thomas, N. Mwanza, and A. Mpoyi (eds) 2010. *Democratic Republic of Congo: Biodiversity and Tropical Forestry Assessment (118/119) Final Report.* Washington DC: USAID.

Shalukoma, C. 2000. News from Kahuzi-Biega. *Gorilla Journal* 21: 3–4.

South, N. and T. Wyatt 2011. Comparing illicit trades in wildlife and drugs: an exploratory study. *Deviant Behaviour* 32: 554–555.

Stevens, M. L. 2007. Iraq and Iran in ecological perspective: the Mesopotamian Marshes and the Hawizeh-Azim Peace Park. (In) S. H. Ali (ed.) *Peace Parks: Conservation and Conflict Resolution: Global Environmental Accord: Strategies for Institutional Innovation.* Cambridge, MA: MIT Press, pp. 313–332.

Sudmeier-Rieux, K., H. Masundire, A. Rizvi, and S. Rietbergen (eds) 2006. *Ecosystems, Livelihoods and Disasters: An Integrated Approach to Disaster Risk Management.* Gland: IUCN.

Swatuk, L. A. 2007. Seeing the forest for the trees: tropical forests, the state and violent conflict in Africa. (In) W. De Jong, D. Donnovan, and K. Abe (eds) *Extreme Conflict and Tropical Forests.* Dordrecht: Springer Netherlands, pp. 93–116.

The Economist 2016. Rejuvenating reefs. *The Economist.* https://www.economist.com/international/2016/02/13/rejuvenating-reefs (accessed 1 May 2023).

The Economist 2020. Into the blue Amazon: Brazil's armed forces and coastal governance. *The Economist.* https://ocean.economist.com/governance/articles/blue-amazon-brazil-armed-forces-and-coastal-governance (accessed 1 May 2023).

The Economist 2021a. Removing carbon dioxide from the air: the world's biggest carbon-removal plant switches on. *The Economist.* https://www.economist.com/science-and-technology/2021/09/18/the-worlds-biggest-carbon-removal-plant-switches-on (accessed 1 May 2023).

The Economist 2021b. Why the war against jihadists in Mali is going badly. *The Economist.* https://www.economist.com/middle-east-and-africa/why-the-war-against-jihadists-in-mali-is-going-badly/21806350 (accessed 1 May 2023).

UNEP 2011. *The Democratic Republic of the Congo: Post-Conflict Environmental Assessment Synthesis for Policy Makers.* Nairobi: United Nations Environment Programme.

UNEP 2012. *Greening the Blue Helmets: Environment, Natural Resources and Peacekeeping Operations.* Nairobi: United Nations Environment Programme.

Vinall, S. and R. Milburn 2021. Countering the illegal wildlife trade. *British Army Review* 181: 94–103.

Wallen, K. E 2018. Global timber trafficking harms forests and costs billions of dollars: here's how to curb it. *The Conversation.* https://theconversation.com/global-timber-trafficking-harms-forests-and-costs-billions-of-dollars-heres-how-to-curb-it-93115 (accessed 1 May 2023).

Warchol, G. L. 2004. The transnational illegal wildlife trade. *Criminal Justice Studies* 17: 57–73.

Watsa and Wildlife Disease Surveillance Group 2020. Rigorous wildlife disease surveillance. *Science* 369(6500): 145–147.

World Bank 2012. *Inclusive Green Growth: The Pathway to Sustainable Development.* Washington DC: The World Bank.

WWF 2022. Stopping the illegal wildlife trade. *World Wildlife Fund.* https://www.wwf.org.uk/what-we-do/stopping-illegal-wildlife-trade (accessed 1 May 2023).

Yamagiwa, J. 2003. Bushmeat poaching and the conservation crisis in Kahuzi-Biega National Park, Democratic Republic of the Congo. *Journal of Sustainable Forestry* 16: 111–130.

# Framings and reflections

# 14

# ECOLOGICAL SECURITY

The new military operational priority for humanitarian and disaster response

*Thammy Evans and Gary Lewis*

## Introduction

Two seminal events took place during the last few years that should force us to think about how to strengthen and shape the military to best serve the world that future generations will inherit. These are the COVID-19 pandemic and the Russian invasion of Ukraine. The COVID-19 pandemic was perhaps the first rapid onset global challenge of the 21st century demonstrating why human beings must learn to think and act as a species in the face of threats that are simultaneous, common, and existential.

The International Panel on Climate Change Assessment Reports 6 (AR6) (IPCC 2021) describes a dramatically increased risk horizon for most of the twelve tipping points in the planet's carbon cycle.[1] Once exceeded, these tipping points will fundamentally produce abrupt, non-linear changes in climate, with run-away heating then occurring. This will be the case even if emissions are subsequently reduced. The result will be a 'hothouse earth' – one in which human efforts to reduce emissions will be increasingly futile. Unexpected consequences will result.

The Ukraine War is a bitter reminder that we are not yet at the stage in human evolution that we can forswear the option to deploy military force. Sadly, the fact of war remains. This will be the case even if, in broad historical terms, inter-state war has been measurably on the decline for some time (Coker 2021; Dyer 2021). History may show that the Ukraine type of war – where one country invades another for ideological and neo-imperial reasons (Snyder 2018) – is an outlier. However, the Ukraine War demonstrates that for the foreseeable future, the West will require an organised, disciplined military to deter invasion. What will really change in the future is the

DOI: 10.4324/9781003377641-18

character of future conflicts. The wars of the future will be very much 'shaped' by climate change and ecological breakdown, its causes, and its consequences (Fant 2022; Goodman 2021; Mathiesen 2022).

In the short term, the Ukraine War's most significant impact has been to reverse the risk of downsizing that NATO had been facing (Stroobants and Vincent 2022). Instead, NATO has been growing. NATO's militaries are also getting more political and financial attention. Additional resourcing provides an opportunity for NATO to become more functionally adept to face potential new threats. NATO's purpose has been to guarantee peace through strength in what is arguably one of the planet's historically bloodiest theatres of conflict: Europe. However, while NATO's central focus must remain on collective defence, its 'threat perspective' must change. This perspective must henceforth include appreciation of transborder threats, instability, and insecurity that are driven by climate change and ecological breakdown which may be actorless (Busby 2020) or weaponised.[2] A significant military role in this context – both within Europe, but especially outside of it – will come in the form of humanitarian and disaster relief (HADR) operations. This will also include the need to support national authorities to deal with run-away infectious diseases.

The Ukraine War has also shown the need for the global community to recommit to a rules-based international order. One which needs to be stronger and more empowered in order to tackle drivers of insecurity as massive and relentless as climate change and ecological breakdown. One which is able to see beyond the parochial solidarities of flag and anthem. One which can help ensure the attainment of the Sustainable Development Goals (SDGs),[3] to which almost all governments have committed themselves. The SDGs provide the most meaningful pathway for enhancing human security in the coming century.

This chapter explores aspects of HADR and Military Aid to the Civilian Authority (MACA) operations as they relate to climate change. The compound risks of food, energy, water, and displacement are described. It is shown that the military's new-found amplified focus on human insecurity provides opportunities to build resilience, lock in first-mover advantage, deliver climate justice, and broaden security-relevant climate policy. In highlighting ground realities and necessitating enhanced ecological literacy amongst security practitioners and policymakers, the chapter proposes that ecological security must become an essential aspect of all future security planning, with approaches centred on mitigation, adaptation, resilience, and regeneration.

## Key dimensions of future HADR-related conflicts

Our future will be beset by conflict. Lots of conflicts will be propelled by the consequences of droughts, heatwaves, floods, storms, and the impact of sea-level rise. Other chapters in this volume set out in detail the climate change/ecological breakdown threats to security. But outlined below are some key dimensions of this nexus as they relate to HADR.

Most future climate/ecological-driven conflicts will take place in what are already known as the 'shatter belt' regions.[4] On our current trajectory, climate change and ecological breakdown will generate greater scarcity and competition for resources. This, in turn, will produce a toxic cocktail of poverty, grinding underdevelopment, and environmental degradation. When combined with state fragility – measured both in poor capacity to cope (effectiveness) and poor state-to-society relationships (legitimacy) – the challenge gets compounded. When environmental risk and state fragility are combined, the consequences are destructive. Countries of particular concern also tend to exhibit the following characteristics: (1) people's livelihoods highly dependent on agriculture or natural resources; (2) trend for low levels of inclusivity (religious, gender, ethnic minorities, youth) in power sharing; (3) lengthy or recent histories of conflict; (4) weapons already present or cached in-country; and (5) the capacity to adapt to climate change impact is low (Mach et al. 2020).

It is important to note that of the world's 25 most 'fragile' states,[5] the UN already has peacekeeping operations in 20.[6] Over half of such peacekeeping operations are in countries ranked 'most exposed' to climate change. Of the UN peacekeeping operations, five out of the six largest are located in countries, 'highly vulnerable to the effects of climate change' (Krampe 2021). Both UN peacekeeping operations and special political missions are increasingly mandated by the UN Security Council to consider and respond to climate-related security risks. These countries account currently for approximately 75% of the UN's field-based special political missions. According to the Notre Dame Gain Index of climate vulnerability, of the 20 countries considered most vulnerable and least prepared to adapt to climate change, at least 12 experienced violent conflict during 2020 (ND-GAIN 2023).

### 'FEWD': key drivers of compound risk

In *All Hell Breaking Loose*, Michael Klare (2019) outlines a useful conceptual ladder of escalation for militaries for when they are called upon to respond to climate/ecological emergencies. The first rung is where they are requested to provide emergency support to (developing) countries. The next rung involves more extended operations, where humanitarian support often needs to be combined with counter-insurgency operations. Military deployments spanning several countries are the subsequent rung. And so on. In our assessment, the higher the level of intervention, the more we move away from 'simple' HADR operations and instead, enter the realm of systemic and relentless threats driven by four essential processes. These fundamental pathways of influence – or drivers of compound risk – that threaten human security are food, energy, water, and displacement (FEWD).

## Food

The combination of conflict, climate change, soil degradation, water scarcity, and pest and disease upticks, such as locusts[7] or avian flu, is already threatening multiple cascading risks on food systems and supplies (see Benton et al., this volume). Both acute (e.g. price hikes and labour shortages) and chronic (e.g. climate change, drought, and water scarcity) shocks to supplies of food, as well as to systems that produce, store, and transport food, are already having a long-term impact that will require years to resolve (Menker 2022). The resulting disasters, complex emergencies, and humanitarian needs will strain national responses in the affected countries. In some cases, this will require calling-in military assistance, both from home and abroad. This is not necessarily a distant reality. The recent Russian blockade of Ukrainian ports, the destruction of its seedstock, the theft of its grain, and the diversion of its agricultural workers into a military response have shown that the weaponisation of the food system can have quick, powerful, and global security implications. During 2022, wheat, barley, and sunflower seeds were, for example, left sitting in warehouses and could not be moved to market via normal channels. This led *The Economist* (2022) to suggest that in order to get much-needed grain out of Ukraine, 'convoys may require armed escorts endorsed by a broad coalition'. Such inflection points can lead quickly to requests for military aid. Responding effectively to them requires an intelligence system that is already cognisant of where those points of strain lie. One which has already considered, in advance, under what conditions, and crucially – when – to take action.

## Energy

The overdue but accelerating global shift to renewable energy in order to mitigate climate change may, perversely, also run the risk of generating conflict. The world needs clean power. But decarbonisation calls for a massive increase in the mining, extraction, and recycling of minerals, and metals like lithium, graphite, and cobalt. Access to critical minerals or so-called 'rare earths' (Klinger 2017) will thus be essential for the digital and decarbonised transitions required in the main greenhouse gas (GHG) emitting countries. There will, inevitably, be a scramble for such resources. These minerals and metals are located all over the world; however, the most easily accessible quantities are found in the tropics, and, worryingly, mainly in states already considered 'fragile'. More than this, they are found in some of the world's most critical ecosystems – the ones that regulate the global climate regime as valuable carbon sinks (both marine and terrestrial). States that control rare earth supply chains, including China and Russia, could become considerably advantaged. There will be a need to anticipate, and de-escalate, such resource competition, including through resource and technology efficiencies. Predatory extraction without regard to consequences on human security and livelihoods will drive riots, food and water insecurity, migration, and the possibly even eruption of threats from 'ungoverned spaces'. These contexts are perhaps particularly likely to require future HADR, MACA, and other military operations.

The global challenge is thus to source such materials in a responsible, resilient, and regenerative way (Elkington 2021). Ways and means that do not feed into a socially and environmentally predatory politics-for-profit extraction cycle. These processes are the very ones that have historically created, and on prevailing trends will continue to create, institutions and processes that are part and parcel of a business model which generates marginalisation, inequality, conflict, and destitution. Effective industrial sector solutions should promote effective environmental and social governance safeguards. There is therefore a need to reconcile climate and biodiversity strategies. This will lessen the impacts of a new scramble for resources that may drive geopolitical tensions in the coming decades (Lazard and Youngs 2021).

To counterbalance the above, it is also worth noting that the expansion of renewable energy options will likely allow countries to increase their energy independence and simultaneously reduce potential sources of conflict over non-renewable energy sources, such as oil and natural gas.

### Water

A recent Carnegie report shows how insecurity, with implications for conflict and transnational organised crime, will likely come from two related phenomena: increasing scarcity of natural resources and the breakdown of the hydrological cycle at local and global levels (Lazard and Youngs 2021). Water scarcity, through green water depletion – in other words losing water from soil and plant moisture, rainfall, and evaporation – has now been reported to have exceeded safe limits (Wang-Erlandsson et al. 2022). The hydrological cycle not only ensures the regulation of global heat dynamics and underpins the carbon cycle, but also the reliable, dependable, and manageable distribution of liquid water across the globe (Sheil 2014; Spracklen 2018). When water becomes scarce, this reduces food yield, the viability of livelihoods, and, ultimately, the ability of the land to sustain human life. Without water, humans are forced to move. Relatedly, Mami Mizutori, the UN Secretary General's Special Representative for Disaster Risk Reduction, has noted, ominously: 'drought is on the verge of becoming the next pandemic and there is no vaccine to cure it. Most of the world will be living with water stress in the next few years. Demand will outstrip supply during certain periods. Drought is a major factor in land degradation and the decline of yields for major crops' (UNDRR 2021a, 2021b).[8] Record temperatures in India and Pakistan, in May 2022, cut wheat production for many farmers by 30–60%, leading India to ban the export of wheat (Johnson 2022). In this way, one threat cascades onto another, and then another, like slow-falling dominoes.

### Displacement

The Internal Displacement Monitoring Centre (IDMC) and the United Nations High Commissioner for Refugees (UNHCR) track forced migration within borders and across borders, respectively. As tracked by UNHCR's Refugee Data Finder (UNHCR 2021), internally displaced people (IDPs), refugees, and asylum seekers totalled 84 m people worldwide by mid-2021. Furthermore, 2020 witnessed a record 30.7 m new IDPs driven to move, often at short notice, by climate and other natural disasters. This was three times more people than those displaced that very same year by conflict (IDMC 2022).[9] Many of these people do return to what is left of their homes, whilst others move on to new homes permanently. Nonetheless, of the total number of IDPs who remained displaced at the end of 2021 (59 m), 10% were still deemed to have been displaced by long-term natural disasters – predominantly due to climate change. At the current rate of increase in natural disasters, the Institute for Economics and Peace assessed in their inaugural Ecological Threat Register that 1.2bn people could be displaced globally due to natural disasters (IEP 2020, 2021) by 2050. In response, the World Bank's updated Groundswell report states that 'decisive collective action could reduce climate migration by as much as 80%' (Clement et al. 2021). With regard to internal versus external forced migration, only the migration origins of IDPs are accurately divided into conflict versus climate disasters in IDMC's graphic-rich annual Global Report on Internal Displacement (GRID). According to the World Economic Forum, 'data on climate refugees … is limited, which is why they're called the "forgotten victims of climate change"' (Ida 2021). Many conflicts and their related displacement, as described elsewhere in this volume, are caused indirectly by drivers of climate and ecological breakdown.

To summarise this section, the key drivers of compound risk can be conceptualised in the acronym FEWD. These drivers will produce a spectrum of invidious governance challenges that will likely call for military activities. Responsible, professional military engagement will be in ever-greater demand to assist civilian authorities. As a consequence, the role of the military to support HADR both at home and abroad will increase. Anticipating these trends will become critical to plan prevention, mitigation, and adaptation to the best degree possible. For this reason, it will become necessary to properly understand the nature of the threat.

## Refocusing the lens: ecological security[10]

We live in the age of the Anthropocene. This means we are now operating, quite literally, in a new world; one we have changed through human impact on the Earth's geology and ecosystems. Our environment – including our security and economic environment – has changed, and it has changed for all of us. Viewed in these terms, approaching international relations exclusively from the parochial perspective of narrowly defined 'national self-interest' is not credible or viable. In the Anthropocene, we can no longer define national self-interest without regard to the condition of the biosphere. In the same way that global financial flows have brought about the need for a Financial Stability Board, we will need to start to look at food systems and the hydrological cycle, for example, in terms of the need for global governance of food and water system stability (Dixson-Declève et al. 2021).

One of the key lessons of the COVID-19 pandemic is that there are existential, over-the-horizon threats which transcend borders powerfully. These can only be resolved if UN member states recalibrate their understanding of what it means for something to be in their 'national self-interest'. There will, of course, be critics who view such a conceptual assault on the traditional, realist perspective as hopelessly naïve. However, in today's world, states can only become strong and succeed if they base their actions on a shared sense of responsibility for the protection of the biosphere. As sweeping as it may sound, this is the manner of thinking that is called for modern global leadership. It must henceforth underpin all aspects of international relations. This is an approach which the original framers of the UN Charter also had in mind in the years leading up to 1945. The logic is even more compelling today. Nations must act collaboratively because their need for self-preservation in a heating world demands it. We recognise that such a recalibration of perspective will not take place overnight, even if the evidence for doing so is already persuasive. Nonetheless, having responsible military leadership that understands and acts upon this proposition, will make a substantial contribution to the necessary – and inevitable – political and security changes already underway.

It is fortunate that a roadmap for acting in a globally responsible way is already in place. This is because we now have a common plan: the 2030 Agenda for the SDGs. In SDG terms, all states should see themselves as 'developing countries' aspiring to progress. The SDGs are founded on the notion that people cannot be assured of their own peace and security while ignorant of the uncertain conditions under which others live. We all inhabit the same planet. The SDGs are 17 in number. All have well-defined targets and indicators which can be and are being, measured (UN 2023). Nothing like this has ever existed before. What is important to appreciate is that underpinning all the other goals are four environmental goals:

- Ensure availability and sustainable management of water and sanitation for all (Goal 6);
- Take urgent action to combat climate change and its impacts (Goal 13);
- Conserve and sustainably use the oceans, seas, and marine resources for sustainable development (Goal 14); and
- Protect, restore, and promote sustainable use of terrestrial ecosystems, sustainably manage forests, combat desertification, and half and reverse land degradation and halt biodiversity loss (Goal 15).

Whilst it may be true that human development is projected to improve in the short term, the world – in general – is not on track to meet the SDGs. For example, as regards the environmental indicators, the UN Environment Programme reports that we are currently only achieving around 20% of the SDG targets, with 10% definitely not being met. A lack of data makes the remainder impossible to assess (UNEP 2021, 2023).

It is in this context that ecological security must become an essential aspect of future security planning. This will require that, while militaries continue to view threats from state-centric adversaries as legitimate priorities, they must broaden their current focus away from so-called 'hard security' to include ecological security approaches that encompass mitigation, adaptation, resilience, and regeneration. Doing so will improve our understanding of the breadth of oncoming scenarios to which we need to respond.[11] We contend that the security community must act now to make up for lost time. It must widen its perspective and engage beyond its traditional comfort zone.

To date, and broadly, military intelligence assets have focused little on climate and ecological security. In terms of adaptation to conflict-related and eco-based security risks, there is a need to collect, share, and incorporate data – both on ecological breakdown threats and the importance of early warning systems – into traditional security horizon-scanning capabilities. This would feed, for example, into the UK's defence aspirations to possess 'a cutting-edge climate threat horizon-scanning capability' (MOD 2021a). The World Meteorological Organisation (WMO) has made a strong case for such anticipation linked to the number of lives that can be saved with such an approach (WMO 2022a; 2022b). Indeed, improved early warning and other protective/adaptive measures, if instituted in time, can invert the pattern that sees a correspondence between the rise in the number of natural disasters and the number of deaths.

Despite this accumulated evidence, recent national assessments, intelligence reports, and political actors, however, 'still downplay [the] threat of climate change' (McKenzie 2021) and other global boundaries to conflict and national security (Weathering Risk 2022). Whether deliberate or not, such a diminishment of relevant data and analysis sadly limits the corresponding 'pull' these issues have on security and defence resources. However, the news is not all bad. A number of clear-sighted intelligence, security, and policy officials have called for the need to 'recast security' (Goodman 2021). In recent decades, traditional state-centric national security perspectives have been slowly tiptoeing towards an emphasis on elevating human security as the basis of national security (see UNDP 1994; MOD 2021a). Moreover, there have been calls for a much greater degree of national intelligence capability to focus on climate and ecological security. Such calls have come from some of the most senior defence and security leaders with insight into climate change and environment insecurity (IMCCS 2020; 2021; MdA, 2022; Nugee 2020). Their reasoning is simple. Without a dedicated intelligence focus on climate change and ecological security geared towards defending global sustainability, our preparedness falls significantly short of our potential to mitigate disasters upstream. The result will inevitably be more resources expended – downstream – on HADR.

It is also worth pointing out that there is a need to move beyond casting all environmentally driven threats as deriving immediately from climate change. This is because climate change is assessed to be only one of the nine 'planetary boundaries' of ecological security.[12] It is for this reason that, in this chapter, we have been consistently referring to threats emanating from 'climate change and ecological breakdown'. We believe that the paucity of linkages to biodiversity is blinkered. Military strategic focus should not selectively fall upon abiotic impacts, or degradation or decarbonisation. It should instead be about increasing regeneration in order to avoid the increase in pathogens and other negative organisms in the first place. As Evans and Schroeder (2021) note, 'Regenerative security has the potential to restore confidence, legitimacy, a sense of rule of law, basic rights, access to justice, resources themselves and, ultimately, trust in people, systems, and environments. Simply put, regenerative security is the regeneration of a state towards the attainment of sustainable positive security'. A further step towards regenerative security approaches would resource and use more MACA to defend sustainability, restore landscapes and seascapes, and so protect nature's own ability to regenerate (King 2022).[13]

The evolution of security concepts from state-centric to human security to ecological security is shown in Figure 14.1. To realise such a trajectory, national security dialogues, instruments, and intelligence assets urgently need to ramp-up their literacy of climate and ecological security (Goodman 2021; MdA 2022; NATO 2021). In addition, a focus on ecological security, we would argue, will more easily allow security and defence responses to be better framed in terms of 'worst-case' and 'likely-case' scenarios. Moreover,

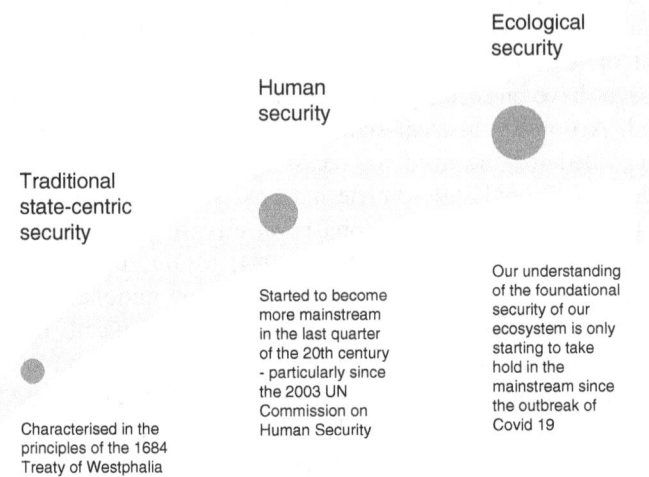

Ecological
security

Human
security

Traditional
state-centric
security

Started to become
more mainstream
in the last quarter
of the 20th century
- particularly since
the 2003 UN
Commission on
Human Security

Our understanding
of the foundational
security of our
ecosystem is only
starting to take
hold in the
mainstream since
the outbreak of
Covid 19

Characterised in the
principles of the 1684
Treaty of Westphalia

**FIGURE 14.1** Evolution of concepts of security.

such an approach would help identify relative degrees of likelihood. If the ecological security framework is adopted, then military planning for the full impact of climate change and ecological breakdown on HADR would – properly – become an 'urgent operational requirement'.[14]

### Seizing the advantage

If defence and security actors start to attain a higher level of ecological awareness and utilise an ecological security framework, four main benefits will result. We discuss each of these in turn.

### *Reducing public unpreparedness and increasing societal resilience*

Making environmental intelligence assessments publicly available – even if only partially – will contribute greatly to a 'whole-of-society' level of preparedness (Homer-Dixon 2021, this volume). This will increase the opportunity to maximise combined solutions to come from military, industry, and society. In most countries, military-actor involvement in awareness raising and shared analysis usually adds a level of gravitas, legitimacy, and credibility. It also opens potential discussion space for direction and guidance to efforts which often otherwise struggle. Militaries have witnessed that this can produce an earnest effort from the civilian authority to respond in kind, and with equal resolve and seriousness. Many examples can be highlighted, including (1) use of the UK National Health Service's (NHS) GoodSam app to find additional volunteers during the COVID-19 pandemic; (2) humanitarian assistance provided by the veteran teams in US' Team Rubicon and the UK's Re:ACT; and (3) focus of emerging decentralised autonomous organisations (DAOs) on sustainable blockchain and regenerative futures (Djordjevic 2021). What such a win–win approach does – simply by getting different societal actors engaged and collaborating earlier – is reduce the net overall eventual burden on defence assets. There is also considerable scope for the military to partner with industry to help bring sustainable green technologies to market. This could assist in disaster response situations, especially when these turn into long-term commitments. Examples here include smart grid solar panels for military installations; electric vehicles, both cars and e-bikes; mini wind generator energy supplies; water evaporation capture; biodome shelters; vertical farming; algal biofuel; and ground or air source heat pumps (Clack and Nugee 2022; Givetash 2019). Additionally, climate impact risk assessments already developed with militaries, e.g. the United States (Miller 2016), the United Kingdom (MOD 2019), and France (Gemenne et al. 2021; IRIS 2021), could be rolled out to catalyse better civil-military, resilience, human security (MOD 2021b), and climate-adaptive regenerative solutions (Carbon Brief 2021).

### Locking-in first-mover advantage

The decision to become positioned upstream of potential disaster-related interventions will provide a first-mover advantage in a multi-polar world where influential powers are vying for hegemony. China is already well positioned, probably more than any other country, on the use of green technology overseas (Lovins in Vidal 2022). Indeed, China has already embedded the concept of 'Ecological Civilisation' into its constitution. Its 'Belt and Road initiative' has already installed significant supply and logistic chains in a number of countries, including many across the 'shatter belt' (above). Furthermore, its military, the People's Liberation Army (PLA), has significant large-scale experience in responding to floods, fire, pandemics, and other disasters at home. Within China, the PLA has even experimented with regenerative landscaping in its severely degraded Loess Plateau region. In short, China's military has considerable deployable experience in regenerative opportunities. China has also used vaccine diplomacy to reach out to countries seeking support in dealing with COVID-19. For countries dealing with ecological crises, the allure of pivoting towards China will continue to be compelling since China possesses global reach, technology, and the political will to respond first. One such example is the recent pivot of the Solomon Islands away from the United States to China through a new security pact. The Solomon Islands have already lost a significant portion of their land to sea-level rise in the last 50 years (Albert 2016). China has a track record of literally building islands.

### Working 'with the grain' on climate justice

With the world urgently in need of regenerative solutions and an environmental justice narrative fast gathering pace, disaster relief MACA options that work 'with the grain' on net zero will help to avoid the inevitable looming legal battles, constraints, and calls for compensation. The UN's effort to clean up its own act in terms of 'greening the blue', i.e. by reducing the carbon footprint of its peacekeeping missions, is an approach worthy of review and replication (UNEP 2012). The Kyoto Protocol on net zero did not include military emissions (Parkinson 2020) or military operations but, increasingly, national militaries and their operations are coming under closer public scrutiny and are not immune from the court of public opinion. The 'rising tide of climate litigation' (Botsford 2021) against private and public sector alike is a sign of the times to come. And while 'ecocide' is not yet a fifth international crime, it is perhaps only a matter of time until it becomes one (Babson 2022). In war, widespread, long-term, and severe damage to the natural environment is already deemed an environmental war crime under the Rome Statute. However, the threshold for military action in response to an environmental war crime is high (Daft 2019). But with environmental degradation now so close to several tipping points, and the impact of military action on already precarious environments and climates resulting in long-term complex humanitarian consequences, the threshold for environmental war crime will likely decrease.

### Security-relevant policy

Climate change policy to date focuses too narrowly on abiotic symptoms, economic implications, and adaptation responses, tempered by public and business tolerance. Broader systemic and intersectional intelligence analysis based on an ecological security policy framework would help to widen perspectives, options, and opportunities for policy and decision makers. However, providing intelligence assessments alone does not in itself lead to sound policies. Militaries have observed that recommendations contained in assessments can run the risk of resulting in superficial, box-checking, and quick-fix 'solutions'. These, of course, are riddled with 'backdraft'-prone, unintended consequences, particularly if the policy-making process is accompanied by interest groups, staff writers, and special advisers who are not ecologically literate.[15] Past efforts to publicise wider ecological analysis have been stymied by adverse narrowly focused interests, resulting in a policy vacuum and delayed societal resilience (Schoonover 2019).

### Advancing the policy spectrum for MACA

The way in which many militaries are preparing themselves for their role in a climate-changed and ecologically threatened world does not yet span the full

Advancing the policy spectrum

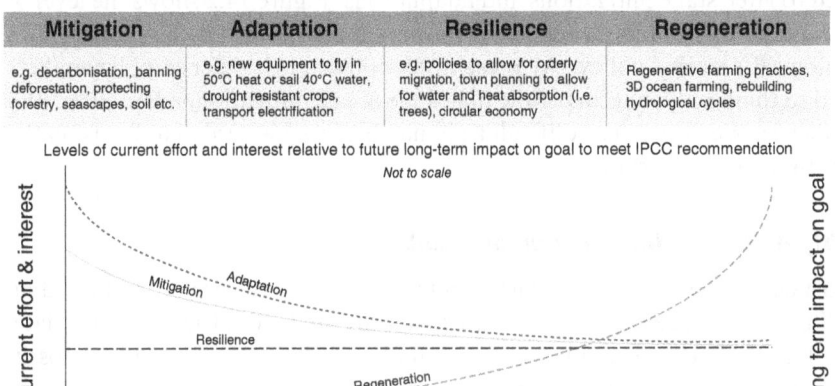

| Mitigation | Adaptation | Resilience | Regeneration |
|---|---|---|---|
| e.g. decarbonisation, banning deforestation, protecting forestry, seascapes, soil etc. | e.g. new equipment to fly in 50°C heat or sail 40°C water, drought resistant crops, transport electrification | e.g. policies to allow for orderly migration, town planning to allow for water and heat absorption (i.e. trees), circular economy | Regenerative farming practices, 3D ocean farming, rebuilding hydrological cycles |

**FIGURE 14.2** Advancing through the mitigation to regeneration policy spectrum.

spectrum of the four crucial stages of an ecological security policy framework. These stages relate to mitigation, adaptation, resilience, and regeneration. This section of the chapter describes these four stages along the mitigation-regeneration policy spectrum. It then looks at how to recalibrate and advance policies to better prevent, manage, and recover from future disasters. To date, some defence forces around the world are developing limited options in mitigation, i.e. avoiding the unmanageable. Most are preparing mainly for climate adaptation, i.e. managing the unavoidable. Few are taking an integrated approach to resilience, i.e. managing what can be managed before it becomes unmanageable. Only an exceptional few are contributing towards regeneration, i.e. pre-empting what will become both unmanageable and unavoidable if left too late (see Figure 14.2).

For this section, we have conducted a literature review drawing upon the very limited body of existing research that focuses on observable military operations (IRIS 2021; Regaud et al. 2022). Together with the reporting of militaries that are implementing climate change and sustainability measures, this shows a broad trend. This is because the current level of effort given over to policy actions along the mitigation-regeneration spectrum tends to focus on securing immediate, tangible benefits in the short term. This happens repeatedly despite the fact that significant evidence demonstrates the supreme advantage of early interventions to reap advantages later down the road (e.g. WMO 2022b). The perfect adage to fit our current situation is that, 'the best time to plant a tree is 20 years ago'. The emphasis must therefore now increasingly focus on down-the-road benefits through regeneration. This is where states can deliver considerably more benefits, including financial benefits, over the medium- to long-term. Conveniently, it will also reduce the need to later-stage mitigations and adaptations. Figure 14.2 shows the level of effort currently put into each of the four main areas of what can be classed as ecological security policy. At present, as noted, most defence policymakers tend to think only in terms of the first three areas: mitigation, adaptation, and resilience. We argue that, difficult as it may be, more is needed to focus on the fourth area: regeneration.

### Mitigation: avoiding the unmanageable

Most decision makers and publics still tend not to see the irony that providing military assistance in fragile and climate-stressed countries has, to date, involved by a very high military carbon footprint. Militaries have a long fossil-fuel-driven logistics tail. Moreover, transporting emergency supplies from one side of the world to the other is carbon intensive. There are also issues when a foreign military outfit draws heavily upon scarce local resources when delivering the requested support. There have, it is true, been some military efforts to experiment with lower-emission platforms, such as tanks, ships, planes, and electric bikes (above). But the fast-follower strategy – one that relies on quickly imitating innovations pioneered by others – can leave MACA flat-footed when working alongside more agile, crowd-sourced, blockchain, and distributed autonomous organisations (DAOs). Low-carbon low-logistics military footprints are becoming more commonplace in some countries, often encouraged by UN peacekeeping operations. The commitment to a net zero footprint for disaster relief would demonstrate relevance and sensitivity. Ironically, though, it may be the Russian invasion of Ukraine, and Europe's need to diminish reliance on Russian oil and gas, that may force a faster switch to a low-footprint military in several countries.

### Adaptation: managing the unavoidable

Several factors contribute to the predominant focus by most militaries on adaptation. Firstly, the reactive nature of most professional and especially expeditionary militaries.[16] Secondly, the threat-based approach adopted by defence analysis that tends to overemphasise worst-case scenarios. Thirdly, the gravitational pull of defence contractors, who are first-up to supply solutions often biased towards their own corporate offerings. These, in turn, drive budgetary and, particularly, platform-led decision making, thereby baking-in certain supply driven responses at a high risk of maladaptation. This can come at the expense of seeking system-led opportunities and options much earlier and positioning to ward off upstream calamity.

One example of heightened adaptation focus is some research undertaken in 2021 by RAND Europe, at the behest of the UK MOD. The key output of which is entitled, 'Crisis Response in a Changing Climate: Implications of Climate Change for UK Defence Logistics in Humanitarian Assistance and Disaster Relief (HADR) and Military Aid to the Civil Authorities (MACA) Operations' (Retter et al. 2021). Despite a recommendation for the UK to move from an 'emergency' to 'resilience' paradigm, the underlying paradigmatic starting point for the research was climate change disruption on ecological integrity rather than ecological disruption on climate integrity. In either scenario, 'worst case' thinking assumed that neither climate change nor its cause of ecological disruption could be avoided. 'Worst case' thinking sees the need to build adapted platforms. In the case of the military, this would drive early commissioning of contracts for new ships that could operate in 40°C water, and better helicopters that could fly in 50°C of dust-laden heat (Retter et al. 2021, pp. 23, 72).

The financial 'winners' from this sort of adaptation-driven approach are inevitably the equipment suppliers. To date, military adaptation efforts have largely been applied to sovereign bases. To the best of our knowledge, only France seems to have considered the broader view beyond looking only at the adaptation required for its military installations. This applies to installations in both metropolitan France and those in its extensive string of overseas departments and territories. Such thinking also extends to France's strong defence advisory relations with francophone Africa in particular. In advance from other similar adaptation vulnerability assessments, the French 'Climate Change Evaluation Methodology for Military Camps' (IRIS 2021) aims to look at both the technical vulnerabilities of camps as well as the socio-natural vulnerabilities in which the installation is situated.

### Resilience: managing what can be managed now

There has been increasing recognition and development in defence forces around the world to integrate the military into cross-government and cross-societal resilience measures at home. The steady rise in abiotic disasters and the recent COVID-19 pandemic has seen most militaries draw lessons from homeland resilience. For most militaries, resilience support lies with the national reserves. In the United States, homeland support has long been the purview of the US National Guard. As Britzky (2022) notes, five years ago, the US National Guard spent roughly 14,000 personnel hours fighting wildfires but by 2021 that had increased to 170,000 hours. Resilience support – in order to fight wildfires – is also now an annual draw on the Australian 2nd Division (Reserves). Germany's creation of the voluntary service, *Dein Jahr für Deutschland* ('Your Year for Germany') has taken place not only in response to the withdrawal of US troops under the Trump administration, but also to prepare for increasing rainfall and flooding in the agricultural plains of Germany (Beste 2021). These plains have been slowly stripped, like elsewhere around the world (see Liu 2007; Rattan and Stewart 2019), of absorbent soil and tree capacity. To try to map and track the incidents of miltary use to combat climate related disasters, the US based Center for Climate and Security launched its Military Response to Climate Hazards (MiRCH) tracker in 2023. Deaths of military personnel from combatting climate hazards are also mentioned.

With the long-predicted increase in resilience tasks, defence readiness is being tested across more of its operating concept, i.e. from warfighting, through other types of military operations and engagement, to protection from natural and human-induced crises at home. Being pinned down at home on resilience tasks will reduce capacity and, in turn, a states' ability to defend its interests or engage abroad in HADR. As such, even the United Kingdom, which has long felt relatively secure within its island perimeter, is starting to look at how to work together with local communities to prepare in advance how to be more resilient (RAND 2021). RAND Europe's study for increasing collaborative resilience is modelled on a number of international examples including the Hampton River military base in Florida, where increased flooding, storms, and sea-level rise was causing the military base to become unworkable for periods of time because of the inability to move key staff in and out to work (see Caves 2021).

### Regeneration: pre-empting what will become unavoidable and unmanageable

In recent years, states in the Global North have had the relative freedoms afforded by a secure home base to prioritise a professional expeditionary army as a military ideal.[17] Authorities in the developing world and Global South, in contrast, and particularly those governing climate-stressed 'fragile' states, have often been obliged to use their disciplined, equipped, and labour-ready military capacity to build, protect, sustain, and develop an ecological foundation to human security and state legitimacy. Within Africa and Latin America, for example, some militaries are having to take an active role in the environmental-climate security dramas already playing out in their countries. This is especially the case where corruption and profit-driven paramilitary interests that have risen up around resource extraction have exceeded the management capacity of civil authorities, such as on the high seas, in illegal logging and mining practices, and in illicit wildlife trade (Regaud et al. 2022). Such activities range from countering illegal wildlife trade (CIWT) in Zambia, Kenya, and Namibia (MOD 2021c); countering illegal logging in Ivory Coast; countering illegal gold mining in Brazil (Digital 2022); countering shipments of toxic waste as well as deterring piracy off the Horn of Africa (EUNAVFOR 2023); and prohibiting, interdicting, and dismantling illegal, unregulated and unreported fishing (IUUF) by distant water fishing

(DWF) (Farrer 2020; Katz 2023). As early as 1982, India formed its first Territorial Army Ecological Task Force by establishing the 127th Infantry Battalion to combat land degradation in recognition of the role of soil erosion in environmental and human security (Jayaram 2015, 2017). In August 2020, Cote d'Ivoire stood up its Brigade spéciale de surveillance et d'intervention (BSSI) (LSI Africa 2021) – consisting of 650 soldiers – to try to combat serious illegal logging in a country which has lost almost all its forestry.

Combatting destructive ecological practices is only one-half of preventing or diminishing potential disaster. So-called 'regenerative security' must entail addressing the risk of aggravated climate impacts. This can be done by building resilience through circular economic models that provide for long-term resource security and human and ecological security (Lazard and Youngs 2021). To date, the use of the military to restore, sustain, develop, and defend ecological sustainability has so far proven relatively uncontentious. More than just local community projects, some projects have occurred on a massive scale with massive ambitions, such as combatting desertification through the planting of an 8,000 km long 'Great Green Wall' across the Sahel from Senegal to Djibouti (GGW 2023; Igini 2022). Success in such undertakings, as seen in the re-terracing, cordoning off, and reforesting of degraded landscapes in China (Groome 2016; Oliver 2018), India (D'Souza 2022; Jayaram 2015; 2017), and Jordan (Jordan 2022), requires considerable physical labour and engineering know-how.

### Effective engagement of foreign MACA in HADR: working with on-the-ground realities

While the Oslo Guidelines (OCHA 2007a; 2007b) call, essentially, for the military to be used only as a 'last resort', this will perforce become ever less the case. Indeed, certain countries, especially in Asia (e.g. China, Pakistan, and India) use the military as a 'first resort'. Given the number of disasters likely to result from climate change and ecological breakdown in the future, it would be unwise to relegate an entity with such capacity (e.g. logistics, labour, and discipline) to only come in at the tail end of HADR. There can be no doubt that in most cases, deployment of the military can bring significant skills and emergency-level equipment to disaster relief situations.[18] Nonetheless, civilian control of such interventions remains essential for reasons of trust, transparency, and legitimacy. It is important therefore that MACA is integrated into an already existing domestic response.

In the Philippines in 2013, the response to Typhoon Haiyan, for example, resulted in 23 foreign militaries providing assistance. This could quickly have overwhelmed a country that had not already prepared for such an influx of such aid.

In our opinion, for external HADR action to have the greatest likelihood of success, the military should demonstrate the following principles: neutrality, impartiality, and independence.[19] However, the military – domestic or foreign – can sometimes be seen as a belligerent or a party to the conflict or post-conflict scenario where a disaster has occurred. For example, the Turkish military was not welcomed or accepted in the aftermath of the Van earthquake, in 2011, by the PKK in southeastern Turkey. Problems related to perceived bias tend to occur more often in complex emergency contexts than in simple natural or human-made emergencies.

There is also a need to distinguish between national MACA and foreign MACA. The following thus responds to situations where militaries are invited to contribute to the MACA in countries requesting support. In the conflicts of the future, the threat environment will be complex, fluid, and rapidly changing. For militaries, in almost every such situation, the objectives of foreign MACA should be twofold: (1) securing the wellbeing of the people in need, and protecting civilians; and (2) securing access to vulnerable populations. For soldiers trying to engage on the ground and deliver HADR in fragile contexts, there will likely be threats of death, injury, and – importantly – kidnapping. Gaining and maintaining access will be critical for competent HADR delivery. Regular and frequent dialogue with local elders, government, military, and religious leaders will be essential. The military will, with the government partner, be responsible for outreach, ideally in a way that is as 'de-Westernised' as possible. Often this should be done in collaboration with the UN Humanitarian Coordinator on the ground, for which there is a tried and tested system in place. Negotiating access will be essential. Often there will be a need for a transparent interposition between the government authorities and opposition leadership – a 'deconfliction' of the local civil/military players. In addition, militaries should plan for remote management of HADR, particularly in contexts where movement is restricted and/or where internationals have already been withdrawn. In such cases, full responsibility for the management of HADR will have already been shifted back to nationals. In this context, the three first principles of humanitarian assistance outlined above should be followed scrupulously.

## Conclusion

In the Anthropocene, states can simply no longer define national self-interest without regard to the condition of the biosphere and its shared protection. A potent mix of our dependence on a globalised economy together with cascading risks from ecological breakdown will produce increasing humanitarian emergencies beyond the current scope of most government services.

As a conclusion, our paper offers the following ten lessons:

- The COVID-19 pandemic has demonstrated that there are powerful, over-the-horizon threats which transcend borders and are existential for human civilisation;
- The Ukraine War has demonstrated that the threat of military force confrontation requires states to maintain an organised and disciplined military, even though the conflicts of the future will be very much shaped by climate change and ecological breakdown;
- Our collective future will be beset by conflict. A lot of it will be caused or propelled by the results of droughts, heatwaves, floods, storms, and the impact of sea-level rise;
- Many future conflicts will take place within or between 'shatter belt' states;
- The fundamental pathways of influence that threaten future human security will be food, energy, water, and displacement (FEWD);
- Opportunities arise as we confront the increase in HADR, including a switch to resilience, locking in first-mover advantage, working with the grain of climate justice, and broadening security-relevant climate policy;
- Ecological security must become an essential aspect of all future security planning, with approaches centred on mitigation, adaptation, resilience, and regeneration;
- The security community must start by increasing its own literacy in ecological security, widening its perspective, developing shared solutions, and engaging beyond its traditional comfort zone;
- For militaries, the objectives of foreign MACA should be to secure the wellbeing of the people in need as well as protect civilians and secure access to vulnerable populations; and
- The primacy of civilian control in disaster relief interventions remains essential for reasons of trust, transparency, and legitimacy, with MACA integrated into existing domestic responses.

## Notes

1 The 12 IPCC-reported tipping-point elements: (1) permafrost thaw; (2) loss of Arctic winter sea ice; (3) loss of West Antarctic ice sheets and shelves; (4) global sea-level rise; (5) ocean acidification weakening ocean carbon sinks; (6) ocean deoxygenation from increased respiration requirements in the oceans; (7) Atlantic Meridional Overturning Circulation (AMOC); (8) global monsoon; (9) Southern MOC; (10) tropical forest; (11) boreal forest; and (12) Antarctic sea ice and polar ice sheets (see Carbon Brief 2021).
2 In the war in Ukraine, for example, concentration risk in the global food system, exacerbated by climate-change-stressed drought, heat, spring frosts, pollination degradation, and wildfires, has been weaponised by Russia who, in 2022, prevented 20 m tons of grain from reaching the international market. As a result,

400 m people more than in previous years were put at risk of starvation (Beasley 2022; see also Winter 2022).

3 The seventeen SDGs followed from the UN Millennium Goals of the year 2000. Together they strive to develop ecological, social, and economic sustainability by 2030 (UN 2023).

4 The shatter belt groups a number of countries under threat. Viewed on a map, these countries are seen to form a broad belt stretching from the Sahel, across the Middle East and North Africa (MENA) region, and then through parts of southern Asia across to the Philippines. Many of these countries have problems of both fractured governance and a variety of endemic environmental issues. These regions also correspond to global priority areas for ecological regeneration; largely because they have witnessed ecological disruption. Often this has also been due to the impact that conflicts, extraction, and exploitation have had on the use of natural resources (see Diehl and Hensel 1994).

5 There is no accepted UN definition of 'fragile' in terms of either state capacity or state legitimacy. There is, however, an existing fragile state index (FSI 2023). The Organisation for Economic Cooperation and Development Assistance Committee (OECD DAC) defines the concept as follows: 'a fragile state is one with weak capacity to carry out the basic functions needed for poverty reduction development and to safeguard the security and human rights of their populations and in its territory it also lacks the ability or political will to develop mutually constructive and reinforcing relations with society' (see Moran et al. 2018).

6 Around two dozen countries are deemed to be most seriously at risk. Of these, more than ten have produced interagency humanitarian appeals for each of the past seven years in succession. As a continent, Africa is already suffering the greatest impact of climate change. Some of the countries most affected are in, and around, the Horn of Africa, where the risk of severe drought has increased from 1-in-7 years to 1-in-2 years.

7 The Horn of Africa locust swarms of 2019-2020 caused 356,000 tons of cereal loss, wasted 197,000 ha of cropland, and 1.35 m ha of pasture-lands in Ethiopia alone (Government of Ethiopia 2020). As a result, one million Ethiopians required food assistance due specifically to damage from locusts, prior to the start of its civil war in late 2020. Around 20.2 m people in Ethiopia, Kenya, Somalia, South Sudan, Uganda, and Tanzania faced severe acute food insecurity, exacerbated by these locust infestations (FAO 2023).

8 Water stress is defined as having fewer than 1,700 cubic metres per person per year. Water scarcity exists when that number drops to below 1,000.

9 According to the International Displacement Monitoring Centre (IDMC): 'Every year, millions of people are forced to flee their homes because of conflict and violence. Disasters and the effects of climate change regularly trigger new and secondary displacement, undermining people's security and wellbeing ... The scale of displacement worldwide is increasing, and most of it is happening within countries' borders' (IDMC 2022). Countries with the highest disaster-driven internal displacements were: Afghanistan (1.1 m people); India (0.93 m); and Pakistan (0.8 m). In 2020, The East Asia and Pacific region saw 30.3% of new displacements, with sub-Saharan Africa accounting for 27.4% (IDMC 2022).

10 Ecological security or 'eco-security' is the presence of ecosystem resilience, now and into the future. As a term found in the lexicon of the scholarship of international relations for decades, it has entered the realm of national politics only relatively recently (see McDonald 2015).

11 These include biotic factors, such as pathogenic, algal, and fungal responses to climatic change; risks to basic economic security at the lower ends of Maslow's

hierarchy of needs; and human resilience mechanisms, such as migration, pressure on governance, and potential of conflict (see Chen 2021; Quiggin et al. 2021).

12 The nine planetary boundaries are climate change; biosphere integrity; land-system change; freshwater use; biogeochemical flows (phosphates and nitrates); ocean acidification; atmospheric loading (not yet quantified); ozone depletion; and novel entities such as plastics pollution (SRC 2022).

13 Resorting to the use of potentially lethal military force to protect key elements of the biosphere from human damage is a dystopia that already exists in some countries, such as Ivory Coast (e.g. illegal logging), the DRC (e.g. gorilla protection), Guyana and Brazil (e.g. illegal gold mining and deforestation). It is now even extending to the oceans to counter illegal unregulated and unreported fishing (IUUF) (see Sinclair 2021). In recognition of the entry of at least climate security if not ecological security into the 'high politics' realm of current defence prioritisation and funding, NATO and ASEAN militaries have taken a number of steps and declarations to increase their awareness and understanding climate security (MOD 2021a; NATO 2021; Rabson 2021). The US DoD's first climate and environmental security table-top exercise (TTX), 'Elliptic Thunder' took place in 2021 (Vergun 2021). Contingency planning and exercises, which often involve working together with local resilience forums either on already developed or in developing together national risk registers, also occur in some countries.

14 Urgent operational requirement (UOR) or needs (UON) are specific military terms used to develop or obtain equipment or capability, which if left unfilled would result in capability gaps potentially leading to loss of life or critical mission failure (CJCSI 2018).

15 The unintended consequences of climate change responses that are not thought through can result in explosive backdraft (see NSB 2017).

16 Expeditionary armies are those organised to achieve military effect abroad, as opposed to developmental armies which confine their ambition to home territory and, explicitly, include societal benefit and economic development in their doctrine. For the latter, see the PLA pre-changes to China's third Defence White Paper of 2015, and the Congolese (article 168), the Gabonese (articles 1–22), and the Chadian (article 194) constitutions.

17 Over the past century, the trend in military focus has been to turn away from developmental roles in defence towards the apex of the 'warrior culture' as an expeditionary military. The apolitical separation of the military from the rest of the executive, its independence from business interests, and the contracting out of non-core warfighting roles have forged what is now seen as the 'professional' army.

18 Key military assets in a disaster response include organisation; discipline; ability to project determination; ability to respond at speed and scale-up quickly; possession of support systems which decrease calls upon civilian resources; and, in certain contexts, an already-established degree of trust.

19 The military should be neutral (e.g. demonstrate no political bias), impartial (e.g. demonstrate no ethnic or religious bias), and independent (e.g. demonstrate, to the extent possible, independence from donors and central or local government).

## References

Albert, S., A. Grinham, B. Gibbes, J. Leon, and J. Church 2016. Sea-level rise has claimed five whole islands in the Pacific. *The Conversation.* https://theconversation. com/sea-level-rise-has-claimed-five-whole-islands-in-the-pacific-first-scientific-evidence-58511 (accessed 1 May 2023).

Babson, K. 2022. Ecocide: The fifth international crime. *The Climate Change Review*. https://www.ucsdclimatereview.org/post/ecocide-the-fifth-international-crime (accessed 1 May 2023).

Beste, A. 2021. Flood protection: let's start with soil. *Agricultural and Rural Convention*. https://www.arc2020.eu/flood-protection-lets-start-with-soil/ (accessed 1 May 2023).

Botsford, P. 2021. The rising tide of climate litigation. *International Bar Association*. https://www.ibanet.org/The-rising-tide-of-climate-litigation (accessed 1 May 2023).

Britzky, H. 2022. Climate change is running the National Guard ragged. *Task & Purpose*. https://taskandpurpose.com/news/climate-change-national-guard/ (accessed 1 May 2023).

Busby, J. 2020. Actorless threats. *Duck of Minerva*. https://www.duckofminerva.com/2021/02/actorless-threats.html (accessed 1 May 2023).

Carbon Brief 2021. The IPCC's sixth assessment report on climate science. *Carbon Brief*. https://www.carbonbrief.org/in-depth-qa-the-ipccs-sixth-assessment-report-on-climate-science (accessed 1 May 2023).

Caves, B., R. Lucas, L. Dewaele, J. Muravska, C. Wragg, T. Spence, Z. Hernandez, A. Knack, and J. Black 2021. Enhancing Defence's Contribution to Societal Resilience in the UK. *RAND Europe*. https://www.rand.org/pubs/research_reports/RRA1113-1.html (accessed 1 May 2023).

Chen, C. 2021. Greening security: the military as a climate game changer. Rajaratnam School of International Studies (RSIS). https://www.rsis.edu.sg/rsis-publication/idss/ip21009-greening-security-the-military-as-a-climate-game-changer/ (accessed 7 September 2023).

CJCSI 2018. Charter of the Joint Requirements Oversight Council (JROC) and implementation of the Joint Capabilities Integration and Development System (JCIDS). *Council of the Joint Chiefs of Staff Instructions*. https://www.acq.osd.mil/asda/jrac/docs/CJCS-Instruction-5123.01H.pdf (accessed 1 May 2023).

Clack, T. and R. Nugee 2022. Climate change is creating security threats around the world – and militaries are responding. *The Conversation*. https://theconversation.com/climate-change-is-creating-security-threats-around-the-world-and-militaries-are-responding-173668 (accessed 1 May 2023).

Clement, V., K. Rigaud, A. de Sherbinin, B. Jones, S. Adamo, J. Schewe, N. Sadiq, and E. Shabahat 2021. Groundswell Part 2: Acting on Internal Climate Migration. *World Bank*. https://www.worldbank.org/en/news/press-release/2021/09/13/climate-change-could-force-216-million-people-to-migrate-within-their-own-countries-by-2050 (accessed 1 May 2023).

Coker, C. 2021. *Why War?* London: Hurst.

Daft, S. 2019. Environmental destruction is (already) a war crime, but is almost impossible to commit, *Justice Info*. https://www.justiceinfo.net/en/43121-environmental-destruction-war-crime-impossible-to-commit.html (accessed 1 May 2023).

Diehl, P. F. and P. R. Hensel 1994. Testing empirical propositions about shatterbelts, 1945–76. *Political Geography* 13(1): 33–51.

Digital, A. B. 2022. Brazilian government crackdown on illegal gold mining activity. *Benzinga*. https://www.benzinga.com/pressreleases/22/01/ab25236704/brazilian-government-crackdown-on-illegal-gold-mining-activity (accessed 1 May 2023).

Dixson-Declève, J. A. Ocamp, and F. Salim 2021. The case for a food systems stability board. *Project Syndicate*. https://www.project-syndicate.org/commentary/the-world-needs-a-food-systems-stability-board-by-sandrine-dixson-decleve-et-al-2021-08 (accessed 1 May 2023).

Djordjevic, I. 2021. The rise of ethical DAOs for a sustainable blockchain future. *Unblock*. https://unblock.net/ethical-dao/ (accessed 1 May 2023).

Dyer, G. 2021. *The Shortest History of War*. London: Old Street Publishing.

D'Souza, E. 2022. Rôle potentiel de l'armée dans la protection de l'environnement: le cas de l'Inde. *FAO*. https://www.fao.org/3/v7850F/v7850f12.htm#:~:text=de %20l'Inde-,R%C3%B4le%20potentiel%20de%20l'arm%C3%A9e%20dans%20la %20protection%20de%20l,le%20cas%20de%20l'Inde&text=L'arm%C3%A9e%20a %20en%20effet,et%20non%20sur%20l'affrontement (accessed 1 May 2023).

Economist 2022. The coming food catastrophe. *The Economist*. https://www.economist. com/leaders/2022/05/19/the-coming-food-catastrophe (accessed 1 May 2023).

Elkington, J. 2021. *Green Swans: The Coming Boon in Regenerative Capitalism*. London: Fast Company Press.

EUNAVFOR (EU Naval Force) 2023. Forces of operation Atalanta. *European Union External Action*. https://eunavfor.eu (accessed 1 May 2023).

Evans, T. and P. Schroeder 2021. Building global climate security. *Chatham House*. https://www.chathamhouse.org/2021/09/building-global-climate-security (accessed 1 May 2023).

Fant, S. 2022. Ukraine: all lithium reserves and minerals in war zone. *Renewable Matter*. https://www.renewablematter.eu/articles/article/ukraine-all-lithium-reserves-and-mineral-resources-in-war-zones (accessed 1 May 2023).

Farrer, S. 2020. Modern slavery and the nexus to illegal unreported and unregulated fishing (IUUF). *Financial Crime News*. https://thefinancialcrimenews.com/illegal-fishing (accessed 1 May 2023).

FAO 2023. Food and Agricultural Organisation. *Locust Watch*. https://www.fao.org/ ag/locusts/en/info/info/index.html (accessed 1 May 2023).

FSI 2023. What does state fragility mean? *Fragile State Index*. https://fragilestatesindex. org/frequently-asked-questions/what-does-state-fragility-mean/ (accessed 1 May 2023).

Government of Ethiopia 2020. Impact of desert locust infestation on household livelihoods and food security in Ethiopia. *Government of Ethiopia: Joint Assessment Findings*. https://www.humanitarianresponse.info/sites/www. humanitarianresponse.info/files/assessments/desert_locust_impact_assessment_ report_for_ethiopia.pdf (accessed 1 May 2020).

Gemenne, F., J. Tasse, S. Kabbej, R. Monange, and F. Babalone 2021. Intégration des enjeux climato-environnementaux par les forces armées. *IRIS*. https://www.iris-france.org/wp-content/uploads/2021/03/RE15-202101-Integration_climat_forces_ armees_defense_climat-rapport-15.pdf (accessed 1 May 2023).

Givetash, L. 2019. Militaries go green, rethink operations in face of climate change. *NBC News*. https://www.nbcnews.com/news/military/militaries-go-green-rethink-operations-face-climate-change-n991651 (accessed 1 May 2023).

GGW 2023. Growing a healthy world wonder. *Great Green Wall*. https://www. greatgreenwall.org/about-great-green-wall/ (accessed 1 May 2023).

Goodman, S. 2021. Humanitarian crises. *Climate Change & (In)Security Project*. https://youtu.be/CODvLytmf9M (accessed 1 May 2023).

Groome, A. 2016. Meet John D. Liu, the Indiana Jones of landscape restoration. *Regeneration International*. https://regenerationinternational.org/2016/03/07/meet-john-d-liu-the-indiana-jones-of-landscape-restoration/ (accessed 1 May 2023).

Homer-Dixon, T. 2021. Closing address. *Climate Change & (In)Security Project.* https://youtu.be/er8GqHMeAn0 (accessed 1 May 2023).

Ida, T. 2021. Climate refugees – the world's forgotten victims. *World Economic Forum.* https://www.weforum.org/agenda/2021/06/climate-refugees-the-world-s-forgotten-victims/ (accessed 1 May 2023).

IDMC 2022. Global Report on internal displacement 2021. *Internal Displacement Monitoring Centre.* https://www.internal-displacement.org/global-report/grid2022/ (accessed 1 May 2023).

IEP 2020. Ecological threat register press release. *Institute for Economics & Peace.* https://www.economicsandpeace.org/wp-content/uploads/2020/09/Ecological-Threat-Register-Press-Release-27.08-FINAL.pdf (accessed 1 May 2023).

IEP 2021. Ecological threat register 2021: understanding ecological threats, resilience and peace. *Institute for Economics & Peace.* https://www.visionofhumanity.org/wp-content/uploads/2021/10/ETR-2021-web-131021.pdf (accessed 1 May 2023).

Igini, M. 2022. Deforestation in Africa. *Earth.org.* https://earth.org/deforestation-in-africa/ (accessed 1 May 2023).

IMCCS 2020. World Climate Security Report 2020. *Center for Climate Security.* https://imccs.org/report2020/ (accessed 1 May 2023).

IMCCS 2021. World Climate Security Report 2021. *Center for Climate Security.* https://imccs.org/the-world-climate-and-security-report-2021/ (accessed 1 May 2023).

IPCC 2021. Assessment report 6: climate change 2022: mitigation of climate change. *International Panel on Climate Change.* https://www.ipcc.ch/reports/ (accessed 1 May 2023).

IRIS (Institut de Relations Internationales et Stratégiques) 2021. Climate change evaluation methodology (CEMC) for military camps. *IRIS.* https://www.iris-france.org/wp-content/uploads/2021/09/ENG_Executive-summary-RE16.pdf (accessed 1 May 2023).

Jayaram, D. 2015. Environmental security, land restoration and the military: a case study of the ecological task forces in India. (In) I. Chabay, M. Frick, and J. Helgeson (eds) *Land Restoration: Reclaiming Landscapes for a Sustainable Future.* Salt Lake City: Academic Press, pp. 163–181

Jayaram, D. 2017. Climate diplomacy and India's ecological task force. *Climate Diplomacy.* https://climate-diplomacy.org/magazine/environment/climate-diplomacy-and-indias-ecological-task-force (accessed 1 May 2023).

Johnson, S. 2022. Pakistan hits 120°F as climate trends drive spring heat wave. *Ars Technica.* https://arstechnica.com/science/2022/05/pakistan-hits-120f-as-climate-trends-drive-spring-heatwave/ (accessed 1 May 2023).

Jordan 2022. The armed forces. *Jordan Government.* http://www.kinghussein.gov.jo/government5.html#:~:text=Tree-Planting%3A%20Military%20involvement%20in%20the%20afforestation%20of,desert%20and%20helped%20to%20restore%20Jordan%E2%80%99s%20green%20cover (accessed 1 May 2023).

Katz, D. 2023 Cooperation, competition and conflict: JADE SPEAR, China, and new strategic art. *NSI.* https://nsiteam.com/cooperation-competition-conflict-jade-spear-china-and-new-strategic-art/ (accessed April 2023).

King, L. 2022. Oceans and their largest inhabitants could be the key to storing our carbon emissions. *The Conversation.* https://theconversation.com/oceans-and-their-largest-inhabitants-could-be-the-key-to-storing-our-carbon-emissions-180901 (accessed 1 May 2023).

Klare, M. 2019. *All Hell Breaking Loose*. Stuttgart: Macmillan.

Klinger, J. 2017. *Rare Earth Frontiers: From Terrestrial Subsoils to Lunar Landscapes*. Ithaca, NY: Cornell University Press.

Krampe, F. 2021. Why United Nations peace keeping operations cannot ignore climate change. *SIPRI*. https://www.sipri.org/commentary/topical-backgrounder/2021/why-united-nations-peace-operations-cannot-ignore-climate-change (accessed 1 May 2023).

Lazard, O. and R. Youngs 2021. The EU and climate security: toward ecological diplomacy. *Carnegie Europe*. https://carnegieeurope.eu/2021/07/12/eu-and-climate-security-toward-ecological-diplomacy-pub-84873 (accessed 1 May 2023).

Liu, J. 2007. Earth's hope: learning to communicate the lessons of the Loess Plateau. *Environmental Education Media Project*. https://www.academia.edu/12899656/Learning_to_Communicate_the_Lessons_of_the_Loess_Plateau (accessed 1 May 2023).

LSI Africa 2021. Côte d'Ivoire: une armée verte contre la déforestation. *LSI Africa*. https://www.lsi-africa.com/fr/actualite-africaine/cote-d-ivoire-environnement-deforestation.html (accessed 1 May 2023).

Mach, K. J., W. N. Adger, H. Buhuang, M. Burke, J. Fearon, C. Field, C. S. Hendrix, C. M. Kraan, J-F. Maystadt, P. Roessler, J. Scheffran, K. A. Schultz, and N. von Uexkull 2020. Directions on research for climate and conflict. *Earth's Future* 8(7). 10.1029/2020EF001532 (accessed 1 May 2023).

Mathiesen, T. 2022. The link between Putin and climate change, *Politico*. https://www.politico.eu/article/link-vladimir-putin-climate-change-russia-ukraine/ (accessed 1 May 2023).

McDonald, M. 2015. Ecological security. *E-International Relations*. https://www.e-ir.info/2015/11/28/ecological-security/ (accessed 1 May 2023).

McKenzie, J. 2021. 'Harrowing' intelligence report still downplays threat of climate change to national security. *The Bulletin*. https://thebulletin.org/2021/11/harrowing-intelligence-report-still-downplays-threat-of-climate-change-to-national-security/ (accessed 1 May 2023).

MdA (Ministère des armées) 2022. *Stratégie climat et défense*. Paris: MdA. https://www.actu-environnement.com/media/pdf/news-39570-Strategie-Climat-Defense.pdf (accessed 1 May 2023).

Menker, S. 2022. Gro's CEO, Sara Menker, briefs the United Nations Security Council: conflict and global food security. *Gro Intelligence*. https://gro-intelligence.com/blog/gro-s-ceo-sara-menker-briefs-the-united-nations-security-council (accessed 1 May 2023).

Miller, R. 2016. Hampton Roads climate impact quantification initiative: baseline assessment of the transportation assets and overview of economic analyses useful in quantifying impacts. *National Academy of Sciences*. https://trid.trb.org/view/1428258 (accessed 1 May 2023).

MOD 2019. Climate impacts risk assessment methodology. *Sustainability & Environmental Appraisal Tools Handbook*. London: Ministry of Defence.

MOD 2021a. Climate change and sustainability strategic approach. *Ministry of Defence*. https://assets.publishing.service.gov.uk/government/uploads/system/uploads/attachment_data/file/973707/20210326_Climate_Change_Sust_Strategy_v1.pdf (accessed 1 May 2023).

MOD 2021b. Human security in defence (JSP985). *Ministry of Defence*. https://www.gov.uk/government/publications/human-security-in-defence-jsp-985 (accessed 1 May 2023).

MOD 2021c. Illegal poachers all tied up by Op CORDED. *Ministry of Defence.* https://www.army.mod.uk/news-and-events/news/2021/06/illegal-poachers-tied-up-by-corded/ (accessed 1 May 2023).

Moran, A., J. Busby, C. Riley, T. G. Smith, R. Kishi, N. Krishnan, C. Wight 2018. The intersection of global fragility and climate risks. *USAID.* https://reliefweb.int/sites/reliefweb.int/files/resources/PA00TBFH.pdf (accessed 1 May 2023).

NATO 2021. NATO Climate Change and Security Action Plan. *NATO.* https://www.nato.int/cps/en/natohq/official_texts_185174.htm (accessed 1 May 2023).

ND-GAIN 2023. Country index. *Notre Dame Global Adaptation Initiative.* https://gain.nd.edu/our-work/country-index/ (accessed 1 May 2023).

NSB 2017. Backdraft. *New Security Beat.* https://www.newsecuritybeat.org/category/backdraft-podcast/ (accessed 1 May 2023).

Nugee, R. 2020. Climate change and sustainability strategic approach. *UK Ministry of Defence.* https://assets.publishing.service.gov.uk/government/uploads/system/uploads/attachment_data/file/973707/20210326_Climate_Change_Sust_Strategy_v1.pdf (accessed 1 May 2023).

OCHA 2007a. Oslo guidelines: guidelines on the use of foreign military and civil defence assets in disaster relief. *UN Office for the Coordination of Humanitarian Affairs.* https://www.unocha.org/publication/oslo-guidelines-use-foreign-military-and-civil-defence-assets-disaster-relief (accessed 1 May 2023).

OCHA 2007b. Global humanitarian overview. *UN Office for the Coordination of Humanitarian Affairs.* https://gho.unocha.org/ (accessed 1 May 2023).

Oliver, L. 2018. China has sent 60,000 soldiers to plant trees. *World Economic Forum.* https://www.weforum.org/agenda/2018/02/china-army-soldiers-plant-trees/ (accessed 1 May 2023).

Parkinson, S. 2020. The carbon boot-print of the military. *Responsible Science.* https://www.sgr.org.uk/resources/carbon-boot-print-military-0 (accessed 1 May 2023).

Quiggin, D., K. de Meyer, L. Hubble-Rose, and A. Froggatt 2021. Climate change risk assessment 2021. *Chatham House.* https://www.chathamhouse.org/sites/default/files/2021-09/2021-09-14-climate-change-risk-assessment-quiggin-et-al.pdf (accessed 1 May 2023).

Rabson, M. 2021. A NATO centre for climate security? Canada and Holland say yes. *Canada's National Observer.* https://www.nationalobserver.com/2021/10/30/latest-news/nato-centre-climate-security-canada-and-holland-say-yes (accessed 1 May 2023).

Rand Europe 2021. Prepare, respond and recover: how UK defence could support societal resilience to crisis events. *RAND Europe.* https://www.rand.org/randeurope/research/projects/supporting-uk-societal-resilience.html (accessed 1 May 2023).

Rattan, L. and B. Stewart 2019. *Soil Degradation and Restoration in Africa.* Boca Raton: CRC Press.

Regaud, N., B. Alex, and F. Gemenne 2022. *La guerre chaud: enjeux stratégiques du changement climatique.* Paris: Presses de Sciences Po.

Retter, L., A. Knack, Z. Hernandez, R. Harris, B. Caves, M. Robson, and N. Adger 2021. Crisis response in a changing climate: implications of climate change for UK Defence logistics in humanitarian assistance and disaster relief (HADR) and military aid to the civil authorities (MACA) operations. *RAND Corporation.* https://www.rand.org/pubs/research_reports/RRA1024-1.html (accessed 1 May 2023).

Schoonover, R. 2019. The White House blocked my report on climate change and national security. *The New York Times.* https://www.nytimes.com/2019/07/30/opinion/trump-climate-change.html (accessed 1 May 2023).

Sheil, D. 2014. How plants water our planet: advances and imperatives. *Trends in Plant Science* 19(4): 209–211.

Sinclair, M. 2021. The national security imperative to tackle illegal, unreported, and unregulated fishing. *Brookings.* https://www.brookings.edu/blog/order-from-chaos/2021/01/25/the-national-security-imperative-to-tackle-illegal-unreported-and-unregulated-fishing/ (accessed 1 May 2023).

Snyder, T. 2018. *The Road to Unfreedom: Russia, Europe, America.* New York: Vintage.

Spracklen, D. 2018. The effects of tropical vegetation on rainfall. *Annual Review of Environment and Resources* 43: 193–218.

SRC 2022. The nine planetary boundaries. *Stockholm Resilience Centre.* https://www.stockholmresilience.org/research/planetary-boundaries/the-nine-planetary-boundaries (accessed 1 May 2023).

Stroobants, J. and E. Vincent 2022. How Vladimir Putin resurrected NATO. *Le Monde.* https://www.lemonde.fr/en/international/article/2022/05/17/how-vladimir-putin-resurrected-nato_5983836_4.html (accessed 1 May 2023).

UN 2023. Do you know all 17 SDGs? *United Nations Department of Economic and Social Affairs.* https://sdgs.un.org/goals (accessed 1 May 2023)

UNDP 1994. Human development report 1994. *United Nations Development Programme.* https://www.undp.org/publications/human-development-report-1994 (accessed 1 May 2023).

UNDP 2020. Human development report 2020: the next frontier: human development and the Anthropocene. *United Nations Development Programme.* https://hdr.undp.org/en/content/human-development-report-2020 (accessed 1 May 2023).

UNDRR 2021a. Drought is on the verge of b coming the next pandemic and there is no vaccine to cure it. *World Humanitarian Forum.* https://www.linkedin.com/pulse/drought-verge-becoming-next-pandemic-vaccine-/ (accessed 1 May 2023).

UNDRR 2021b. Global assessment report special report on drought 2021. *United Nations Office for Disaster Risk Reduction.* https://www.undrr.org/publication/gar-special-report-drought-2021 (accessed 1 May 2023).

UNEP 2012. Greening the blue helmets: environment, natural resources and UN peacekeeping operations. *United Nations Environment Programme.* https://operationalsupport.un.org/sites/default/files/unep_greening_blue_helmets_0.pdf (accessed 1 May 2023).

UNEP 2021. Measuring progress: environment and the SDGs. *United Nations Environment Programme.* https://www.unep.org/resources/publication/measuring-progress-environment-and-sdgs (accessed 1 May 2023).

UNEP 2023. World environment situation room. *United Nations Environment Programme.* https://wesr.unep.org/ (accessed 1 May 2023).

UNHCR 2021. Refugee data finder. *United Nations High Commissioner for Refugees.* https://www.unhcr.org/refugee-statistics/ (accessed 1 May 2023).

Vergun, D. 2021. DOD exercise highlights need to address climate change, its impacts. *Department of Defense.* https://www.defense.gov/News/News-Stories/Article/Article/2596591/dod-exercise-highlights-need-to-address-climate-change-its-impacts/ (accessed 1 May 2023).

Vidal, J. 2022. Energy efficiency guru Amory Lovins: 'It's the largest, cheapest, safest, cleanest way to address this crisis'. *The Guardian.* www.theguardian.com/environment/2022/mar/26/amory-lovins-energy-efficiency-interview-cheapest-safest-cleanest-crisis (accessed 1 May 2023).

Wang-Erlandsson, L., A. Tobian, R. J. van der Ent, I. Fetzer, S. te Wierik, M. Porkka, A. Staal, F. Jaramillo, H. Dahlmann, C. Singh, P. Geve, D. Gerten, P. W. Keys, T. Gleeson, S. E. Cornell, W. Steffen, X. Bai, and J. Rockstorm 2022. A planetary boundary for green water. *Nature Reviews: Earth & Environment* 3: 380–392.

Weathering Risk 2022. Joint statement on climate change and impact in IPCC report. *Weathering Risk.* https://weatheringrisk.org/en/event/IPCC-Climate-Conflict-Joint-Statement (accessed 1 May 2023).

Winter, E. 2022. The role of the environmental war crime in the Russian invasion of Ukraine. *Jurist.* https://www.jurist.org/commentary/2022/05/elliot-winter-ukraine-conflict-environmental-war-crime/ (accessed 1 May 2023).

WMO 2022a. United in science: we are heading in the wrong direction. *World Meteorological Organization.* https://public.wmo.int/en/media/press-release/united-science-we-are-heading-wrong-direction (accessed 1 May 2023).

WMO 2022b. Early warning systems save millions of lives. *World Meteorological Organization.* https://library.wmo.int/doc_num.php?explnum_id=7560 (accessed 1 May 2023).

# 15

# THE HYPERTHREAT AND POLITICO-MILITARY RESPONSE

## Outcomes from a military appreciation of entangled security

*Elizabeth Boulton*

## Introduction

In January 2023, as the Ukraine–Russian war seemed it might morph into a 'the West' versus 'the rest' stand-off, French philosopher Emmanuel Todd declared that World War III (WWIII) – albeit in a different form from prior world wars – had begun (Devecchio 2023). Related to this, 2022 saw 'the largest number of violent conflicts since 1946' (UN 2022, p. 58). 2022 was also the year in which scientists assessed that the aim to limit global warming to 1.5°C, necessary to forestall dangerous climate change, was now out of reach without extraordinary measures (Armstrong McKay et al. 2022; UNEP 2022; Warszawski et al. 2021). The three main forms of greenhouse gases (GHG) (carbon dioxide; methane and nitrous oxide) reached 'record highs' in 2021 (WMO 2022), while it was calculated that planned fossil fuel projects would emit twice as much carbon as feasible under a 1.5°C carbon budget (Kühne et al. 2022; SEI 2021). Meanwhile, Earth's biodiversity was 'declining faster than at any time in human history' and required 'transformational' change to halt or slow its deterioration (IPBES 2019).

All of this follows decades of war in the Middle East; a global pandemic; years of peaceful civil disobedience by climate activists; the substantive failures of the last two Conference of the Parties (COP26 and 27) to the United Nations Framework Convention on Climate Change (UNFCCC); rolling and increasingly frequent extreme weather events across the globe; and backsliding on sustainable development goals (UN 2022) – even before the pandemic (UN DESA 2018). The breakdown in planetary, human, and state security is no longer questioned, while the niche field of climate security now sits within broader research on existential threat (Ord 2020); collapsology

DOI: 10.4324/9781003377641-19

(Servigne and Stevens 2020); global catastrophic risk (GCR) (Beard et al. 2021; Richards et al. 2021; Sears 2020); or polycrisis (Homer-Dixon et al. 2021). Thus, there has been no shortage of efforts to describe future risks and problems, however, what still eludes humanity is the capacity to respond effectively to converging risks. The scale of security failure must be confronted squarely. First, the initiating objective of the UN, to prevent major war, has not been achieved. Second, the probable death of planetary life has been announced, with this occurring in the way in which a person who appears healthy can be told they have a terminal illness.

Both outcomes arose in a subtle way. The outcomes crept up upon humanity; there was no single discernible event, rather a small line after small line was crossed, one opportunity to forge a new path thwarted after another. Yet the almost imperceptible ways both thresholds were crossed denies the significance of what has occurred. It is in this context, like a doctor who hesitantly advises the terminally ill patient there may be one outlier treatment option after all, that this chapter presents an alternate approach to security. It involves framing the climate and environment crisis neither as a scientific or economic policy issue nor as one existential risk among many (the list approach) and not as a 'threat multiplier' (Goodman 2007) but rather as the primary threat – and moreover, as an entirely new type of foe called a 'hyperthreat' (Boulton 2018).

The hyperthreat construct acts as an analytical hinge; it facilitated the application of traditional military-style threat analysis and strategic planning tools to the problem (Boulton 2022a). In turn, this created space to explore and reimagine politico-military response, leading to 'PLAN E' – which is a climate and ecologically centred grand security strategy (Boulton 2022b). The intention of this chapter is to provide a broad introduction to the whole hyperthreat, entangled security, and PLAN E approach. It explains the conceptual background, highlights key analytical insights that emerged, briefly introduces PLAN E, and then discusses the research implications. As a prelude, the chapter briefly reviews prior literature in the field and explains how hyperthreat research attempts to address identified shortfalls in knowledge.

## Incoherence of climate security discourse

In 1995, the short-lived UN Commission on Global Governance suggested that 'the concept of global security must be broadened … to include the security of people and the security of the planet' (Commission on Global Governance 1995, p. 108). Since then, there has been expansive research on human and planetary security. This includes arguments that the meaning of security relates to context and 'survival, urgency and emergency' (Trombetta 2008, p. 588). Alternate concepts have been developed such as the human, gender, and environmental security (HUGE) approach (Oswald Spring 2009) or ecological security (McDonald 2018). These developments have yet to fundamentally alter humanity's security posture.

Climate-security literature also features many calls for action, such as arguments for 'a profound restructuring of international politics and order that can assure the planet's survival' (Burkeet al. 2016, p. 522) or that 'traditional notions of security need a rapid overhaul' and to focus upon decarbonisation and making flourishing ecosystems (Dalby 2020). Climate security studies have been criticised for being fragmented and incoherent, due to siloed policy communities and institutions (Mobjörk et al. 2016). Related to this, Dellmuth et al. (2018) find there is a lack of theoretical or empirical research on how to integrate the varied climate-security approaches that are found across different inter-governmental organisations (IGO) and assess their effectiveness. They write:

'[T]here is little evidence that climate change has been coherently securitized across IGOs, and scholars' debate whether we are witnessing a "failed securitization" of climate change or a "climatization" of specific security-related issues such as defense, migration, and development'.

(Dellmuth et al. 2018, p. 4)

Likewise, in overviewing practice, Busby (2021) identifies a patchwork of measures including the UN 'climate security mechanism', which seeks to mainstream climate security considerations into general planning and integrated risk assessment procedures. Yet he laments the lack of any systemic approach or solution, 'the discussion of threats and policy initiatives begs the question of what to do' (Busby 2021, p. 190).

A feature of prior climate security discourse is persistent fear and suspicion of a securitisation approach (Paglia 2018), which is often associated with top-down, simplistic, and draconian measures that would erode human rights (Bostrom 2013, p. 27). Accordingly, military input is limited to providing advice about security impacts and assisting with disaster response. In a report for the UN Security Council (UNSC), an 'informal expert group' of military advisers on climate security suggest their future role could be providing 'fine-grained, contextualized analysis' to UN agencies and in helping persuade decision makers on the need for action. They warn that if the UNSC fails to respond to the climate crisis, it will 'appear out of touch with fundamental threats to international peace and security – and human survival' (UNSC 2021, pp. 16–17).

Also relevant to climate security and existential threat discourse is research on climate action which finds that 'doom and gloom' narratives and top-down solutions may be counter-productive, while 'bottom up' transformative 'win-win' narratives are better at motivating climate action (Hinkel et al. 2020).

The research presented here responds to gaps in prior research in several ways. Related to problems with effective conceptualisation of the problem, the research's primary focus is on 'deep frames' – or entrenched worldviews – and developing a new theory of threat (the hyperthreat notion) and philosophical framework (entangled security), that better matched the security challenges of the 21st century. Additionally, instead of calls for transformational or emergency response, it develops a new grand strategy for a crisis response. Thus, the works sit in the creative space, rather than the problem description space.

## Conceptual development

The research began with the question: why was climate and environmental change (CEC) often described as a type of threat – existential, catastrophic, or an emergency – yet not responded to with the same prioritisation, resourcing, and speed as traditional military threats or crises, such as the Global Financial Crisis? Drawing upon over a decade of research in climate communication and behaviour change, it was deduced that humanity's ineffective response was primarily due to 'deep framing' barriers. The term 'deep frames' comes from neuro and cognitive science research; it refers to complex networks of neuron pathways, built from infancy, which 'hold' a person's guiding worldview, identity, and values, and which influence decision making – mostly at the subconscious level. As cognitive scientist George Lakoff (2010) has explained, decisions are mostly unconscious (98%), require emotion, use the 'logic' of frames, metaphors, and narratives, are physical (in brain circuitry) and vary considerably.

To use the language of military-style threat assessment, it was assessed that humanity's 'centre of gravity' for effective response was the 'deep framing' problem. Accordingly, this analytical activity narrowed its focus to one sole objective: the hunt for a new deep frame or way of conceiving threat and security in the Anthropocene. To do this, it explored a range of emerging Anthropocene-relevant epistemologies and ontologies; interpreted these through security theory; and tested them with real-world planetary, human, and state factors to develop a new transdisciplinary theoretical approach. Some aspects of this process are described below.

## From hyperobject to hyperthreat

Eco-philosopher Timothy Morton's concept of global warming of a hyperobject (Morton 2013) was found to address many of the deep-framing barriers identified by cognitive scientists and climate policy analysts (Boulton 2016). Instead of presenting global warming as a series of statistics, Morton materialises it, turning it into a single phenomenon, a new 'thing' – a baffling, strange type of monster – with five distinct characteristics. As Morton (2013) explains, the hyperobject moves like fog; it is infused through everything yet cannot be seen in its entirety. It operates indirectly – through other objects – and over such vast scales of time and space that it disrupts the human ability to link cause and effect. Morton argues that the hyperobject is now the dominant actor and major shaping influence on Earth. In turn, humans are existentially demoted; thus, humanity's new ontology involves a profound loss of agency.

Along with other relevant theoretical approaches, such as just war theory (Fotion 2007), slow violence (Nixon 2011), and eco-theology (Francis 2015), Morton's hyperobject notion contributed to the concept of CEC as a hyperthreat (Boulton 2018). In contrast to Morton's neutral hyperobject, the hyperthreat framing emphasises that the climate and ecological crisis presents a new form of violence. It includes environmental issues, explores the new nature of conscious hostile intent, and proposes humans' still have some agency. In the end, the hyperthreat of CEC is defined as having warlike destructive capabilities that are so diffuse that it is hard to see the enormity of the destruction coherently nor who is responsible for its hostile actions. It defies existing human thought and institutional constructs.

## Entangled security

At its simplest level, 'entangled security' refers to the inherent interconnectedness of planetary, human, and state security. However, as a theoretical approach, it draws upon a wider body of research in ecofeminism, new materialism, care ethics, and quantum theory (Figure 15.1). Generally, these fields view that 'entanglement' represents the true nature of existence, as opposed to a hierarchical structure or a conception of the human being as self-reliant and autonomous. In the context of the hyperthreat, there is the realisation that almost all forms of life on Earth face a shared, entangled threat. The nature of entanglement is best illuminated by returning to the idea's roots.

Donna Haraway recalls the way humans' physical existence is fundamentally entangled with nature; for example, humans eat food which then 'become' the human. However, her more complex work, encapsulated by the 'meeting well' term, considers the interconnected affective (emotional and sensory) and ethical dimensions of relations between human and non-human in an entangled exitance. For example, she discusses the interaction between animals in a laboratory test and their testers. She wonders if the tester has any empathic response to hearing an animal wince in pain. She reflects upon the nature of this meeting for both participants, and what failure to resonate empathetically to another creature's pain means – at an ethical level. Haraway (2008) uses this, and similar scenarios, to suggest that accurate knowledge and sound decision making in an entangled existence requires understanding of the affective dimensions of interaction. Haraway thus posits a new approach to violence which, rather than ignoring, masking, or

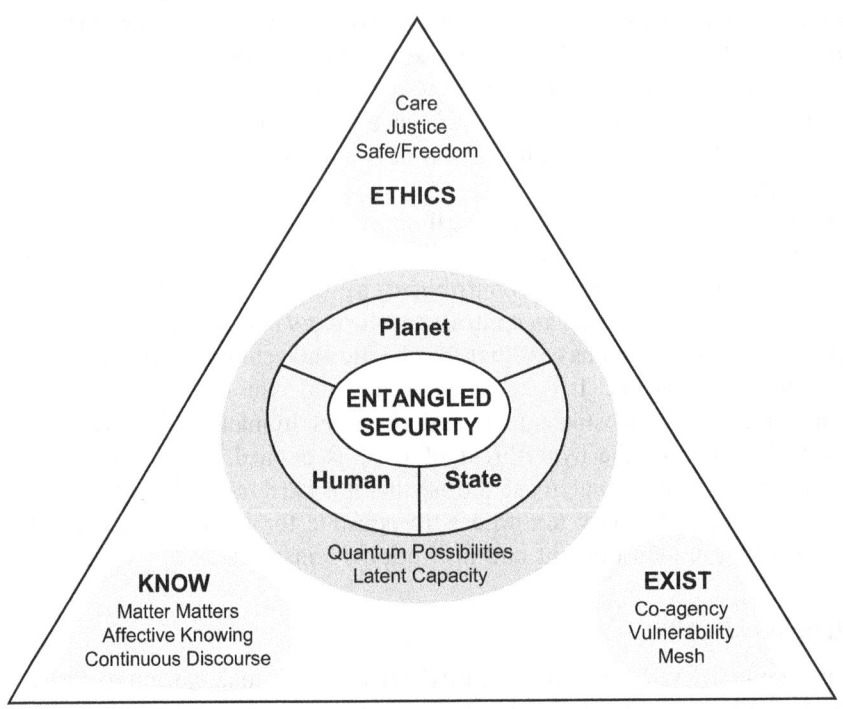

**FIGURE 15.1** Entangled security conceptual compass.

minimising it, involves the opposite: actively seeking acute awareness of the harm caused and being fully accountable for how one's actions may impact the other. This is a new form of ethics and realism, which directly contrasts with some traditional military approaches where efforts are made to decrease soldier empathy for their foe (Grossman 1995).

Karen Barad's 'agential realism' theory draws upon quantum physics to argue that the nature of existence is inherently dynamic, entangled, and subject to abrupt change. For example, at the atomic level, matter – human or otherwise – cannot avoid impacting and colliding with other matter and thus inflicting 'agential cuts' – that is, leaving a mark on the 'other'. If existence is like this, across her body of work, Barad then explores the far-reaching implications it holds for ethics or notions such as justice, time, and agency. Pertinent to questions about framing and the hyperthreat, and building upon Bohr's two-split experiment (Bohr 1963), agential realism also explores entanglement between matter and meaning. Barad proposes that meaning emerges through intra-active conversation between matter (Barad 2007, p. 183). Effective navigation of an entangled reality, Barad posits, requires 'continuous discourse' between humans and all other matter. This idea, on the necessity of ongoing and wide dialogue between all forms of matter to see 'reality' as clearly as possible, informs the (to be described) development of a new 'tribal discourse' analytical method and PLAN E's 'hyper-conversations' as a mechanism for ongoing meaning making.

Agential realism has great significance for the framing of climate and environmental threats. It suggests that to be truthful, meaningful, and relevant, words such as 'threat' or 'security' must be part of the iterative reconfiguration of matter and meaning elicited by CEC. For example, if GHG (or forest clearing, or plastic pollution in oceans) are harmful to other forms of life, then this must be captured within a discourse that is alive to the realities of material reconfiguring of life on Earth. In other words, it simply cannot be that while matter is undergoing seismic changes and having substantial agency and intra-active impacts that old notions of threat or security remain fixed to material conditions of a pre-climate change era.

## The hyperthreat in context

The hyperthreat and entangled security notions can be applied to 'real world' considerations. An activity called a 'tribal discourse' provided a vehicle in which to better understand how the hyperthreat interacts in an entangled security context. In military terms, the tribal discourse also functions as a type of 'friendly forces' analysis. As shown in Figure 15.2, humans are grouped, not as nation-states or according to socio-economic or cultural identity, but in terms of their actions in relation to the hyperthreat.

### *Planetary security tribes*

Considering planetary security, scientists struggled with bridging the science-to-policy gap and, while global citizens showed great potential, at a global level, at

**FIGURE 15.2** Tribal discourse.

the global level they still lacked the impact needed to contain the hyperthreat. Of grave concern, however, and pertinent to a threat inquiry, was the circumstance of the earth protectors who face an increasingly sophisticated, well-resourced, and sometimes militarised human threat. Between 1996 and 2016, there were persistent annual growth rates in global environmental crime (Nellemannet al.2016, p. 39). With an estimated market value of between roughly US$90 and US$276 bn per annum (May 2017, p. xi), environmental crime is perpetuated by corporations, corrupted officials, and transnational criminal and terrorist networks (UNEP 2018, p. ix).

An increasingly quasi-military style of operations has, in turn, led to green militarisation (Lunstrum 2014). Environmental crimes have cascading negative impacts: illegal logging and deforestation reduce carbon sinks, while crime networks undermine nation-states governance capacity and legitimate income (UNEP 2018). Although occasional progress is made (Nellemann et al. 2014, p. 9), overall, the literature indicates that front-line agencies are overwhelmed. For example, proponents of Botswana's controversial 'shoot to kill' policy to address rhinoceros poaching argue that such approaches must be understood in the context of all other measures failing (Mogomotsi and Madigele 2017). As CEC impacts worsen, wilderness becomes rarer, but also more lucrative. Correlated with degrading human security, environmental crime becomes a perverse new form of employment. Conceptually, new battalions of environment criminals are being raised, with greater technological and military capability, who are effectively aligned with the hyperthreat; that is, they increase its destructive power.

### Human security tribes

As global insecurity increases, disturbingly, the capacity of human helpers (i.e. the aid and development sector and the human protection regime) is also in decline (Rüttinger et al. 2015; Cruz 2017; Bellamy and McLoughlin 2018). When combined with the prospect of harsher hyperthreat impacts to come, this creates a downward spiral, called the helper-hyperthreat bind (Figure 15.3). Analysing potential consequences, one question is how a lack of 'help' may be perceived by the most vulnerable. At best, non-helping might be accepted as a non-interventionalist strategy. However, there is also the risk that non-helping is perceived or manifests as a 'leave them to suffer' strategy, or worse, graduates to a 'leave them to their fate' strategy. If it became evident that this was occurring, in turn, it would likely legitimise a sense of grievance towards the developed world which could manifest in destructive ways or be exploited by malevolent agents. A 'leave them to suffer/die' strategy could also create a permissive environment for the worst forms of human behaviour to emerge, such as genocide, slavery, and other forms of

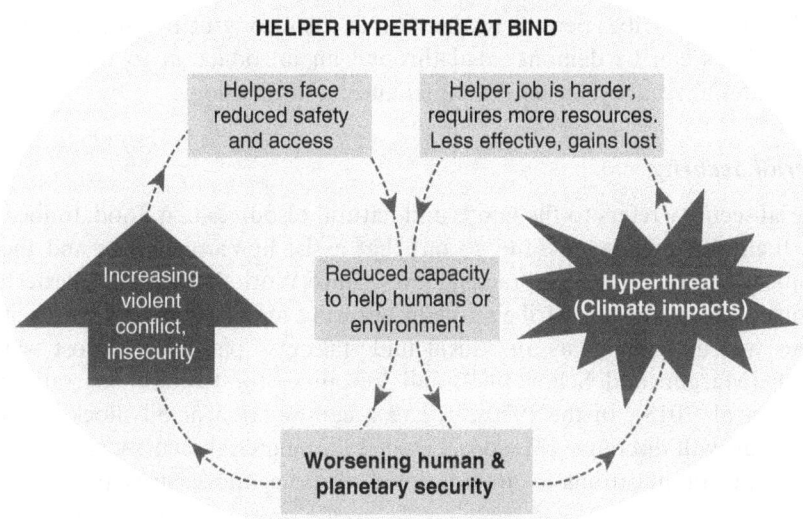

**FIGURE 15.3**   The helper-hyperthreat bind.

abuse. There are already early indicators of links between the onset of the hyperthreat and increasing slavery (Brown et al. 2021). However, another possibility is that the current structural quagmire of international organisations, humanitarianism, and the delivery of aid could be interpreted as an opportunity to reorganise the systems of help, long criticised as constituting new forms of colonialism (Barnett 2011).

### *State security: malign fighters*

The concerns of state security tribes are a prospective quasi-partnership between the malign fighters (non-state actors with intent to harm others) and the hyperthreat. Two examples help explain. Firstly, analysis of non-state armed groups in the Lake Chad region, Syria, Afghanistan, and Guatemala concludes that global warming (the hyperthreat) is contributing to creating an environment in which malign fighters can thrive (Nett and Rüttinger 2016, p. 55). Secondly, the trend of terror groups integrating control of environmental resources into their tactics, such as water (The Stabilisation Network 2017) or, less successfully, oil (Hansen-Lewis and Shapiro 2015) are 'early warnings' of a type of tactic that might be developed further over the next few decades. Outside of conflict zones, malevolent control or disruption of environmental resources can occur through the targeting of critical infrastructure.

## Misaligned state security tribes

The discourse on tribes permits recognition of the incongruent stance of state security. This can be demonstrated through an introduction to three concepts: material security; systems maintenance; and dual-logic.

### Material Security

Material security refers to the goods and natural resources (e.g. food, timber, steel, fuel, minerals, paper, and so on) that assist human societies and the nation-state to function at a practical level. Since World War II, for Western Nations at least, the state tribe's role in ensuring material security for their citizens was understood as an ethical undertaking – part of the 'post war rebuild' (Marglin and Schor 1990) and 'resources for freedom' narratives (Warde et al. 2018). In the 1970s, due to a combination of oil shocks, and limits to growth discourse (Meadows et al. 1972) material security started to have greater ramifications for international relations and security policy.

### Systems maintenance

A generic term to describe the security sector's role in material security is the 'system-maintenance' construct (Stokes 2007). Stokes explains that as global supply chains became more vulnerable to disruptions, there was an increased 'global commons' argument to use tools of force, like the US Central Intelligence Agency (CIA) or military, to 'maintain the system'. This phenomenon has also been described as a modern-era *Lebensraum* strategy (Smith 2003, p. 265), while others argue that logic also infuses stabilisation operations (Morrissey 2017). Thus, under systems maintenance logic, protecting the hyperthreat is regarded as a necessary and dutiful service to the nation.

While systems maintenance may be enacted in a way which is not deleterious to others, this is not always the case. Stokes (2007) finds systems maintenance approaches have led to human rights abuses and more authoritarian regimes. The problem also needs to be considered alongside resource war literature (Sankara and Anderson 1988; Kaldor *et al.* 2007; Klare 2012) and related testimony from a so-called 'economic hit man' (Perkins 2016). The link between the 2003 Iraq war and oil, officially denied (Chilcot 2016) but best understood through Colgan's nuanced analysis (Colgan 2020), especially his causal pathways framework (Colgan 2013), is highly significant to hyperthreat deliberations. Through the systems maintenance prism, the 2003 Iraq War can be viewed as a war waged in support of the hyperthreat. The incoherence of this US-driven security strategy is that, ironically, at the same time this occurred, 'global citizen' was developing new ways to achieve material security for its citizens (e.g. eco-innovations, zero-emission technologies, etc.).

Most concerning to PLAN E is that despite new awareness of the hyperthreat and the increasingly counter-productive effects of systems maintenance (above), the state tribe trajectory remains largely unchanged. It has been reported that resource eagles are set to invest US$4.9 tn on exploration and extraction of new fossil fuel resources between 2020 and 2030. However, it is also reported that this scale of activity can not occur if global warming is to be limited to 1.5°C (Global Witness 2019). The current build-up in global annual military spending, which exceeded US$2 tn in 2021 (Lopes da Silva *et al.* 2021), reflects the 'uniforms' preparing for multiple conflict scenarios, many of which have systems maintenance dimensions. For example, the South China Sea dispute relates to control of shipping lanes and rich fisheries, but also access to natural gas and crude oil resources (Thuy and Welfield 2019; Zhong and White 2017 pp. 17–19).

### Dual-Logic

The above-discussed incongruency extends Stoke's 'dual-logic' to a climate context. 'Business-as-usual' approaches to material security, which impose a systems maintenance burden upon security agencies, pose two threats to humanity: (1) intensification of hyperthreat power and (2) increased likelihood of geopolitical security destabilisation and conflict. Thus, in contrast with prior climate-security literature that considers whether CEC may lead to violent conflict, dual-logic suggests that instead, the failure to rapidly transition to an ecologically sustainable pathways when first prominently identified to the global community as a type of security issue in the early 1990s (IPCC 1990), may already have had security impacts. Overall, positioning the hyperthreat as the most significant threat reveals that state and security tribes – humanities most powerful groups – are inherently misaligned with their *raison d'être* – protecting their human and non-human populations.

Research on climate denialism (Supran and Oreskes 2017; Oreskes and Conway 2010; InfluenceMap 2019; Brulle 2018) highlights how state tribes can be hypnotised by the hyperthreat. Through systems maintenance security operations, the uniforms facilitate hyperthreat growth. In undertaking disaster response roles, increasingly, uniforms also find themselves cleaning up after the hyperthreat. Security agencies do not re-orientate towards this new foe; rather they remain its ally. Yet if this strange, incoherent situation could be reversed, if humanity could reclaim and re-orientate its state tribes, the current balance of probabilities, which currently lie with a hyperthreat victory and a *Hothouse Earth* (Steffen *et al.* 2018) outcome, could be recast.

### New ways to think about threat

The ways in which 'threat' is commonly understood – at the 'deep framing' level (subconsciously held, entrenched worldviews) – are mismatched to the ways in which threat (violence, destruction, and harm) manifests in the Anthropocene. New ways of understanding threat are now required which, in turn, can inform the character of a new threat posture.

'Harm-doing' defies conventional expectations of what a threat or what 'harm' looks like. Akin to the description of the 'banality of evil' (Arendt 1963), with the hyperthreat, there is the problem that those making the most harmful decisions do not look like a stereotypical enemy or threat. There is also the problem that such decision makers exist on a spectrum, there are: those who unconsciously participate in harmful decision making; those who do so knowingly yet sit within the law; and those who consciously conduct harmful and illegal activity but do so with virtual impunity due to a lack of institutional incapacity to address such harm. Another factor could be that key decision makers who are consciously or inadvertently aiding the hyper-threat may hold sanctioned authority, trust, and power (see Milgram 1963). Such decision makers may be a CEO or government official who wears a smart suit, has a confident manner, and appears earnest, reliable, and honourable; a person who is most likely male, with a sense of gravitas and whose narratives orientate around making contributions to society. Despite rhetoric and appearances, however, the decision making could be devastatingly destructive and threatening to many people and forms of life. This could be the greatest challenge to defeating the hyperthreat: the awkwardness of confronting wrongdoing when it appears as proper, with all the symbols of societal authority and validation. There is an incapacity to see threat when it is not dressed like a threat.

For modern Western nations, a particular challenge may be – in official policy statements at least – that it perceives itself as the 'good' protagonist, dedicated to upholding a 'liberal rules-based order'. The 'threat' is predominantly characterised as some form of 'other' which exists not within but 'over there'. Accordingly, in the era of CEC, a most confronting idea is that this time the threat not only looks like oneself but is also within one's own society. This threat analysis presents a very difficult narrative conundrum, which, it is argued here, constitutes a highly significant aspect of the problem of finding a pathway to safe earth. Accordingly, there is a need to find a way to discuss and resolve these issues which, ideally, does not dehumanise people nor rupture trust and cohesiveness within society. In the entangled security context of the 21st century, it is proposed this is achieved by shifting focus towards neutral identification of harm-doing. Threat is not conceived as an 'identity' (an individual or group), but instead focus shifts towards 'actions' or 'activities' that will harm others (including matter) or which will degrade planetary-human-state security. This requires new or re-oriented institutions, laws, and investigative and intelligence capabilities, as proposed in PLAN E.

**Result: Plan E**

How could the hyperthreat be contained and countered? To develop ideas, strategic planning methods have been deployed including SWOT analysis (Mintzberg 2007); principles of war analysis (Australian Army 2017, pp. 16–18); principles of entangled security analysis; and general operational art (McKercher 1996). Real-options analysis is a method for decision making in environments of deep uncertainty (Whitten *et al.* 2012) and its approach of actively identifying outlier possibilities can also be also utilised (see Figure 15.4).

The resultant PLAN E is a climate and eco-centred security strategy (Boulton 2022a, 2022b). The mission is to create a safe path to safe earth. In practical terms, this means limiting global warming to as close to 1.5°C as possible, arresting the sixth extinction event, and maximising safety for all forms of planetary life. It assumes a popular mandate and that a global environmental peace treaty has been achieved. It is envisioned as a civilian led, whole-of-society, layered mobilisation – as distinct from militarisation. PLAN E's operational design revolves around targeting the hyperthreat's key enablers; that is, addressing its invisibility and unknowability; the 'off-the leash' nature of the threat, and humanity's passivity and hesitancy. It involves raising new capabilities and creating new institutional structures, as shown in an initial concept for a Hyper-Response Force (HRF) (Figure 15.5). The proposed structure offers orchestration logic rather than command and control. It is anticipated that HRF would be highly contextualised and vary across local to global scales.

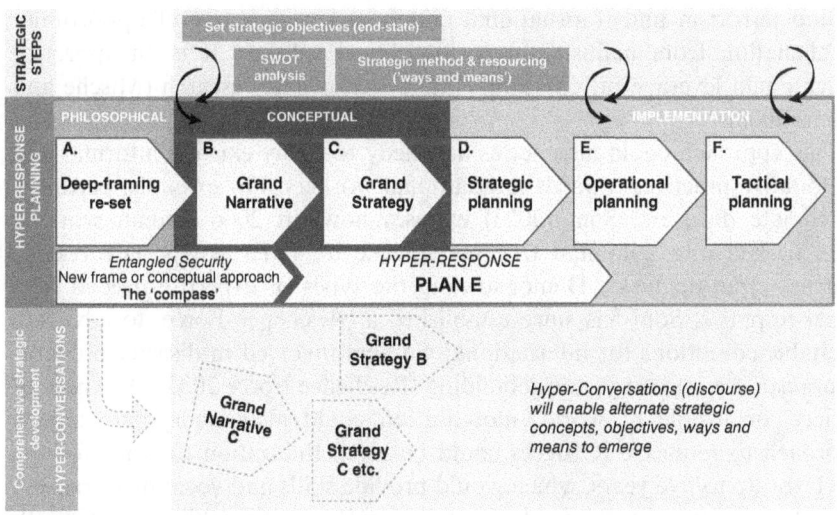

**FIGURE 15.4** Strategic planning and the hyperthreat.

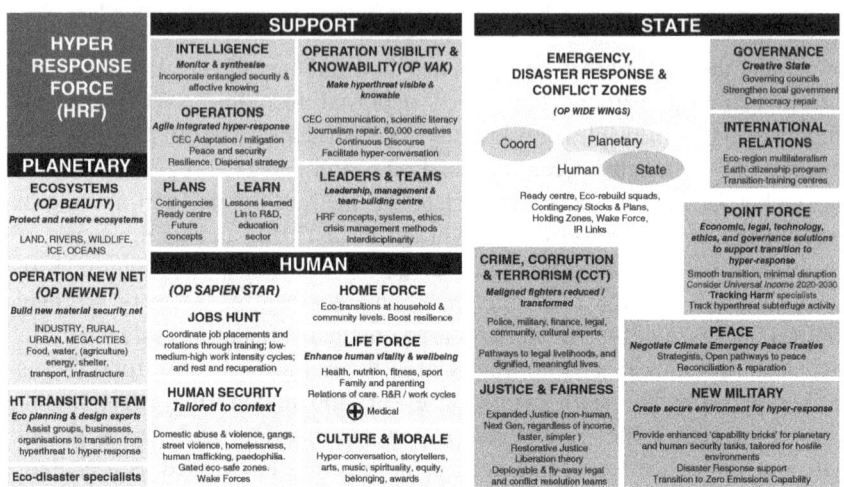

**FIGURE 15.5** Hyper response force.

In the context of a fast-advancing hyperthreat, where there is an enormous amount of work to be undertaken in a short amount of time, Earth's enormous human population is an asset if it can be leveraged towards a hyper-response. Thus, PLAN E envisions an enormous planetary clean-up and rehabilitation project, employing and training billions of people. With sophisticated consultation and honest, fair engagement, it is possible that a hyper-response, which supersedes old colonial style approaches to aid and development, could concurrently address some of the underpinning drivers behind terrorism and transnational crime, such as scarcity of opportunity or alienation from mainstream socio-political spheres. It is an approach which could leverage off environmental peacekeeping research (Mische and Harris 2008).

This approach could also act as a remedy to many existing informal and exploitative practices towards international workers. For example, pertinent to climate disasters, Soni (2023) exposes how, in 2006, Indian workers were tricked into a human tracking scheme to undertake post-Hurricane Katrina clean-up tasks. Demonstrating the types of expertise needed in a hyper-response, Soni has since established a 'Resilience Force' to advocate equitable conditions for international workers involved in disaster preparation, response, recovery, and rebuilding (Resilience Force 2023). A 'whole of society' or 'whole of world' mobilisation could also incorporate a new approach to refugees. Refugees could be given the option to work for the HRF for up to five years, which would provide skills and vocational training but also create a pathway to a new form of citizenship called 'earth citizenship'.

A novel part of PLAN E is a 'bottom-up' and increasingly localised response to threat. This contrasts with approaches that see climate and environmental issues as security issues and advocate a top-down militarised solution (Paglia 2018). Accordingly, PLAN E facilitates ongoing society-wide 'hyper-conversations' to assist people develop their own ways of framing and responding to the 'mission' in ways that make sense given their social, cultural, and environmental context. Local capacity for rapid sense making – especially of meteorological and other environmental intelligence – and design of local hyperthreat response will become increasingly important as the hyperthreat's main body forces arrive and circumstances become more dangerous, fluid, and erratic.

The proposed shift of security resourcing and capabilities to local levels applies at the geopolitical level as well. Instead of being aligned around human-designed political boundaries, multilateralism could align with ecological or climate boundaries, facilitating care of ecosystems and disaster response. Eco-multilateralism may also provide a buffer against fears and risks associated with centralised global power, while also affirming the ethics of neighbourliness and interdependence rather than rivalry.

## Discussion: informing politico-military response

PLAN E is the 'good news' of a hyperthreat framing; it promotes choice and problem solving. The latter being one of humanity's most critical survival attributes. However, as touched upon through the systems maintenance discussion (above), the hyperthreat lens also presents disturbing paradoxes, and the implications of this conceptual instability need to be further drawn out.

### Erasing reality

The first unstable aspect of the current and dominant Western approach to defence and security is that it is disconnected from the physical realities of climate and ecological science. At the strategic level, it is not realistic to consider dangerous climate change as merely another factor which bears upon the security environment. In NATO's (2022) *Strategic Concept*, for example, climate change is listed last. The 'last on the list', 'add on', and even 'threat multiplier' framings deny the irreversible nature of dangerous climate change and species extinctions. Such language convenes insignificance and distort the truth of the hyperthreat's lethality.

Present-day military planning rarely accounts for the effect 21st-century military technologies would have on fragile ecosystems (Ceballos et al. 2020; IPBES 2019). For example, discussion about potential conflict in the South China Sea often includes naval battles, submarine warfare, and sea mines as though there were no fisheries or marine ecosystems in these oceans

(Zhong and White 2017). Aside from its intrinsic worth, such marine life is also important to Asia-Pacific people's food supply and livelihoods. There is a lack of logic in fighting over dwindling resources when doing so might destroy those same resources. In general, in most conflicts, the casualty toll of non-human life is rarely recorded or noted as significant. Similar erasure occurs with global military GHG emissions which are difficult to calculate given the reporting of military emissions to the UNFCCC is voluntary. However, best estimates suggest annual military GHG emissions are enormous – between 1% and 5% of total global emissions (Rajaeifar et al. 2022). This is an extraordinary amount to emit from global climate mitigation planning.

The most significant omission is the consideration of the timeframes for effective responses to climate and ecological issues. The Paris Agreement orientates around the need for the world to undertake dramatic GHG emissions cuts from 2020 onwards; the nature of the climate system means this cannot be put 'on hold'. Yet the dominant security concern remains preparation for, or conduct of, kinetic warfare between 2020 and 2050, with its likely high financial, GHG, and environmental costs, and high demands for engineering, technology, and other human ingenuity and talent. Thus, if current trends in warfare continue, this will derail capacity for effective transformational response to the planetary crisis. In short, it is not feasible to 'fight' the hyperthreat and major warfare at the same time.

### Threat posture

The second unstable aspect relates to threat posture which, it is argued, is currently oriented around defunct logic. This relates to the deep-framing problem. Fossil fuels were once regarded as good. They were seen as essential for military victory (Goralski and Freeburg 2021), and the security and flourishing of the state. Under the systems maintenance logic (above), the role of security forces in protecting resource extraction activities and their supply chains – or supporting 'resource eagles' – has been regarded historically as a necessary and dutiful service to the nation. Due to global warming, this situation has been turned upside down. The industrial-era system that the foreign affairs and defence sectors have supported (Denniss and Behm 2021) and often worked with as partners, now threatens all forms of planetary life, including *Homo sapiens*. Thus, what once made nations, and especially the West, secure, victorious, powerful, and safe, now does the opposite – if timeframes beyond a decade are used.

This circumstance introduces a new form of Mutually Assured Destruction (MAD). The danger of continued warfare, especially that which has underlying resource war dimensions, is that humanity misses the opportunity to save planetary life. A scenario helps to explain. Imagine tensions between 'the

West' and 'the rest' escalate into a 'hot' WWIII or simmer as a long, expensive, and resource-intensive battle of attrition. Whichever side 'wins' militarily would then face an enraged hyperthreat – a descent into a 'hothouse Earth' scenario – which would destroy their security anyhow. Thus, in 2023, with an encroaching hyperthreat, there is no longer any viable pathway to long-term security through warfare. A further deduction is that if security strategy and the defence sector do not pivot towards the hyperthreat, it is unlikely that the civil sector will be able to achieve Paris Agreement ambitions for a habitable planet.

### Purpose of the defence sector

The third unstable aspect relates to the question of whether a nation-state's security strategy is diverging too far from its fundamental purpose. A foundational theoretical concept is that security strategy serves political objectives (Clausewitz 1832, p. 119). In a democracy, it could be assumed that 'political' relates to the need and wishes of the populace. This logic can be applied to the hyperthreat. What if the Paris Agreement and various climate emergency declarations are accepted as representing the will of the global populace? If these establish that global warming and other ecological breakdowns as a significant and urgent threat, then it follows that security strategy must re-orientate to support the larger political objective. Additionally, if security decisions, such as the decision to go to war, are separated from political processes, debate, and interrogation (Chilcot 2016), then a risk arises that such decisions are out of alignment with the security needs and preferences of the broader population. Through paying taxes, the broader civil population funds defence forces and veteran rehabilitation and care. Furthermore, the populace provides or 'are' the personnel who serve in defence forces.

If a nation's governance systems have been even partly appropriated by powerful financial interests, such as through regulatory capture (Dal Bó 2006); lobbyists (Brulle 2018); business and economic actors (Butler 2016); or oligarchical influence (Beetham 2011), then this risks security forces also being appropriated. Corporate capture, or similar, of governance could mean humanity's security assets inadvertently end up working for elites and against the populace's security interests. For security forces to be used to undermine the security of their own people presents a profound moral conflict. It also has implications for the legitimacy of the nation-state structure. Accordingly, at the time in which the hyperthreat vanguard has arrived, there is an urgent need to review the politico-military relationship to ensure it is aligned with the needs of the populace. Furthermore, given the nature of the hyperthreat, it may only be a security mission which has the authority to override those imperilling planetary security.

PLAN E offers an alternative to the current continuation of resource war trajectories and state-sponsored systems maintenance support to industrial-era forms of energy, geopolitical, and corporate power. The security sector abruptly turns its systems maintenance support away from the fossil fuel and extractive resource sector and back to its own people. It supports a different systems maintenance mission: the protection of Earth's planetary life system.

## Conclusion

As a framing device, the hyperthreat highlights the need for security strategy to be philosophically re-anchored from the modern, industrial era to the present time which faces global warming and ecological collapse. There are two key problems with extant approaches to defence and security: strategic disconnection from physical reality and increasing divergence from the public's *bone fide* security needs and preferences. A threat lens also forces confrontation with one of the most difficult aspects of the problem – the complex idea of socially and legally sanctioned harm-doing. However, the hyperthreat lens also opens the door to utterly re-imagining what threat response and security forces look like in the 21st century.

## References

Arendt, H. 1963. *Eichmann in Jerusalem: A Report on the Banality of Evil.* London: Penguin.
Armstrong McKay, D. I., A. Staal, J. F. Abrams, R. Winkelmann, B. Sakshewski, S. Loriani, I. Fetzer, S. E. Cornell, J. Rockstrom, and T. M. Lenton 2022. Exceeding 1.5C global warming could trigger multiple climate tipping points. *Science* 377(6611). 10.1126/science.abn7950.
Australian Army 2017. *Land Warfare Doctrine 1: The Fundamentals of Land Power.* Canberra: Commonwealth of Australia.
Barad, K. 2007. *Meeting the Universe Halfway: Quantum Physics and the Entanglement of Matter and Meaning.* Durham, NC: Duke University Press.
Barnett, M. 2011. *Empire of Humanity: A History of Humanitarianism.* Ithaca, NY: Cornell University Press.
Beard, S. J., L. Holt, S. Avin, A. Tzachor, L. Kemp, P. Torres, and H. Belfield 2021 Assessing climate change's contribution to global catastrophic risk. *Futures* 127. 10.1016/j.futures.2020.102673.
Beetham, D. 2011. *Unelected Oligarchy: Corporate and Financial Dominance in Britain's Democracy.* London: London School of Economics and Political Science.
Bellamy, A. J. and S. McLoughlin 2018. *Rethinking Humanitarian Intervention.* London: Macmillan.
Bohr, N. 1963. *Essays 1958–1962 on Atomic Physics and Human Knowledge.* Oxford: Oxbow Press.
Bostrom, N. 2013. Existential risk prevention as global priority. *Global Policy* 4(1): 15–31.

Boulton, E. 2016. Climate change as a 'hyperobject': a critical review of Timothy Morton's reframing narrative. *Climate Change* 7(5): 772–785.

Boulton, E. G. 2018. Climate Change as a hyperthreat. (In) S. Pearson, J. L. Holloway and R. Thackway (eds) *Australian Contributions to Strategic and Military Geography*. London: Springer, pp. 69–90.

Boulton, E. G. 2022a. An introduction to PLAN E: grand strategy for the twenty-first-century era of entangled security and hyperthreats. *Expeditions with MCUP* 2022(1): 1–91.

Boulton, E. G. 2022b. PLAN E: a grand strategy for the twenty-first-century era of entangled security and hyperthreats. *Journal of Advanced Military Studies* 13(1): 92–128.

Brown, D., D. S. Boyd, K. Brickell, C. D. Ives, N. Natarajan, and L. Parsons 2021. Modern slavery, environmental degradation and climate change: fisheries, field, forests and factories. *Environment and Planning E: Nature and Space* 4(2): 191–207.

Brulle, R. J. 2018. The climate lobby: a sectoral analysis of lobbying spending on climate change in the USA, 2000 to 2016. *Climatic Change* 149(3): 289–303.

Burke, A., S. Fishel, A. Mitchell, S. Dalby, and D. J. Levine 2016. Planet politics: a manifesto from the end of IR. *Millennium* 44(3): 499–523.

Busby, J. W. 2021. Beyond internal conflict: the emergent practice of climate security. *Journal of Peace Research* 58(1): 186–194.

Butler, S. D. 2016. *War is a Racket: The Antiwar Classic by America's Most Decorated Soldier*. New York, NJ: Simon and Schuster.

Ceballos, G., P. R. Ehrlich, and P. H. Raven 2020. Vertebrates on the brink as indicators of biological annihilation and the sixth mass extinction. *Proceedings of the National Academy of Sciences of the United States of America* 117(24): 13596–13602.

Chilcot, S. J. 2016. *The Report of the Iraq Inquiry: Executive Summary*. London: House of Commons.

Clausewitz, C. 1832. *On War*. London: Penguin.

Colgan, J. D. 2013. Fueling the fire: pathways from oil to war. *International Security* 38(2): 147–180.

Colgan, J. D. 2020. Oil and security: The necessity of political economy. *Journal of Global Security Studies* 6(1). 10.1093/jogss/ogaa008.

Commission on Global Governance 1995. *Our Global Neighbourhood: The Report of the Commission on Global Governance*. New York, NJ: United Nations.

Cruz, C. A. 2017. *Improving Security of United Nations Peacekeepers: We Need to Change the Way We are Doing Business*. New York, NJ: United Nations.

Dal Bó, E. 2006. Regulatory capture: a review. *Oxford Review of Economic Policy* 22(2): 203–225.

Dalby, S. 2020. National security in a rapidly changing world. *Balsillie Papers* 3(3). https://balsilliepapers.ca/wp-content/uploads/2020/10/Balsillie-Paper-Dalby.pdf.

Dellmuth, L. M., M.-T. Gustafsson, N. Bremberg, and M. Mobjork 2018. Intergovernmental organizations and climate security: advancing the research agenda. *Wiley Interdisciplinary Reviews: Climate Change* 9(1). 10.1002/wcc.496.

Denniss, R. and A. Behm 2021. Double game: how Australian diplomacy protects fossil fuels. *Australian Foreign Affairs* 12: 20.

Devecchio, A. 2023. Emmanuel Todd: La Troisième Guerre mondiale a commencé. *Le Figaro*. https://www.lefigaro.fr/vox/monde/emmanuel-todd-la-troisieme-guerre-mondiale-a-commence-20230112#Echobox=1673601494-1 (accessed 1 May 2023).

Fotion, N. 2007. *War and Ethics: A New Just War Theory*. London: Continuum.

Francis, P. 2015. Encyclical letter: Laudato Si' of the Holy Father Francis on care for our common home. http://w2.vatican.va/content/francesco/en/encyclicals/documents/papa-francesco_20150524_enciclica-laudato-si.html (accessed 1 May 2023).

Global Witness 2019. *Overexposed: How the IPCC's 1.5 Degree Report Demonstrates the Risks of Overinvestment in Oil and Gas*. London: Global Witness.

Goodman, S. 2007. *National Security and the Threat of Climate Change*. Washington, DC: The CNA Corporation.

Goralski, R. and R. W. Freeburg 2021. *Oil and War: How the Deadly Struggle for Fuel in WWII Meant Victory or Defeat*. Quantico, VA: Marine Corps University Press.

Grossman, D. 1995. *On killing: The Psychological Cost of Learning to Kill in War and Society*. New York, NJ: Bay Back Books.

Hansen-Lewis, J. and J. N. Shapiro 2015. Understanding the Daesh economy. *Perspectives on Terrorism* 9(4): 1–14.

Haraway, D. 2008. *When Species Meet*. Minnesota, MN: University of Minnesota Press.

Hinkel, J., D. Mangalagiu, A. Bisaro, and J. D. Tabara 2020. Transformative narratives for climate action. *Climatic Change* 160(4): 495–506.

Homer-Dixon, T., O. Renn, J. Rockstrom, J. F. Donges, and S. Janzwood 2021. A call for an international research program on the risk of a global polycrisis. *SSRN*. 10.2139/ssrn.4058592 (accessed 1 May 2023).

Influence Map 2019. *Big Oil's Real Agenda on Climate Change*. London, UK: InfluenceMap.

IPBES 2019. *Summary for Policymakers of the Global Assessment Report on Biodiversity and Ecosystem Services of the Intergovernmental Science-Policy Platform on Biodiversity and Ecosystem Services*. Bonn: IPBES.

IPCC 1990. *Climate Change: The IPCC Scientific Assessment (Contribution of Working Group I to the First Assessment Report of the Intergovernmental Panel on Climate Change)*. London: Intergovernmental Panel on Climate Change.

Kaldor, M., T. L. Karl, and Y. Said 2007. *Oil Wars*. London: Pluto Press.

Klare, M. 2012. *The Race for What's Left: The Global Scramble for the World's Last Resources*. Basingstoke: Macmillan.

Kühne, K., N. Bartsch, R. D. Tate, J. Higson, and A. Habet 2022. 'Carbon bombs': mapping key fossil fuel projects. *Energy Policy* 166: 112950. 10.1016/j.enpol.2022.112950.

Lakoff, G. 2010. Why it matters how we frame the environment. *Environmental Communication* 4(1): 70–81.

Lopes da Silva, D., X. Liang, A. Marksteiner, L. Beraud-Sudreau, and N. Tian 2021. *Trends in World Military Expenditure*. Sweden: Stockholm International Peace Research Institute.

Lunstrum, E. 2014. Green militarization: anti-poaching efforts and the spatial contours of Kruger National Park. *Annals of the Association of American Geographers* 104(4): 816–832.

Marglin, S. A. and J. B. Schor 1990. *The Golden Age of Capitalism: Reinterpreting the Postwar Experience*. Oxford: Oxford University Press.

May, C. 2017. *Transnational Crime and the Developing World*. Washington, DC: Global Financial Integrity.

McDonald, M. 2018. Climate change and security: towards ecological security? *International Theory* 10(2): 153–180.

McKercher, B. J. C. and M. A. Hennessy (eds) 1996. *The Operational Art: Developments in the Theories of War*. London: Praeger.

Meadows, D. H., D. L. Meadows, J. Randers, and W. W. Behrens 1972. *The Limits to Growth: A Report to the Club of Rome*. Washington, DC: Potomac Associates.

Milgram, S. 1963. Behavioral study of obedience. *The Journal of Abnormal and Social Psychology* 67(4): 371–378.

Mintzberg, H. 2007. *Tracking Strategies: Toward a General Theory*. Oxford: Oxford University Press.

Mische, P. and I. Harris 2008. Environmental peacemaking, peacekeeping, and peacebuilding. *Encyclopedia of Peace Education*: 1–9.

Mobjörk, M., L. M. Dellmuth, N. Bremberg, M-T. Gustafsson, H. Sonnsjo, and S. van Baalen 2016. *Climate-Related Security Risks: Towards an Integrated Approach*. Sweden: SIPRI.

Mogomotsi, G. E. and P. K. Madigele 2017. Live by the gun, die by the gun: Botswana's 'shoot-to-kill'policy as an anti-poaching strategy. *South African Crime Quarterly* 60: 51–59.

Morrissey, J. 2017. *The Long War: CENTCOM, Grand Strategy, and Global Security*. Athens, GA: University of Georgia Press.

Morton, T. 2013. *Hyperobjects: Philosophy and Ecology after the End of the World*. Minneapolis: University of Minnesota Press.

NATO 2022. *NATO 2022 Strategic Concept*. Belgium: NATO.

Nellemann, C., R. Henriksen, A. Kreilhunter, D. Stewart, M. Kotsovou, P. Raxter, E. Mrema, and S. Barrat (eds) 2016. *The Rise of Environmental Crime: A Growing Threat to Natural Resources, Peace, Development and Security. A UNEP-INTER-POL Rapid Response Assessment*. Kenya: UNEP.

Nellemann, C., R. Henriksen, P. Raxter, N. Ash, and E. Mrema (eds) 2014. *The Environmental Crime Crisis: Threats to Sustainable Development from Illegal Exploitation and Trade in Wildlife and Forest Resources*. The Hague: United Nations Environment Programme.

Nett, K. N. and L. Rüttinger 2016. *Insurgency, Terrorism and Organised Crime in a Warming Climate: Analysing the Links Between Climate Change and Non-State Armed Groups*. Berlin: Climate Diplomacy.

Nixon, R. 2011. *Slow Violence and the Environmentalism of the Poor*. Cambridge, MA: Harvard University Press.

Ord, T. 2020. *The Precipice: Existential Risk and the Future of Humanity*. London: Hachette Books.

Oreskes, N. and E. M. Conway 2010. *Merchants of Doubt*. New York, NJ: Bloomsbury.

Oswald Spring, Ú. 2009. A HUGE gender security approach: towards human, gender, and environmental security. (In) H. G. Brauch, U. O. Spring, J. Grin and C. Mesjasz (eds) *Facing Global Environmental Change: Environmental, Human, Energy, Food, Health and Water Security Concepts*. London: Springer, pp. 1157–1181.

Paglia, E. 2018. The socio-scientific construction of global climate crisis. *Geopolitics* 23(1): 96–123.

Perkins, J. 2016. *The New Confessions of an Economic Hit Man*. London: Berrett-Koehler Publishers.

Rajaeifar, M. A., O. Belcher, S. Parkinson, B. Neimark, D. Weir, K. Ashworth, R. Larbi, and O. Heidrich 2022. Decarbonize the military: mandate emissions reporting. *Nature* 611(7934): 29–32.

Resilience Force 2023. We are the resilience workforce. *Resilience Force*. https://resilienceforce.org/ (accessed 1 May 2023).

Richards, C., R. Lupton, and J. Allwood 2021. Re-framing the threat of global warming: an empirical causal loop diagram of climate change, food insecurity and societal collapse. *Climatic Change* 164(3): 1–19.

Rüttinger, L., G. Stang, D. Smith, D. Tanzler, and J. Vivekananda 2015. *A New Climate for Peace: Taking Action on Climate and Fragility Risks*. Paris: G7.

Sankara, T. and S. Anderson 1988. *Thomas Sankara Speaks: The Burkina Faso Revolution, 1983–87*. New York, NJ: Pathfinder Press.

Sears, N. A. 2020. International politics in the age of existential threats. *Journal of Global Security Studies* 6(3): 1–20.

SEI 2021. *The Production Gap Report 2021*. Stockholm: Stockholm Environment Institute.

Servigne, P. and R. Stevens 2020. *How Everything Can Collapse: A Manual for Our Times*. London: John Wiley & Sons.

Smith, N. 2003. After the american lebensraum: 'empire', empire, and globalization. *Interventions* 5(2): 249–270.

Soni, S. 2023. *The Great Escape: A True Story of Forced Labor and Immigrant Dreams in America*. New York, NJ: Algonquin Books.

Steffen, W., J. Rockstrom, K. Richardson, T. M. Lenton, C. Folke, D. Liverman, C. P. Summerhayes, A. D. Barnosky, S. E. Cornell, M. Crucifix, J. F. Donges, I. Fetzer, S. Ladee, M. Scheffer, R. Winkelman, and H. J. Schellnhuber 2018. Trajectories of the Earth system in the Anthropocene. *Proceedings of the National Academy of Sciences of the United States of America* 115(33): 8252–8259.

Stokes, D. 2007. Blood for oil? Global capital, counter-insurgency and the dual logic of American energy security. *Review of International Studies* 33(2): 245–264.

Supran, G. and N. Oreskes 2017. Assessing ExxonMobil's climate change communications (1977–2014). *Environmental Research Letters* 12(8). 10.1088/1748-9326/aa815f.

The Stabilisation Network 2017. *TSN Update: Daesh Withholding Medicine and Medical Care from Raqqawis*. Dubai: The Stablisation Network.

Thuy, T. T. and J. B. Welfield 2019. *Building a Normative Order in the South China Sea: Evolving Disputes, Expanding Options*. London: Edward Elgar Publishing.

Trombetta, M. J. 2008. Environmental security and climate change: analysing the discourse. *Cambridge Review of International Affairs* 21(4): 585–602.

UN 2022. *The Sustainable Development Goals Report*. New York, NJ: United Nations.

UN DESA 2018. *The Sustainable Development Goals Report 2018*. New York, NJ: UN Department of Economic and Social Affairs.

UNEP 2018. *The State of Knowledge of Crimes that have Serious Impacts on the Environment*. Nairobi: United Nations Environment Programme.

UNEP 2022. *Emissions Gap Report*. Nairobi: United Nations Environment Programme.

UNSC 2021. *The UN Security Council and Climate Change*. New York, NJ: United Nations Security Council.

Warde, P., L. Robin, and S. Sörlin 2018. *The Environment: A History of the Idea.* Baltimore, NJ: John Hopkins University Press.

Warszawski, L., E. Kreiger, T. M. Lenton, O. Gaffney, D. Jacob, D. Klingenfeld, R. Koide, M. M. Costa, D. Messner, N. Nakicenovic, H. J. Schellnhuber, P. Schlosser, K. Takeuchi, S. van der Leeuw, G. Whiteman, and J. Rockstrom 2021. All options, not silver bullets, needed to limit global warming to 1.5°C: a scenario appraisal. *Environmental Research Letters* 16(6). 10.1088/1748-9326/abfeec.

Whitten, S. M., G. Hertzler, and S. Strunz 2012. How real options and ecological resilience thinking can assist in environmental risk management. *Journal of Risk Research* 15(3): 331–346.

WMO 2022. *Greenhouse Gas Bulletin, No.18: The State of Greenhouse Gases in the Atmosphere Based on Global Observations through 2021.* Geneva: World Meteorological Organization.

Zhong, H. and M. White 2017. South China Sea: its importance for shipping, trade, energy and fisheries. *Asia-Pacific Journal of Ocean Law and Policy* 2(1): 9–24.

# 16

# A REFLECTION ON 30 YEARS OF CLIMATE AND CONFLICT

*Thomas Homer-Dixon*

In the late 1980s, during my doctoral work at the Massachusetts Institute of Technology, I examined the causes of violent conflict through the theoretical lenses of social psychology and group-identity defence. I focused on 'we/they' divisions between groups and the most extreme forms of dehumanisation in conflict. Towards the end of my dissertation work, I considered how environmental stresses such as water scarcity, deforestation, and the degradation of agricultural land could lead to such conflict dynamics.

Several years later, in 1991, this research culminated in an article published in the journal *International Security* titled, 'On the threshold: environmental changes as causes of acute conflict' (Homer-Dixon 1991). There, I presented an analytical framework and a set of hypotheses for investigation about how environmental change might produce violent conflict. The central feature of this article was a causal path analysis of the relationships between various kinds of environmental stress and violent conflict. I suggested that it was unlikely that people were going to fight directly over environmental resources such as water, forests, or soil, but rather that degradation, over-consumption, and demand-driven scarcity of these environmental resources would result in a series of cascading consequences for agricultural production, economic wellbeing, and social cohesion within societies. Furthermore, it would be those intermediate consequences – or 'intermediate social effects' as I came to call them – that would ultimately influence the probability of violence.

As such, I felt that it was very important to trace out these causal pathways in detail to specify the relationship between the main environmental problems, their second-order environmental effects, and the social effects that might ultimately lead to violent conflict. As a first step, I developed diagrams (see Figures 16.1 and 16.2), which identify multiple causal pathways between

DOI: 10.4324/9781003377641-20

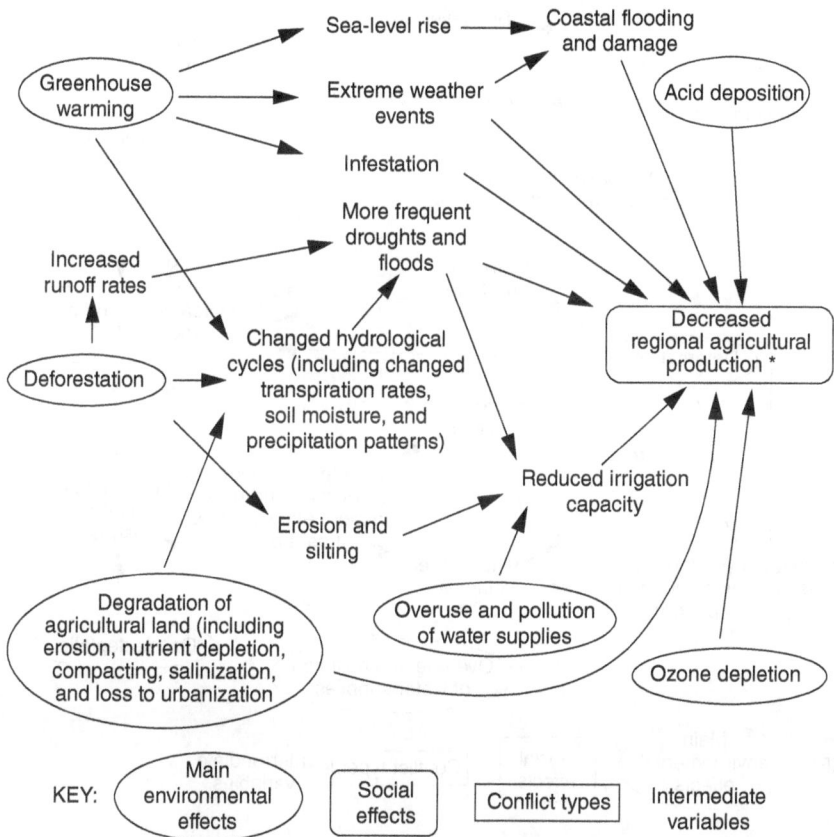

**FIGURE 16.1** Possible effects of environmental change on agricultural production.

'greenhouse warming' (now more commonly referred to as climate change), decreased regional agricultural production, and decreased economic productivity.

A relatively direct pathway identified in Figure 16.1 leads from greenhouse warming to increased infestation by pests that destroy crops – an outcome that decreases regional agricultural production. A less direct pathway proceeds from greenhouse warming to extreme weather events (for instance, larger storms) and, in turn, to greater coastal flooding that then decreases agricultural production. (Much of the world's most productive agricultural land is in coastal regions that are disproportionately affected by increasingly severe storms.) An even more indirect pathway involves multiple steps. An example is the pathway from greenhouse warming to changes in hydrological cycles (the cycle of water between the atmosphere and the landscape) to more

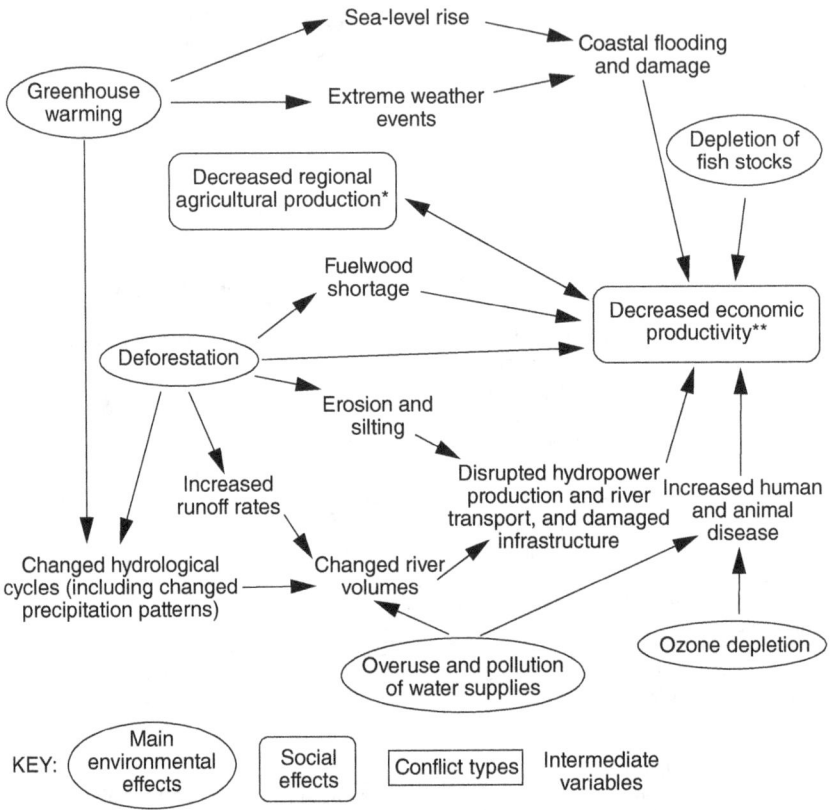

**FIGURE 16.2** Possible effects of environmental change on economic productivity.

frequent droughts and floods, to reductions in irrigation capacity, and finally
to decreased agricultural production. This pathway analysis shows, first, that
climate change can produce multiple negative effects that occur simultaneously, and second, that climate change interacts with other environmental
challenges. For example, in many regions in the world, climate change is
worsening the impacts on agricultural yields of ongoing deforestation, degradation of agricultural land, and the over-consumption and pollution of
water supplies.

Figure 16.1 then fits within Figure 16.2 as a subset of that latter figure. This
integration of the two figures was a useful way to make extremely complex,
multi-level, and intersecting causal pathways easy to understand visually.
Figure 16.2 shows that the same environmental factors, including greenhouse
warming, decrease economic productivity, and another critically important
intermediate social effect. But these environmental factors operate through
pathways that are additional to those identified in Figure 16.1; for example,

warming's impacts on hydrological cycles and river volumes also disrupt hydro-power production, which affects economic output.

I developed this path analysis in the early 1990s, and 30 years of subsequent evidence largely supports the identification of effects and pathways in these figures. From our perspective today, the key conclusion is that climate change is a powerful amplifying factor across many different dimensions; and the impact of the whole set of interacting factors is far more than the sum of their individual impacts (Homer-Dixon 1991).

Building on this initial analysis in the mid-1990s, my research group at the University of Toronto, which became known as the 'Toronto Group', used empirical case studies to illustrate the various causal pathways from environmental scarcity, through their intermediate social effects, to specific types of violent conflict. We pursued case studies in Haiti, Rwanda, South Africa, Israel/Palestine, India, and the Philippines. The results of this research were published in a second article, 'Environmental scarcities and violent conflict: evidence from cases' again in *International Security*, in 1994 (Homer-Dixon 1994). In this second article, I was clear that environmental stresses, including global warming, were generally not sufficient in themselves to produce direct conflict between states, or so-called 'resource wars'. Rather, I argued that they tended to interact and exacerbate intergroup conflicts that have a strong 'we/they' identity dynamic and to increase groups' motivation to engage in conflict through a heightened sense of relative deprivation (a feeling that just expectations are not being satisfied).

The culmination of my work in this area was my book, *Environment, Scarcity and Violence,* published by Princeton University Press in 1999 (Homer-Dixon 1999). The book synthesised a decade of research and analysis of how environmental stress might affect violent civil and interstate conflict. Using the conceptual frameworks and pathway analysis in my original *International Security* articles as starting points, I argued that the deleterious impacts of environmental stress on agricultural production and economic productivity could precipitate three main types of conflicts, which I labelled 'simple-scarcity', 'group-identity', and 'relative-deprivation' conflicts.

Simple-scarcity conflicts occur when people, groups, or countries fight over resources at the sub-national or international levels. Historical examples include Japan's efforts to secure coal, oil, and minerals in China and South-East Asia during World War II and Hitler's advance to seize the Caucasian oil fields that were halted at Stalingrad. Group-identity conflicts occur when environmental scarcities deepen existing we/they cleavages between identity groups, and these deepened cleavages then engender violence. Such conflicts often arise from the displacement of people who have been negatively affected by environmental problems – for instance, falling agricultural yields in their home region – and who then come into conflict with local populations when they move to new regions. When different ethnic and cultural groups are propelled together under stressful circumstances, intergroup hostility usually increases. Relative-deprivation conflicts arise when the scarcity of environmental resources weakens local and national economies, such as when lower food production raises unemployment. When individuals (or groups) perceive that they are not getting what they deserve – that is, when a widening gap develops between what people think they ought to get and what they actually receive – they can experience a deep sense of frustration and grievance, which can, in turn, motivate violence towards those believed to be the source of the problem.

As I carried out this work in the 1990s, I understood that a true test of these hypotheses would not occur for decades, since it would take that long for sufficient field evidence to accumulate to either falsify or support my arguments. My time horizon extended out 40 years, to 2030 or so. I anticipated that by then environmental stress would exacerbate, and at times initiate, violent conflict. Now that we are approaching 2030, I believe my analysis has been substantially vindicated (IEP 2021, 2022).

Let's now turn to the security implications of climate change more specifically. Records at Mauna Loa Observatory illustrate the level of carbon dioxide concentration in the atmosphere for thirty years on either side of my initial research into the links between environmental stress and violence (see GML 2023). For example, the $CO_2$ mole fraction (ppm) was 316 in 1958, 354.45 in 1990, and 424 in 2023. Strikingly, the curve does not show any diminution over time; it remains essentially an exponential upward trend. As a father of two teenagers, this graph keeps me awake at night. Humanity has so far to go to turn this curve downwards to reduce the effects of carbon dioxide on the heating of the planet. Even during periods where humanity actually cut carbon dioxide emissions considerably – during the first year of the COVID pandemic, in particular – the overall warming trend was not even slightly abated.

In terms of surface temperature, Figure 16.3 takes us back to the end of the last ice age, 11,300 years ago. From then until now, the total variation of temperature on the surface of the planet has been around 0.7°C. During the last 2,000 years, a period when humans established the foundations of modern civilisation – its major cities and ports, its agricultural zones, and its transportation networks – temperature varied no more than 0.5°C. Warming to date has already significantly shifted us out of this historical range. Without a large reduction in greenhouse gas emissions, we are set on a near-vertical trajectory of 2–3°C of warming within this century.

For context, Earth's average surface temperature is currently around 13.9°C. An increase to nearly 16.0°C in this incredibly short timeframe – a blink of an eye in geological terms – will shred almost all of the planet's ecological systems, because they will not have time to adapt. It will also irrevocably damage the agricultural, forest, and fishery systems on which our

## PAST AND PROJECTED TEMPERATURE CHANGES AT EARTH'S SURFACE

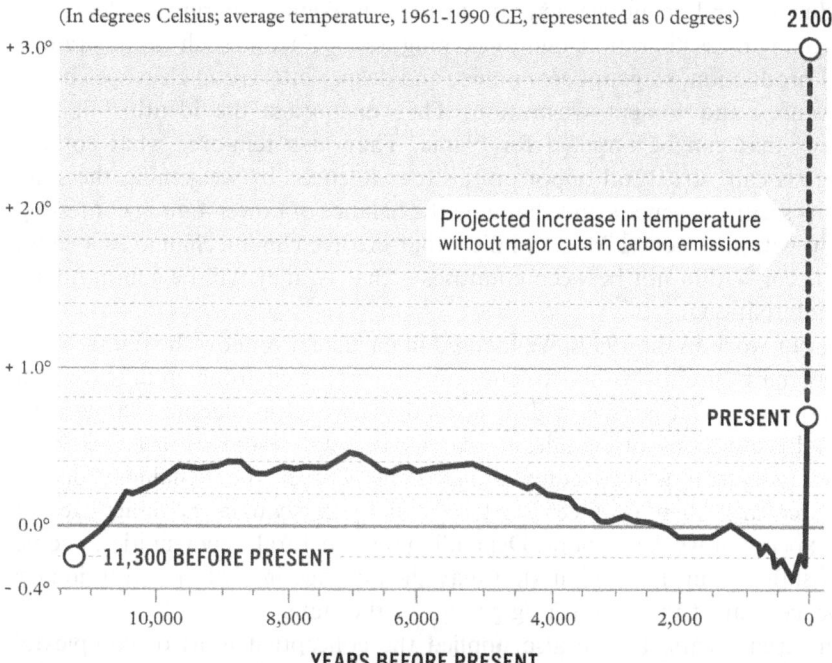

(In degrees Celsius; average temperature, 1961-1990 CE, represented as 0 degrees)    **2100**

**FIGURE 16.3**  Past and present temperature changes at Earth's surface.

societies depend for their wellbeing. With such warming, wide swaths of the tropics will become too hot during summer months for people to work outside; and global food systems will almost certainly be unable to support the projected human population of 9–10bn. The impacts on economic wellbeing and growth will make it difficult, if not impossible, to sustain anything resembling modern liberal democracy.

As I emphasised in my work in the 1990s, this warming can contribute to conflict through multiple pathways, both direct and indirect, by *interacting with additional stressors* that already exist within societies, including institutional weakness and intergroup cleavages. Societies that are already vulnerable are being affected first and worst. But the influence of climate change on conflict is often obscure, because causal pathways are so complicated. Even where climate change is a powerful influence, causal complexity can confound analysis.

Still, I believe our research in the 1990s was prescient. As we anticipated, climate change is now causing around the world extreme droughts and floods, extensive mortality of forests, more and larger storms, rising seas, urban heat emergencies, and widespread water scarcity. Together, these effects can have severe impacts, when they combine with other social stresses, on agricultural production and economic activity, in turn deepening intergroup cleavages, provoking mass migration, and weakening states. Factors such as constrained food production, stagnant economies, and deeper intergroup cleavages boost grievances and foment frustration. They encourage the identification of groups that can be targeted for blame. They also reinforce what conflict theorists call 'structural opportunities for violence', by weakening the state and its security agencies and changing the balance of power among contesting group within societies. Most conflicts that involve climate change as a cause will occur within not between countries – that is, they will be sub-national, not international.

In our work in the 1990s, we focused in particular on how the rent-seeking behaviour of groups within societies can exacerbate environmental scarcity, a phenomenon that is now evident around the world as climate change worsens. In situations of significant scarcity of vital resources, such as fresh water, powerful groups commandeer control over the remainder of the resource and then profiteer ('extract rents', in economists' language) at the expense of weaker groups. Depending on structural opportunities, people who suffer from this exploitation may then engage in civil or revolutionary violence against the profiteering groups or the state.

In recent years, I have also applied the conceptual tools of complexity science to understand better such conflicts' causal dynamics and the capacity of societies to adapt to the effects of climate change without social turmoil. An especially useful concept here is that of the 'tipping event'. Complex systems such as Earth's climate and human societies have the capacity to flip from one state to another, often in entirely unanticipated ways.

We can illustrate tipping events with an 'energy landscape' visual metaphor (Figure 16.4). This landscape involves a ball on a surface; the surface contains lots of dips or 'basins of attraction'. The ball represents the system in question, and it can be pushed from one dip or basin to another. Basins of attraction are akin to the system's zones of stability or equilibrium; that is, they are places where less energy input or 'work' is required to keep the system stable. Over time, because of larger changes in the system's surrounding context, basins can become deeper or shallower. If a basin becomes

Tipping Event

**FIGURE 16.4** Tipping event.

shallower, a sudden shock or external perturbation can more easily flip or 'tip' the system from that basin to another. The shallowing of a basin is analogous to a loss of resilience in the system.

Evidence suggests that Earth's climate system might be becoming less resilient to future perturbation and so more likely to generate cascades of tipping events. A key paper published in 2019 elaborates on this idea. Timothy Lenton and his co-authors write that 'the clearest emergency would be if we were approaching a global cascade of tipping points that led to a new, less habitable, "hothouse" climate state … We argue that cascading effects are common [and] examples are starting to be observed' (Lenton et al. 2019, p. 594).

Lenton's team identify several components of Earth's climate system that could exhibit tipping behaviour in the future and then suggest possible causal links between these elements that could produce tipping cascades. One example is the link between the loss of the Arctic Ocean's sea ice and a slowdown of the Atlantic Meridional Overturning Circulation (Caesar et al. 2018), the natural process of marine heat and salt transportation in the north Atlantic Ocean. This shift could, in turn, destabilise the West African monsoon, triggering drought in Africa's Sahel region; it could also dry the Amazon and create a build-up of heat in the Southern Ocean, accelerating ice loss in the Antarctic (Lenton et al. 2019, p. 594). In sum, our climate system may be on the precipice of a series of sharp changes that would have enormous socio-economic consequences – producing, in turn, societal tipping events and subsequent violence.

In this context, perhaps the most pressing security issue is the prospect of a sharp rise in food scarcity. Lack of ready access to food has enormous socio-economic and emotional consequences. Systems scientists, such as Franziska Gaupp, have considered how climate change is increasing the probability of 'simultaneous breadbasket failures' (Gaupp et al. 2020). Her team 'combine region-specific data on agricultural production with spatial statistics of climatic extremes to quantify the changing risk of low production for the major food-producing regions (breadbaskets) over time' (Gaupp et al. 2020, p. 54). Gaupp's team quantified the risks of regional production failures for staples, such as wheat, maze, and soybean, and showed that the risk of simultaneous global breadbasket failure is rising (Gaupp 2020). Simultaneous food shocks could happen as soon as this year, not at some indefinite point in the future, when most people, including many policy experts, assume severe climate problems will arise. The impacts on global economic and political stability would be staggering. Even the relatively minor food shock that coincided with the 2008/09 financial crisis exacerbated civil instability in almost 60 countries (Brinkman and Hendrix 2011).

In conclusion, after 30 years of debate over the links between climate change and violent conflict, we can say definitively that climate change is now a powerful factor influencing economic growth and agricultural production around the world; it will increasingly contribute to societal tipping events, where economic stagnation and food insecurity produce sharp outbreaks of social instability and violence. Climate change, once considered merely a security 'threat multiplier', is now a direct threat to global security in itself.

## References

Brinkman, H.-J. and C. S. Hendrix 2011. Food insecurity and violent conflict: causes, consequences, and addressing the challenges. *World Food Programme*. https://www.researchgate.net/publication/267450250_Food_Insecurity_and_Violent_Conflict_Causes_Consequences_and_Addressing_the_Challenges (accessed 1 May 2023).

Caesar, L., S., Rahmstorf, A., Robinson, G., Feulner and V., Saba 2018. Observed fingerprint of a weakening Atlantic Ocean overturning circulation. *Nature* 556: 191–196.

Gaupp, F., J. Hall, S. Hochrainer-Stigler, and S. Dadson 2020. Changing risks of simultaneous global breadbasket failure. *Nature Climate Change* 10: 54–57.

GML 2023. Mauna Loa Baseline Observatory. *Global Monitoring Laboratory*. https://gml.noaa.gov/obop/mlo/ (accessed 13 June 2023).

Homer-Dixon, T. 1991. On the threshold: environmental changes as causes of acute conflict. *International Security* 16(2): 76–116.

Homer-Dixon, T. 1994. Environmental scarcities and violent conflict: evidence from cases. *International Security* 19(1): 5–40.

Homer-Dixon, T. 1999. *Environment, Scarcity and Violence*. Princeton, NJ: Princeton University Press.

IEP (Institute for Economics & Peace) 2021. *Ecological Threat Report 2021: Understanding Ecological Threats, Resilience, and Peace*. http://visionofhumanity.org/resources (accessed 1 May 2023).

IEP (Institute for Economics & Peace) 2022. *Ecological Threat Report 2021: Understanding Ecological Threats, Resilience, and Peace*. http://visionofhumanity.org/resources (accessed 1 May 2023).

Lenton, T. M., J. Rockstrom, O. Gaffney, S. Rahmstorf, K. Richardson, W. Steffen, and H. J. Schellnhuber 2019. Climate tipping points: too risky to bet against. *Nature* 575(7784): 592–595.

# INDEX

Note: Page numbers in *italic* indicate figures, and page numbers followed by 'n' refer to notes.